Introduction to Nonlinear Optics

Since the early days of nonlinear optics in the 1960s, the field has expanded dramatically, and is now a vast and vibrant field with countless technological applications. Providing a gentle introduction to the principles of the subject, this textbook is ideal for graduate students starting their research in this burgeoning area.

After basic ideas have been outlined, the book offers a thorough analysis of second harmonic generation and related second-order processes, before moving on to third-order effects, the nonlinear optics of short optical pulses, and coherent effects such as electromagnetically-induced transparency. A simplified treatment of high harmonic generation is presented at the end. More advanced topics, such as the linear and nonlinear optics of crystals, the tensor nature of the nonlinear coefficients, and their quantum mechanical representation, are confined to specialist chapters so that readers can focus on basic principles before tackling these more difficult aspects of the subject.

Geoffrey New is Professor of Nonlinear Optics with the Quantum Optics and Laser Science Group in the Blackett Laboratory, Imperial College London. A highly regarded laser physicist and a Fellow of the Optical Society of America, he has spent nearly half a century researching nonlinear optics and laser physics.

Introduction to Nonlinear Optics

GEOFFREY NEW

Quantum Optics & Laser Science Group

Blackett Laboratory

Imperial College London

CAMBRIDGE
UNIVERSITY PRESS

CAMBRIDGE
UNIVERSITY PRESS

University Printing House, Cambridge CB2 8BS, United Kingdom

One Liberty Plaza, 20th Floor, New York, NY 10006, USA

477 Williamstown Road, Port Melbourne, VIC 3207, Australia

314-321, 3rd Floor, Plot 3, Splendor Forum, Jasola District Centre, New Delhi - 110025, India

79 Anson Road, #06-04/06, Singapore 079906

Cambridge University Press is part of the University of Cambridge.

It furthers the University's mission by disseminating knowledge in the pursuit of
education, learning and research at the highest international levels of excellence.

www.cambridge.org
Information on this title: www.cambridge.org/9780521877015

© G. New 2011

First published 2011

A catalogue record for this publication is available from the British Library

Library of Congress Cataloging in Publication data
New, Geoffrey (Geoffrey H. C.)
Introduction to Nonlinear Optics / Geoffrey New.
p.   cm.
Includes bibliographical references and index.
ISBN 978-0-521-87701-5 (Hardback)
1. Nonlinear optics. I. Title.
QC446.2.N475 2011
535′.2–dc22   2010051428

ISBN 978-0-521-87701-5 Hardback
ISBN 978-1-107-42448-7 Paperback

Contents

Preface

I set out to write this book in the firm belief that a truly introductory text on nonlinear optics was not only needed, but would also be quite easy to write. Over the years, I have frequently been asked by new graduate students to recommend an introductory book on nonlinear optics, but have found myself at a loss. There are of course a number of truly excellent books on the subject – Robert Boyd's *Nonlinear Optics*, now in its 3rd edition [1], is particularly noteworthy – but none of them seems to me to provide the gentle lead-in that the absolute beginner would appreciate.

In the event, I found it a lot harder to maintain an introductory flavour than I had expected. I quickly discovered that there are aspects of the subject that are hard to write about at all without going into depth. One of my aims at the outset was to cover as much of the subject as possible without getting bogged down in crystallography, the tensor structure of the nonlinear coefficients, and the massive perturbation theory formulae that result when one tries to calculate the coefficients quantum mechanically. This at least I largely managed to achieve in the final outcome. As far as possible, I have fenced off the 'difficult' bits of the subject, so that six of the ten chapters are virtually 'tensor-free'. Fortunately, many key aspects of nonlinear optics (the problem of phase matching, for example) can be understood without knowing how to determine the size of the associated coefficient from first principles.

In the process of writing, I also discovered several areas of the subject that I had not properly understood myself. It is of course always said that the best way to learn a subject is to teach it, so perhaps I should not have been too surprised. There were also things that I had simply forgotten. There was one 'senior moment' when, after struggling with a thorny problem for several hours, I found the answer in a book which cited a paper by Ward and New!

As to the subject itself, I have sometimes teased research students by telling them that nonlinear optics was all done in the 1960s. This statement is of course manifestly ridiculous as it stands – what have all those thousands of papers on nonlinear optics published in the last 40 years been about then? On the other hand, the statement that *all the basic principles* of nonlinear optics were established in the 1960s, does embody an element of truth. There really are very few fundamental concepts underlying the nonlinear optics of today that were not known by 1970. What has happened is not so much that new principles have been established, but rather that the old principles have been exploited in new ways, in new materials, in new combinations, in new environments, and on new time and distance scales. Technology has advanced so that many possibilities that could only be dreamt of in the 1960s can now be realised experimentally and, more than that, may even be intrinsic to commercial products. This point is elaborated in Section 1.12.

The book itself has a modular rather than a serial structure. Consequently, some chapters stand on their own feet, and most readers will probably not want to read the ten chapters in sequence. Beginners should certainly start with Chapter 1, but a reader wanting to get up to speed on crystal optics could read Chapter 3 on its own, while someone interested in dispersion could jump in at Chapter 6. The purpose of both these chapters is to support those that follow. On the other hand, readers who are not interested in crystallography should avoid Chapters 3 and 4, while those who are not concerned with the tensor nature of the nonlinear coefficients should certainly skip Chapter 4 and maybe the start of Chapter 5 too. And anyone who is allergic to quantum mechanics should be able to survive the entire book apart from Chapter 8 and parts of Chapter 9. Of course, some people may be looking for more detailed information, which they may be able to find in one of the nine appendices.

The following table shows what the reader can expect in each chapter; ticks in brackets imply a small amount of material.

Chapter	Tensors	Crystallography	Quantum Mechanics
1	×	×	×
2	×	×	×
3	(✓)	✓	×
4	✓	✓	×
5	(✓)	(✓)	×
6	×	×	×
7	×	×	×
8	✓	(✓)	✓
9	(✓)	×	✓
10	×	×	×

As the table suggests, the chapters have a varied flavour. Some are fairly specialist in nature, while others contain mostly bookwork material. Chapter 1 serves as a gentle introduction. The polarisation is expanded in the traditional way as a power series in the electric field (treated as a scalar) with expansion coefficients that are regarded as constants. A range of potential nonlinear interactions is explored in this way. The problem of phase mismatch emerges naturally from the discussion. Although in one sense the procedure amounts to little more than cranking a mathematical handle, a surprisingly large number of important nonlinear processes can be discovered in this way. At the end of the chapter, attention is drawn to the shortcomings of the simple approach adopted, and the necessary remedies are outlined. The chapter ends with a brief overview of the entire field.

Chapter 2 starts with a detailed analysis of second harmonic generation, based on Maxwell's equations and proceeding via the coupled-wave equations. The fields and polarisations are now complex numbers, but their vectorial nature is still not taken into account. The discussion broadens out into sum and difference frequency generation, optical parametric amplification and optical parametric oscillators. The chapter ends with a treatment of harmonic generation in focused beams.

In Chapter 3, real nonlinear media are encountered for the first time. The chapter focuses on the linear optics of crystals, and contains a fairly detailed treatment of birefringence in uniaxial crystals. The principle behind quarter- and half-wave plates is outlined. A section on biaxial media is included at the end.

Chapter 4 deals with second-order nonlinear effects in crystals, and there is now no way to avoid defining the nonlinear coefficients with all their tensor trappings. Though I have tried hard to keep it simple (as much material as possible has been relegated to Appendix C), the treatment that remains is still rather indigestible. Permutation symmetries are outlined, the contracted suffix notation for the nonlinear coefficients is introduced, and the Kleinmann symmetry condition is explained. The structure of the nonlinear coefficients for the three most important crystal classes is highlighted, and examples are given to show how the results are applied in some typical nonlinear interactions.

Chapter 5 is devoted to a selection of third-order nonlinear processes. Although the coverage is restricted to amorphous media, there are still places (especially early in the chapter) where the tensor structure of the coefficients in unavoidable. The later parts of the chapter deal with stimulated Raman scattering, acousto-optic interactions, and stimulated Brillouin scattering. A section on nonlinear optical phase conjugation is included at the end.

Chapter 6 focuses on dispersion. Although basically a linear effect, dispersion plays a crucial role in many nonlinear processes, and the material covered here therefore provides essential background for Chapter 7. After a general introduction, the evolution of a Gaussian pulse in a dispersive medium is analysed in detail, and this leads on naturally to a discussion of chirping, and pulse compression. A hand-waving argument then enables a rudimentary pulse propagation equation to be deduced, which provides an opportunity to introduce the local time transformation.

Nonlinear optical interactions involving short optical pulses are treated in Chapter 7. The choice of examples is somewhat arbitrary, but the selection covers quite a wide range, and includes self-phase modulation, nonlinear pulse compression, optical solitons, second harmonic generation in dispersive media, optical parametric chirped pulse amplification (OPCPA), pulse diagnostics, and the phase stabilisation of few-cycle optical pulses.

Up to this point in the book, the strength of the nonlinear interactions has been represented by the value of the appropriate nonlinear coefficient or, more exactly, the set of numbers contained in the non-zero tensor elements for the process in question. Chapter 8 focuses on the quantum mechanical origin of the coefficients. The perturbation theory formulae that emerge are large and cumbersome and, to this extent, they may leave the reader not much the wiser! However, every effort has been made to simplify things by taking special cases, and by highlighting dominant terms in the equations, the structure of which is relatively simple. The discussion continues in Chapter 9, which deals with resonant effects including self-induced transparency (SIT) and electromagnetically-induced transparency (EIT).

No book on nonlinear optics would be complete without a mention of high harmonic generation (HHG), and the quest for attosecond pulses. The treatment of these topics in Chapter 10 is essentially classical, and certainly simple. So if the promise in the title of an introductory treatment to the subject has not always been fulfilled in some of the intermediate chapters, at least it is kept in the last chapter.

Some more detailed material is included in the appendices, the last of which contains values of useful constants in nonlinear optics.

The varied nature of the material covered in the book has affected how the reference list has been compiled. A comprehensive list of sources would be almost as long as the book itself, so I have had to adopt a selective policy on which papers to refer to. Broadly speaking, references are of three types: historical papers describing things done for the first time, books and recent journal articles where an aspect of the subject is reviewed, and (in a few cases) papers on topics of particular current interest. To find further sources, it is always worth using the Internet. The information available there may not always be of impeccable quality, but one will almost always be led to an authoritative source within a few minutes.

There are bound to be errors of one sort or another in any book of this type. I have of course made every effort to avoid them, but I doubt if anyone has ever written a physics book without getting a few things wrong. Colleagues who read the book in draft picked up quite a few glitches, but I am sure there are more. If you find something that you think is wrong, please e-mail me on g.new@imperial.ac.uk, so that I can build up a corrections file.

Geoff New
Le Buisson de Cadouin
France
July 2010

Acknowledgements

Many people contributed in many different ways to this book, some during the actual writing, and others on a much longer time-scale.

I want first of all to express my thanks and appreciation to two people who had a strong influence on my early research. Professor John Ward introduced me to nonlinear optics in Oxford in 1964, and has been a good friend ever since, while the late Professor Dan Bradley was a dynamic leader and a constant source of inspiration and support over the following 15 years or so, first at Queen's University Belfast, and later at Imperial College London.

Over the years, I have benefited greatly from interactions with colleagues, collaborators, research associates and graduate students, through which my understanding of nonlinear optics has been expanded. I mention particularly Professor John Elgin, Dr. Graham McDonald, Dr. Paul Kinsler, Dr. Ian Ross, Dr. Mick Shaw, Professor Wilson Sibbett, Professor Derryck Reid, Professor Sabino Chavez-Cerda, Dr Jesus Rogel-Salazar, Dr. Harris Tsangaris, Dr. Phil Bates, Dr. Jonathan Tyrrell, and Dr. Sam Radnor.

I am especially grateful for the help and advice on specific aspects of the book given by Dr. Luke Chipperfield, Professor Ian Walmsley, Professor Majid Ebrahim-Zadeh, Dr. Louise Hirst, Professor Jon Marangos, Dr. Stefan Scheel, Dr. Stefan Buhmann, and Professor Mike Damzen.

I owe a special debt of gratitude to three colleagues in the Blackett Laboratory at Imperial College. Neal Powell in the audio-visual unit drew almost all the non-graphical figures, and improved several of the others as well. He also made a number of late changes at very short notice, and altogether could not have been more helpful. Professor John Tisch read Chapter 10 in detail, made numerous constructive comments that greatly improved the text, and drew my attention to a number of key references. He also provided two figures for that chapter, and one for Chapter 7 as well. Lastly, Professor Richard Thompson somehow found the time to read most of the book in draft, and even said that he enjoyed it! I adopted almost all of his invaluable suggestions and am very grateful for his help and support.

Finally, I would like to acknowledge the Leverhulme Trust, which provided financial support in the form of an Emeritus Fellowship during the period in which the book was being written, and to thank Cambridge University Press for putting up with countless missed deadlines for delivery of the final text.

Introduction

1.1 Nonlinearity in physics

The response of real physical systems is never exactly proportional to stimulus, which is a way of saying that we live in an inherently nonlinear world. The deviation from linearity may be very slight, especially if the stimulus is weak; the assumption of linearity will then be an excellent approximation and probably the only route to an analytical solution. But that does not change the fact that linearity is an idealisation.

A simple example of the linear approximation occurs in elementary mechanics where one assumes that restoring force is proportional to displacement from equilibrium (Hooke's law). This leads to the equation of motion

$$m\ddot{x} = -s_1 x \tag{1.1}$$

where m is the mass and s_1 is the Hooke's law constant (the stiffness of the system). Equation (1.1) has the simple harmonic motion solution

$$x = A\cos\{\omega_0 t + \phi\} \tag{1.2}$$

where $\omega_0 = \sqrt{s_1/m}$ is the angular frequency of oscillation, and A and ϕ are fixed by the initial conditions. Equation (1.1) is usually an excellent approximation when the amplitude A of oscillations is small. But, since real systems are never perfectly linear, their oscillations will never be precisely (co)sinusoidal, and will contain harmonics of ω_0.

Nonlinearity can be incorporated into Eq. (1.1) by including additional terms on the right-hand side to represent the fact that the restoring force is no longer linear in the displacement. For example, it might be appropriate to write

$$m\ddot{x} = -s_1 x + s_2 x^2 = -(s_1 - s_2 x)x \tag{1.3}$$

where the second form highlights the fact that the nonlinear term makes the stiffness dependent on amplitude. Notice that the inclusion of a term in x^2 makes the stiffness asymmetrical in the displacement; if $s_2 > 0$, the net stiffness is lower for positive x and higher for negative x.

Since Eq. (1.3) has no analytical solution, a numerical solution is generally called for. However, the role of the extra term in introducing harmonics can be appreciated by regarding it as a perturbation, with Eq. (1.2) as the zeroth-order solution. To the first order of approximation, the system will be subject to the additional force term

$$s_2 x^2 = s_2 A^2 \cos^2\{\omega_0 t + \phi\} = \tfrac{1}{2} s_2 A^2 [1 + \cos 2\{\omega_0 t + \phi\}], \tag{1.4}$$

and this will lead (through Eq. 1.3) to a second harmonic component in the motion. In principle, the process can be repeated to higher orders of approximation, although the terms quickly become negligible if the nonlinearity is weak.

In many situations in life, one seeks to minimise the effects of nonlinearity. Manufacturers of audio equipment are for example keen to advertise the lowest possible figures for 'harmonic distortion'. But in other circumstances, nonlinearity can be put to good use, and this book is about how it can be exploited to spectacular effect in optical physics.

1.2 The early history of nonlinear optics

In optics, one is interested in the response of atoms and molecules to applied electromagnetic (EM) fields. The interaction of light and matter is of course governed by the Schrödinger equation, which is linear in the wave function but nonlinear in the response of the wave function to perturbations. Despite this, optics proceeded quite successfully for many years on the assumption that the response of optical materials was linear in the applied electric field E. If P is the polarisation of the medium (i.e. the dipole moment per unit volume),[1] one writes

$$P = \varepsilon_0 \chi^{(1)} E \qquad (1.5)$$

where $\chi^{(1)}$ is the linear susceptibility. So if $E = A \cos \omega t$, the consequence is that $P = \varepsilon_0 \chi^{(1)} A \cos \omega t$. It also follows from Eq. (1.5) that the electric displacement is

$$D = \varepsilon_0 E + P = \varepsilon_0 (1 + \chi^{(1)}) E = \varepsilon_0 \varepsilon E \qquad (1.6)$$

where $\varepsilon = 1 + \chi^{(1)}$ is the relative dielectric constant. As we will see later, ε is the square of the refractive index, $n = \sqrt{1 + \chi^{(1)}}$.

Equation (1.5) served as a good approximation for so long because the electric field strengths that scientists were able to deploy in those early years were far weaker than the fields inside atoms and molecules; the perturbations were therefore very small. It was not until the 1870s that the Rev. John Kerr, a lecturer at the Free Church Training College in Glasgow, UK, demonstrated that the refractive index of a number of solids and liquids is slightly changed by the application of a strong DC field [2]. This phenomenon, now known as the DC Kerr effect,[2] was the first nonlinear optical effect to be observed.

Two decades later, in the 1890s, Friedrich Pockels at the University of Göttingen studied a related process known today as the Pockels effect [3]. The Kerr effect and the Pockels effect differ in two respects. In the Kerr effect, the refractive index change is proportional to the *square* of the applied DC field, whereas in the Pockels effect, the change is directly

---

[1] The word 'polarisation' has two meanings in optics. It refers (as here) to the dipole moment per unit volume (in coulombs m^{-2}), and would be represented by the vector **P** in a more rigorous treatment. But the word is also used to describe the *polarisation of light* where it refers to the direction of the fields in the transverse EM wave. The direction (or plane) of polarisation is normally taken to be that of the electric field in the wave.

[2] Sometimes it is just called the Kerr effect, but it must be distinguished from the optical (or AC) Kerr effect.

proportional to the field.[3] Secondly, whereas the Kerr effect is observable in liquids and amorphous solids, the Pockels effect occurs only in crystalline materials that lack a centre of symmetry. This vital distinction and the reason behind it will be discussed in detail later.

1.3 Optical second harmonic generation

There now followed a long gap in the history of nonlinear optics. Further progress had to wait for a source of strong *optical frequency* fields to become available, in other words for the invention of the laser in 1960. With the arrival in the laboratory of the ruby laser, nonlinear optics underwent a second birth, and it has been flourishing ever since. Indeed, since lasers themselves are inherently nonlinear devices, one could even argue that laser physics is itself a compartment within the wider field of nonlinear optics. Traditionally, however, the field of nonlinear optics is taken to exclude lasers themselves, which is fortunate insofar as this book would otherwise be much longer.

The first nonlinear optics experiment of the laser era was performed in 1961 by a team led by the late Peter Franken at the University of Michigan in Ann Arbor [4]. As shown in Fig. 1.1, a ruby laser was focused into a slab of crystalline quartz to discover if the nonlinear response of the medium to the intense optical frequency radiation at 694.3 nm was strong enough to create a detectable second harmonic component at a wavelength of 347.15 nm.[4]

The way to think about this experiment is diagrammed in Fig. 1.2. Consider the response of the electrons in the quartz to the stimulus of the optical frequency electric field of the laser beam written as $E = A \cos \omega t$. The displacement of the electrons creates a dipole moment p per atom,[5] or $P = Np$ per unit volume where N is the atomic number density.

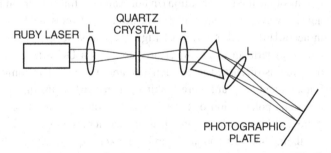

Fig. 1.1   Schematic diagram of the first second harmonic generation experiment by Peter Franken's group at the University of Michigan in 1961 [4].

---

[3] The Pockels effect is also known as the electro-optic effect, or sometimes (confusingly) as the *linear* electro-optic effect because the index change is linearly proportional to the DC field.

[4] The vacuum wavelength of the second harmonic component is of course half that of the fundamental.

[5] One could also write $p = \varepsilon_0 \alpha E$ where α is the atomic polarisibility, in which case $\chi^{(1)} = N\alpha$.

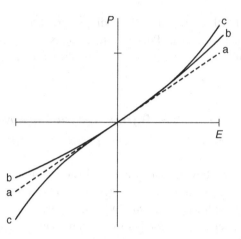

Fig. 1.2 Linear and nonlinear response of polarisation P to applied electric field E: (a) linear case (dotted), (b) quadratic nonlinearity, (c) cubic nonlinearity.

We allow for the possibility of a nonlinear response by writing

$$P = \varepsilon_0(\chi^{(1)}E + \chi^{(2)}E^2 + \cdots). \tag{1.7}$$

For $\chi^{(2)} > 0$, the dependence of P on E is represented by curve b in Fig. 1.2, while in the absence of the nonlinear term, Eq. (1.7) reverts to Eq. (1.5), and the linear relationship between E and P represented by the dotted straight line a in the figure is recovered.

If $E = A\cos\omega t$ is substituted into Eq. (1.7), the polarisation now reads

$$P = \varepsilon_0(\underbrace{\chi^{(1)}A\cos\omega t}_{\text{linear term}} + \underbrace{\tfrac{1}{2}\chi^{(2)}A^2(1 + \cos 2\omega t)}_{\text{nonlinear term}}). \tag{1.8}$$

The analogy between Eqs. (1.7)–(1.8) and Eqs. (1.3)–(1.4) is obvious. Figure 1.3 shows graphs of the total polarisation (in bold) and the linear term on its own (bold dotted); notice how the waveform is stretched on positive half-cycles and flattened on negative half-cycles, in accordance with curve b of Fig. 1.2. The solid grey line in Fig. 1.3 representing the nonlinear term in Eq. (1.8) clearly contains a DC offset (dotted grey), the significance of which is discussed in Section 1.6, and a second harmonic component, which is what Franken and his team were looking for in 1961. A useful picture is to regard the second harmonic polarisation of the nonlinear medium as an optical frequency antenna. Just as radio and TV signals are broadcast by accelerating charges in the transmitter aerial, so the oscillating nonlinear polarisation in Franken's quartz crystal should radiate energy at twice the frequency of the incident laser field. The key question in 1961 was simply: will the radiation from the second harmonic antenna be strong enough for photons at 347.15 nm to be observed experimentally?

The University of Michigan experiment was a huge success to the extent that *second harmonic generation* (SHG) was detected. The down side was that the harmonic intensity was extremely weak, so weak in fact that the photographic plate reproduced in the 15 August 1961 issue of *Physical Review Letters* [4] appears to be totally blank! In fact, the

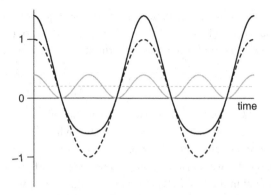

Fig. 1.3 Response (bold line) of a system exhibiting a quadratic nonlinearity to a (co)sinusoidal stimulus (dotted line). The difference between the two (grey line) contains both second harmonic and DC frequency components.

energy conversion efficiency from fundamental to harmonic was about 1 part in 10^8, a clear demonstration that SHG was a real effect although, at this minuscule efficiency, it was clearly a curiosity rather than a practical means of generating ultraviolet (UV) light.

1.4 Phase matching

Why was the efficiency so small? The reason is easy to understand as soon as one includes the spatial dependence of the fields in the equations. Assuming that the ruby laser field is a plane wave propagating in the z-direction, we write

$$E_1 = A_1 \cos\{\omega t - k_1 z\} \tag{1.9}$$

where the angular wave number of the fundamental beam is $k_1 = n_1 \omega / c$, and n_1 is the refractive index.[6] The second harmonic term in the polarisation is therefore

$$P_2 = \tfrac{1}{2} \varepsilon_0 \chi^{(2)} A_1^2 \cos\{2\omega t - 2k_1 z\} \tag{1.10}$$

which seems fine at first sight. But now compare the space-time dependence in Eq. (1.10) with that of a freely propagating field at 2ω namely

$$E_2 = A_2 \cos\{2\omega t - k_2 z\} \tag{1.11}$$

where $k_2 = n_2 2\omega / c$ by analogy with k_1. Notice that the arguments of the cosines in Eqs (1.10) and (1.11) are different unless $2k_1 = k_2$, which is only true if the refractive indices n_1 and n_2 are the same. But dispersion ensures that the indices are usually *not* the

---

[6] Here and throughout the book, the default space-time dependence of an optical field is $(\omega t - kz)$, where the parameter k is called the *angular wave number*, by analogy with angular frequency $\omega = 2\pi \nu$. However, many authors use $(kz - \omega t)$ instead of $(\omega t - kz)$. The choice is immaterial under a cosine but, with complex exponentials, it leads to a difference in sign that permeates into many subsequent equations. See Appendix A for more discussion.

same. The severity of the mismatch is easily quantified by asking over what distance the cosine terms in Eqs (1.10) and (1.11) get π radians out of step. This distance, known as the *coherence length* for the second harmonic process, is easily shown to be[7]

$$L_{\mathrm{coh}} = \frac{\pi}{|\Delta k|} = \frac{\lambda}{4\,|n_2 - n_1|} \qquad (1.12)$$

where $\Delta k = k_2 - 2k_1$ is called the mismatch parameter. For typical optical materials, L_{coh} is a few tens of microns[8] (μm) after which, as we will show in the next chapter, the SHG process goes into reverse, and energy is converted from the harmonic back into the fundamental wave. Hence, only a few microns of the 1 mm quartz sample contributed to the signal in the first SHG experiment, and the rest was redundant.

The phase-matching problem was soon solved. Within a few months of the University of Michigan experiment, researchers at the Ford Motor Company's laboratories, around 30 miles from Ann Arbor in Dearborn, Michigan, had exploited birefringence in a KDP crystal[9] to keep the ω and 2ω waves in step with each other [5]. They did this by finding a particular direction of propagation in KDP for which the refractive index of the ordinary wave at ω was the same as that of the extraordinary wave at 2ω. The process is called *phase matching* or sometimes *birefringent phase matching* (BPM) to distinguish it from other methods of achieving the same end; see Chapter 2. Under phase-matched conditions, the SHG conversion efficiency jumped to tens of percent with careful experimental adjustment, and from that moment on, nonlinear optics stopped being a curiosity and became a practical proposition.

Phase matching is of vital importance in many nonlinear processes. The key principle to grasp is that the direction of energy flow between fundamental and harmonic waves is determined by the relative phase between the nonlinear polarisation and the harmonic field. If the two can be held in step by making the fundamental and harmonic refractive indices the same, the energy will keep flowing in the same direction over a long distance. Otherwise it will cycle backwards and forwards between the fundamental and the harmonic, reversing direction more frequently the more severe the phase mismatch.

1.5 Symmetry considerations

Before moving on, it is worth considering some other key features of the SHG process. First of all, notice that the quadratic term in Eq. (1.6), represented by curve b in Fig. 1.2, implies that the medium responds differently according to the direction of the electric field. A positive field creates a slightly greater (positive) response than that of curve a, whereas a negative field produces a slightly smaller response than in the linear case. This implies that

---

[7] Sadly, there is no generally agreed definition of 'coherence length' in nonlinear optics, and definitions with 1, 2, π, and 2π in the numerator of Eq. (1.12) can be found in different textbooks. The definition in Eq. (1.12) is the one used almost universally by researchers in the 1960s, and I see no reason to change it.

[8] 'Micron' is still widely used for micrometre in colloquial usage.

[9] KDP stands for potassium dihydrogen phosphate.

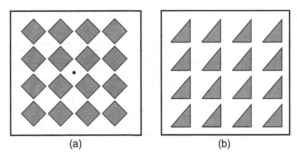

Fig. 1.4   (a) Centrosymmetric pattern and (b) non-centrosymmetric pattern.

SHG (and all other nonlinear processes that depend on $\chi^{(2)}$) can occur only in media that are structurally different in one direction from the opposite direction. Media of this kind are variously said to *lack inversion symmetry*, to *lack a centre of symmetry*, or to be *non-centrosymmetric*. In simple terms, one could say that they have an inherent 'one-wayness', a property that only crystalline materials can possess. Fortunately, many non-centrosymmetric media also exhibit double refraction (birefringence), and so are candidates for birefringent phase matching.

A more detailed discussion of inversion symmetry is given in Chapter 3, but the simple schematic pictures shown in Fig. 1.4 may be helpful at this stage. Imagine you are looking into a crystal along the line of the laser beam. The structure shown on the left possesses inversion symmetry. It can be flipped horizontally ($x \rightarrow -x$) or vertically ($y \rightarrow -y$) and the pattern is unchanged; the centre of symmetry at the mid-point is marked with a dot. It is therefore impossible in principle for this medium to exhibit the asymmetrical characteristics shown in Fig. 1.3. The structure on the right is different because, in this case, a flip in either direction results in a different pattern. So left is different from right, up is different from down, and the response in Fig. 1.3 is no longer forbidden.

A simple mathematical proof of the principle runs as follows [6]. In one dimension, consider the term $P_x = \varepsilon_0 \chi^{(n)} E_x^n$. Under inversion symmetry, P_x must change sign if E_x changes sign. But, when n is even, E_x^n is unchanged if $E_x \rightarrow -E_x$, and it follows that $\chi^{(2)}$, $\chi^{(4)}$, etc. must be zero in this case. No such restriction applies for odd n, so processes involving $\chi^{(3)}$ for example (see Section 1.10) can occur in centrosymmetric media.

1.6 Optical rectification

After the successful demonstration of second harmonic generation, the nonlinear optics bandwagon began to roll, and a multitude of other nonlinear processes were discovered in a 'gold rush' period in the mid-1960s. One such process was *optical rectification*, which has already appeared in the polarisation in Eq. (1.8) above [7]. This is probably the simplest nonlinear optical process both to visualise and to observe. To detect optical rectification, all one has to do is to pass a laser beam through a non-centrosymmetric crystal located

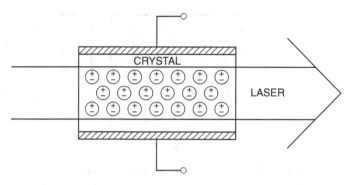

Fig. 1.5 Schematic diagram of optical rectification. Positive and negative charges in the nonlinear medium are displaced, causing a potential difference between the plates and current to flow in an external circuit.

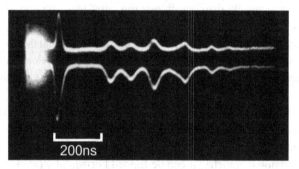

Fig. 1.6 Optical rectification signal (top trace) and laser monitor (lower trace).

between a pair of capacitor plates, as shown in Fig. 1.5. As suggested in the figure, the asymmetric response of the material to the field of the laser displaces the centre of gravity of the positive and negative charges in the medium, creating a DC polarisation in the medium. Dipoles oscillating at the *optical* frequency are of course created in all materials, and these are represented by the linear polarisation of Eq. (1.5). What is shown in Fig. 1.5 is rather the DC offset of Fig. 1.3, and its effect is to induce a potential difference between the plates, which allows a rectified component of the optical frequency field to be detected; see Problem 1.2.

An example of an optical rectification (OR) signal is shown in the dual-beam oscilloscope record of Fig. 1.6, where the upper and lower traces are the OR and the laser monitor signals, respectively. In fact, so accurately does the OR signal track the laser intensity that it is impossible to tell them apart from the record itself.[10] Since the OR process also leaves the laser field essentially unchanged, it was considered as a possible basis for laser power measurement in the early days of the subject. But other technologies prevailed in the end.

Notice that the phase-matching issue does not arise in the case of optical rectification, because the DC term in the polarisation (analogous to the second harmonic contribution in

---

[10] The fact that one trace is the negative of the other is a trivial result of the polarity of the electrical connections.

Eq. 1.10) is

$$P_{dc} = \tfrac{1}{2}\varepsilon_0 \chi^{(2)} A_1^2 \tag{1.13}$$

which has no space-time dependence.

1.7 The Pockels effect

The *Pockels effect* [3] relates to the change in the refractive index of a non-centrosymmetric medium when a DC electric field is applied. The effect can be discovered within the framework of Eq. (1.7) if a DC term is included in Eq. (1.9) so that

$$E = A_0 + A_1 \cos\{\omega t - k_1 z\}. \tag{1.14}$$

Substituting this form into Eq. (1.6) produces a new nonlinear term in the polarisation with the same space-time dependence as the linear polarisation. When the two are combined, the net contribution at frequency ω is

$$P = \varepsilon_0(\chi^{(1)} + 2\chi^{(2)} A_0) A_1 \cos\{\omega t - k_1 z\} + \cdots \tag{1.15}$$

The new term is proportional to the DC field A_0, and is in phase with the optical frequency field provided $\chi^{(2)}$ is real. No phase-matching considerations arise since, as for optical rectification, phase matching is automatically ensured. It is clear from Eq. (1.15) that the linear susceptibility has been modified by $2\chi^{(2)} A_0$, and the refractive index has correspondingly changed from $n = \sqrt{1 + \chi^{(1)}}$ to $n = \sqrt{1 + \chi^{(1)} + 2\chi^{(2)} A_0}$.

The Pockels effect allows the polarisation properties of light to be controlled electrically and, in combination with a polarising beam splitter, is routinely used to create an optical switch. A more detailed discussion of a Pockels cell can be found in Chapter 4.

1.8 Sum frequency generation

The catalogue of second-order nonlinear processes can be extended further by considering the process of sum frequency generation (SFG), a possibility that emerges if the incident electric field contains two different frequency components namely

$$E = A_1 \cos\{\omega_1 t - \mathbf{k_1}.\mathbf{r}\} + A_2 \cos\{\omega_2 t - \mathbf{k_2}.\mathbf{r}\}. \tag{1.16}$$

The angular wave numbers have now been written as vectors to allow for the possibility that the waves at ω_1 and ω_2 are non-collinear (i.e. travelling in different directions).[11] Inserting Eq. (1.16) into the polarisation expansion of Eq. (1.7) yields a term of the form

$$P = \varepsilon_0(\cdots + \chi^{(2)} A_1 A_2 \cos\{\omega_3 t - (\mathbf{k_1} + \mathbf{k_2}).\mathbf{r}\} + \cdots) \tag{1.17}$$

---

[11] When written as a vector, the angular wave number is variously called the wave vector, the propagation vector or simply the **k**-vector.

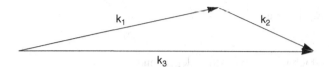

Wave vector triangle for phase-matching non-collinear beams.

where $\omega_3 = \omega_1 + \omega_2$. It must be stressed that the subscripts 1–3 are now merely labels, their only numerical significance being that ω_3 is the highest frequency of the set. In this notation, second harmonic generation is the special case where $\omega_1 = \omega_2 = \omega_3/2$.

If the polarisation wave of Eq. (1.17) is to match the space-time dependence of a freely propagating field at the sum frequency ω_3 and space-time dependence $\cos\{\omega_3 t - \mathbf{k}_3.\mathbf{r}\}$, the phase-matching condition to be satisfied is

$$\mathbf{k_1} + \mathbf{k_2} = \mathbf{k_3} \tag{1.18}$$

which is represented by the vector triangle of Fig. 1.7. For collinear beams, this equation becomes

$$n_1\omega_1 + n_2\omega_2 = n_3\omega_3 \tag{1.19}$$

which is more complicated than for SHG where phase matching simply requires that the fundamental and harmonic refractive indices are the same. The term 'phase matching' is to be preferred to 'index matching' for this reason. In media that exhibit normal dispersion,[12] the right-hand side of Eq. (1.19) will always be larger than the left-hand side (see Problem 1.2). This means that adjusting the angle between $\mathbf{k}_1$ and $\mathbf{k}_2$ is not sufficient on its own to fix the phase-mismatch problem, although it may still be useful for adjusting the phase-matching condition. An example of the use of non-collinear beams for increasing the bandwidth of a nonlinear interaction can be found in Chapter 7.

At the photon level, one can regard the SFG process as the annihilation of photons $\hbar\omega_1$ and $\hbar\omega_2$, and the creation of a photon $\hbar\omega_3$, so that

$$\hbar\omega_1 + \hbar\omega_2 \rightleftharpoons \hbar\omega_3 \tag{1.20}$$

which is a statement of conservation of energy. This equation has been written to suggest that the process may go into reverse, which will happen for certain phase relationships between the three fields.

Notice that multiplying all terms in Eq. (1.18) by $\hbar$ yields

$$\hbar\mathbf{k_1} + \hbar\mathbf{k_2} = \hbar\mathbf{k_3}. \tag{1.21}$$

Since momentum is $\mathbf{p} = \hbar\mathbf{k}$, this equation demonstrates that the phase-matching condition for the interacting waves is equivalent to the momentum-matching condition for the interacting photons.

---

[12] Normal dispersion means that the refractive index rises with frequency (or falls with wavelength).

1.9 Difference frequency generation and optical parametric amplification

We now consider a further possibility. Suppose that the relationship $\omega_1 + \omega_2 = \omega_3$ still applies, but that the two waves entering the nonlinear medium are ω_2 and ω_3 ($> \omega_2$) rather than ω_1 and ω_2. In this case, the relevant term in the polarisation is

$$P = \varepsilon_0(\cdots + \chi^{(2)} A_3 A_2 \cos\{\omega_1 t - (\mathbf{k_3} - \mathbf{k_2}).\mathbf{r}\} + \cdots). \tag{1.22}$$

This represents the process known as difference frequency generation (DFG) in which ω_1 is generated through the interaction of ω_3 and ω_2.[13] Equations (1.18)–(1.21) are all unchanged in the DFG case, although it perhaps makes more sense to write Eq. (1.20) the other way round, namely

$$\hbar\omega_3 \leftrightharpoons \hbar\omega_1 + \hbar\omega_2. \tag{1.23}$$

This equation highlights an important feature of DFG. Since the difference frequency component grows as photons at ω_3 divide into photons at ω_1 and ω_2, it follows that the other input beam at ω_2 grows at the same time, indeed it gains a photon for every photon created at ω_1.

A further scenario now presents itself. What would happen if the sole input wave were the one at ω_3? Since there is always some background noise covering the optical spectrum, might this be able to provide seed signals at ω_1 and ω_2 to get the DFG process off the ground? If so, how would an ω_3 photon know how to divide its energy between the two smaller photons? After all, there is an infinite number of ways to break a pencil into two parts, so which of the infinite set of possibilities would the photon choose?

The quick answer to these questions is that the process does indeed work as suggested, and that the split that most closely satisfies the phase-matching condition is the one that prevails. If you like, you can think of the photons trying out all options at once, with the most efficient being the one that wins. The process under discussion is called *optical parametric amplification* (OPA), and it can be performed in an optical cavity to create an *optical parametric oscillator* (OPO) [8]. OPOs offer a route to generating tunable coherent radiation because the output frequency can be varied by controlling the phase matching. As an alternative source of coherent radiation, the OPO should be regarded as a close relative of the laser in which the fundamental physical process is optical parametric amplification rather than stimulated emission.

Frequency-mixing processes are conveniently represented diagrammatically as shown in Fig. 1.8, where the first three frames depict SHG, SFG, and DFG/OPA, respectively. Up-arrows indicate waves that are (ideally) being depleted, and down-arrows those that are being enhanced. But unless the phase-matching conditions are satisfied, these directions apply only to the first (and subsequent odd-numbered) coherence lengths, and are reversed

---

[13] Of course, the difference frequency $(\omega_2 - \omega_1)$ would also have appeared in Eq. (1.17) had that equation been written out in full. The focus on $(\omega_3 - \omega_2)$ is to emphasise the fact that difference frequency generation is sum frequency generation in reverse.

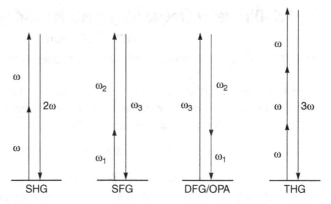

Diagrammatic representation of various nonlinear processes including second harmonic generation (SHG), sum frequency generation (SFG), difference frequency generation (DFG), optical parametric amplification (OPA), and third harmonic generation (THG).

in the others. More sophisticated versions of these diagrams featuring virtual energy levels will be introduced in Chapter 8.

1.10 Third-order processes

A number of important *third-order processes* can be identified within the framework used in this chapter if we include a cubic term in Eq. (1.7), in which case the equation becomes

$$P = \varepsilon_0(\chi^{(1)}E + \chi^{(2)}E^2 + \chi^{(3)}E^3 + \cdots). \tag{1.24}$$

The kind of nonlinear response represented by the new term is typified by curve c in Fig.1.2. Notice that, unlike the quadratic term, the way the response depends on the magnitude of E is independent of the sign of E, so the behaviour is symmetrical between positive and negative fields. The consequence is that $\chi^{(3)}$ is non-zero in centrosymmetric materials in which $\chi^{(2)} = 0$ by symmetry, so an ordinary block of glass (and liquids and gases too) exhibit processes arising from the third-order nonlinearity.

What new processes does the cubic term give rise to? If one inserts Eq. (1.14) into Eq. (1.24), one obtains a number of new polarisation components including

$$\begin{aligned}
P = \varepsilon_0(&\cdots + 3\chi^{(3)}(A_0^2 + \tfrac{1}{4}A_1^2)A_1 \cos\{\omega t - k_1 z\} \\
&+ \tfrac{3}{2}\chi^{(3)}A_0 A_1^2 \cos\{2\omega t - 2k_1 z\} + \tfrac{1}{4}\chi^{(3)}A_1^3 \cos\{3\omega t - 3k_1 z\} + \cdots).
\end{aligned} \tag{1.25}$$

As we discovered in Section 1.6, terms in $\cos \omega t$ represent refractive index changes, and indeed the *DC Kerr effect* is represented by the term in A_0^2 in Eq. (1.25). However, there is clearly now a further index change based on the following term, which involves $A_1^2 = 2I/c\varepsilon_0 n$ where I is the intensity (i.e. the power per unit area) of the field at ω. This term arises from the optical-frequency Kerr effect (sometimes called the AC Kerr effect), and it is the origin of the *intensity-dependent refractive index*, a process in which a beam of light changes its own refractive index. It is quite easy to show from Eq. (1.25) that the index change is[14]

$$\Delta n \simeq \left(\frac{3\chi^{(3)}}{4c\varepsilon_0 n^2} \right) I. \tag{1.26}$$

This effect underlies the potentially damaging effect of self-focusing, as well as the temporal effects of self-phase modulation (SPM) and self-steepening. Self-phase modulation plays a crucial role in a variety of pulse compression processes, as well as in optical soliton formation.

A generalised version of the intensity-dependent refractive index can be discovered if one inserts the two-frequency field of Eq. (1.16) into Eq. (1.24). In this case, the polarisation includes the terms

$$P = \varepsilon_0(\cdots + \left(\tfrac{3}{2}\chi^{(3)} A_1^2 \right) A_2 \cos\{\omega_2 t - \mathbf{k}_2.\mathbf{r}\} + \left(\tfrac{3}{2}\chi^{(3)} A_2^2 \right) A_1 \cos\{\omega_1 t - \mathbf{k}_1.\mathbf{r}\}). \tag{1.27}$$

The first term represents a change in the refractive index at ω_2 associated with the intensity of the wave at ω_1, while in the second term the roles are reversed. This process in which one beam modifies the refractive index of another is known as cross-phase modulation (XPM). The case of the intensity-dependent refractive index mentioned earlier is the special case where a single beam modifies its own propagation environment.

The two final terms in Eq. (1.25) represent *DC-induced second harmonic generation* and *third harmonic generation,* respectively. SHG is of course normally forbidden in centrosymmetric materials, but an applied DC field lifts the symmetry and enables the process to occur. Third harmonic generation (THG; see the right-hand frame of Fig. 1.8) is the third-order analogue of second harmonic generation. Unlike SHG, it can occur in a centrosymmetric medium, although the larger frequency difference between the fundamental and the harmonic means that birefringent phase matching is harder to achieve.

1.11 Theoretical foundations

We now consider the strengths and weaknesses of the simplified treatment given in this introductory chapter. By expanding the polarisation in powers of the electric field (see Eq. 1.24), we have certainly succeeded in discovering quite a number of nonlinear optical phenomena, all of which exist, and most of which are in the mainstream of current technology. The vital importance of achieving phase matching (in processes where it is not automatic) has also emerged, and the entire treatment has been simple and straightforward. It is in fact remarkable how much activity in nonlinear optics today has its roots in phenomena that have featured in this chapter.

---

[14] It is common to write $\Delta n = n_2 I$ where n_2 is known as the nonlinear refractive index; see Section 5.2.6. We have not done that here to avoid notational confusion with the refractive index at frequency ω_2.

The approach we have adopted is, however, deficient in a number of respects, and one or two false impressions have been given. Firstly, the fundamental origin of optical nonlinearity has never even been mentioned nor, in consequence, has any proper justification been given for writing the polarisation as a power series in the electric field to begin with. Indeed, plugging different electric field combinations into the expansion, and finding the resulting contributions to the polarisation feels more like a mathematical exercise than a serious exercise in nonlinear physics.

The key point is that, so far, all the physics at the atomic and molecular level has been hidden inside the nonlinear coefficients $\chi^{(n)}$. To do things properly, one needs to study the origin of optical nonlinearities quantum mechanically. Perturbation theory can normally be used for this purpose provided the applied electric fields are weak compared to those within the nonlinear material. The expansion of the polarisation in powers of the electric field then arises naturally as the theory is taken to successive orders. Elaborate quantum mechanical expressions for the nonlinear coefficients emerge from the analysis (take a quick glance at Chapter 8!), and numerical values can be calculated, at least in principle. One discovers that the coefficient for a given process depends on the frequencies of all the participating fields,[15] so the implication in this chapter that all nonlinear processes of a given order share a single nonlinear coefficient (i.e. that all second-order interactions depend on the same $\chi^{(2)}$), turns out to be incorrect. For example, the nonlinear coefficient governing optical rectification is not the same as the one controlling second harmonic generation, nor (at third order) is the coefficient of the DC Kerr effect the same as that for the optical Kerr effect.

Things become more complicated under resonant conditions, in other words when one or more of the field frequencies approaches transition frequencies in the nonlinear medium. Resonance does not of itself necessarily invalidate the perturbation theory approach. In some circumstances, the nonlinear coefficients simply become complex, and a range of resonant nonlinear interactions (such as two-photon absorption and stimulated scattering) then appear. However, perturbation theory certainly fails in the case of coherent interactions (such as self-induced transparency) where relatively strong fields act on time-scales that are shorter than the damping times. These issues are discussed in Chapters 5 and 9. Resonance or no resonance, perturbation theory invariably fails if the field strength is sufficiently high, an obvious example being high harmonic generation, which is based on the brute force ionisation of an atom and the subsequent recollision of the extracted electron with the parent ion. The topic is treated in Chapter 10.

Other important issues that have so far been ignored include the vector nature of the field and polarisation, along with all structural aspects of the nonlinear media, many of which are crystalline and probably birefringent too. The parameters in Eq. (1.24) were all written as scalars, whereas the polarisation and the field are in fact vectors, and the nonlinear coefficients are actually tensors. So the equation should really be written

$$\mathbf{P} = \varepsilon_0(\chi^{(1)}\mathbf{E} + \chi^{(2)}\mathbf{EE} + \chi^{(3)}\mathbf{EEE} + \cdots) \qquad (1.28)$$

---

[15] Some deeper questions about the nature of optical susceptibilities, and whether the equations for P as a function of E should be interpreted in the time domain or the frequency domain, are considered in Appendix B.

where $\boldsymbol{\chi}^{(1)}$, $\boldsymbol{\chi}^{(2)}$ and $\boldsymbol{\chi}^{(3)}$ are second-, third-, and fourth-rank tensors, respectively. Readers who are daunted by tensors will be reassured to know that, just as a simple vector (a first-rank tensor) can be written as a column matrix containing its Cartesian components, so a second-rank tensor can be written as a two-dimensional matrix and, in general, an nth-rank tensor written as an n-dimensional matrix. In matrix form, Eq. (1.28) duly becomes

$$P_i = \varepsilon_0 \left(\sum_j \chi_{ij}^{(1)} E_j + \sum_{jk} \chi_{ijk}^{(2)} E_j E_k + \sum_{jkl} \chi_{ijkl}^{(3)} E_j E_k E_l \right) \qquad (i,j,k,l = x,y,z).$$

$$(1.29)$$

Notice that Eq. (1.29) appears to suggest that the linear susceptibility $\chi_{ij}^{(1)}$ has $3^2 = 9$ components, the second-order nonlinear coefficient $\chi_{ijk}^{(2)}$ has $3^3 = 27$ components, and the third-order coefficient $\chi_{ijkl}^{(3)}$ has $3^4 = 81$. Fortunately, the situation is not as bad as it seems. At worst, it turns out that there are three independent components of $\chi_{ij}^{(1)}$, while in many common nonlinear crystals, most elements of $\chi_{ijk}^{(2)}$ are zero. Again, third-order processes are most often studied in isotropic media,[16] where most of the 81 potential coefficients are zero by symmetry. These issues are treated in more detail in Chapters 3, 4 and 5.

1.12 The growth of nonlinear optics

It was suggested in the Preface that most of the foundations of nonlinear optics were laid in the 1960s. It is certainly a salutary experience to re-read the seminal 1962 paper of Armstrong, Bloembergen, Ducuing and Pershan (ABDP) [9] and see the depth and breadth of its coverage of the subject at such an early date. By the end of that decade, most of the basic principles of the field as we know it today were established. However, knowing the principles is one thing; exploiting them is quite another. Lasers themselves were still in their infancy in the 1960s, and the story of the last half-century has been of laser physics and nonlinear optics growing in parallel, each fertilising the other, and with technological advances always making yesterday's dreams today's reality. The basic themes have constantly been re-orchestrated using superior materials, in new combinations and environments, and especially on increasingly challenging time and distance scales.

Nowhere has progress been more crucial than in materials technology where developments in ultrahigh purity crystal growth and the fabrication of complex semiconductor heterostructures, 'designer' nonlinear media, photonic crystals, and quantum-confined materials have been astonishing. Consider, for example, the early history of the optical parametric oscillator (or OPO). The possibility of generating tunable coherent light using OPOs was understood as early as 1962, and the first OPO was demonstrated in 1965 [10].

---

[16] The distinction between structural and optical isotropy is addressed in Chapter 5.

But for the next 20 years, OPO research proceeded only slowly, primarily because the non-linear materials required to create viable devices were not available. But by the 1980s, the necessary fabrication techniques had been developed, and sophisticated OPO sources based on high-grade crystals are now in widespread use.

A similar story can be told about quasi-phase matching (QPM). Suggested as a phase-matching technique in ABDP (see Fig. 10 in [9]), QPM could not be implemented reliably for many years, because it was not possible at that time to grow thin-layered stacks of nonlinear crystals with alternating orientation. But once periodic poling techniques were perfected [11], QPM quickly became standard, to the point where, today, some researchers regard birefringent phase matching (BPM) as outdated.

Of the nonlinear optical phenomena discovered in this chapter, almost all the frequency-mixing processes are now commonplace. It turns out that a remarkably large number of nonlinear optical techniques are based in one way or another on the optical Kerr effect, which appeared in Eq. (1.25). As mentioned in Section 1.10, this leads to self-phase modulation in which a beam or pulse of light changes its own local refractive index. This effect is the physical basis of self-focusing, spatial and temporal soliton formation, optical bistable devices, as well as standard techniques for optical pulse compression and laser mode-locking.

As noted already, cross-fertilisation between nonlinear optics and laser physics has been an important feature of developments in both areas; indeed the two fields are so closely related, that it is sometimes hard to draw a clear line between them. The technique known as Kerr-lens mode-locking (KLM) is an example of this synergy [12]. The spatial properties of a laser beam are altered by the optical Kerr effect because the refractive index near the beam centre, where the intensity is greatest, is increased relative to that at the periphery. If the effect becomes strong enough to overcome diffractive beam spreading, catastrophic self-focusing can occur. But at a weaker level, the effect on an optical pulse can be exploited to remarkable effect. If it can be arranged that, at an aperture placed somewhere in the beam path, the beam size decreases as the intensity increases, the transmission through the aperture will be greatest at the peak of the pulse. But this is precisely the recipe for a passive mode-locking device, and this is basically how KLM mode-locking works. The technique is widely used today and, in Ti:sapphire lasers (among others), it enables pulses shorter than 10 fs to be generated. Of course, a modest amount of energy compressed into 10 fs corresponds to high peak power, and this is just what is needed to excite optical nonlinearities. In one scenario, after further compression into the sub-10 fs regime (using other nonlinear optical techniques), the conditions are met for high harmonic generation (HHG) in which a wide range of harmonics is generated in a gas jet serving as the nonlinear medium. Detailed analysis (see Chapter 10) indicates that a train of ultra-broadband pulses is generated with a periodicity that is twice the optical carrier period. Under optimum conditions, the signal emerges as a train of attosecond pulses,[17] or possibly as a single pulse. At the time of writing, this particular problem lies at the research frontier.

---

[17] 1 attosecond (1 as) $= 10^{-18}$ s.

The principle behind high harmonic generation is radically different from anything envisaged in the 1960s, and this process certainly involves new principles. They may have been called 'the swinging sixties', but it is very doubtful if anyone in that decade had the idea of tearing an electron out of an atom in an intense optical field, and then using the field to drive the electron back to collide with the parent ion [13]. Of course, Eq. (1.24) is a series expansion, and series expansions always break down when signal strengths are high. It would therefore not be surprising to find that other strong-field effects involve new principles too. The foundations for many strong-field coherent effects in optics were in fact laid in the classic papers of McCall and Hahn on self-induced transparency (SIT) [14], and the effects studied there can certainly not be accommodated within Eq. (1.24). However, some of the principles and much of the terminology and mathematical notation in McCall and Hahn's work were inherited from nuclear magnetic resonance, which was discovered by Isidor Rabi in 1938.[18] The terms 'Rabi frequency' and 'Rabi flopping' are now commonplace in optics (see Chapter 9). Coherent nonlinear optical processes studied more recently include electromagnetically-induced transparency (EIT) [15,16], lasing without inversion (LWI) [17] and slow light [18,19]. Both EIT and LWI arise when there are three atomic levels coupled by two EM fields, one strong and the other weak, and the strong field transition modifies the properties of the weak field transition. Rabi flopping on the strong transition is crucial in both cases, as it is in SIT, but the way in which the process is exploited is novel.

Most of the topics mentioned in the foregoing discussion will be treated in more detail later: OPOs in Chapter 2, self-focusing in Chapter 5, pulse compression and solitons in Chapter 7, SIT and EIT in Chapter 9, and high harmonic generation and attosecond pulses in Chapter 10. But nonlinear optics is a huge subject, and no book, let alone an introductory text like this one, can include everything of importance in the field.

Problems

1.1 Consider second harmonic generation in lithium niobate for a fundamental field whose (vacuum) wavelength is 1.064 μm. If the effective refractive indices are 2.2339 and 2.2294 for the fundamental and second harmonic fields, respectively, find the coherence length.

1.2 Assuming that the DC polarisation in Fig. 1.5 is uniform throughout the crystal, and that edge effects are neglected, show that the open circuit voltage between the plates is $V_{open} = P_{dc}d/\varepsilon\varepsilon_0$, where d is the plate separation and ε is the DC dielectric constant of the crystal. What is the polarity of the induced voltage? Determine the magnitudes and directions of the vectors $\mathbf{P}$, $\mathbf{E}$ and $\mathbf{D}$ in the crystal.

1.3 Prove Eq. (1.17).

1.4 Show that Eq. (1.19) can never be satisfied in a normally dispersive medium (one where the refractive index rises monotonically with frequency).

1.5 A phase-matching configuration is possible in beta-barium borate (BBO) in which two separate non-collinear beams at 1.064 μm generate a second harmonic beam at

---

[18] Rabi received a Nobel Prize for this achievement in 1944.

0.532 μm. If the effective refractive indices at the two wavelengths are 1.65500 and 1.55490, respectively, find the angle between the two fundamental beams.

1.6 What are the dimensions of the three coefficients in Eq. (1.25)?

1.7 (a) Show that the refractive index change implied at the end of Section 1.7 is $\Delta n \cong \chi^{(2)} A_0 / n$.

   (b) Prove Eq. (1.26).

Frequency mixing

2.1 Introduction and preview

In this chapter, we will consider several of the basic frequency-mixing processes of nonlinear optics. The simplest is second harmonic generation (SHG), and we will take this as our basic example. In SHG, a second harmonic wave at 2ω grows at the expense of the fundamental wave at ω. As we will discover, whether energy flows from ω to 2ω or vice versa depends on the phase relationship between the second harmonic field and the nonlinear polarisation at 2ω. Maintaining the optimal phase relationship is therefore of crucial importance if efficient frequency conversion is to be achieved.

The SHG process is governed by a pair of coupled differential equations, and their derivation will be our first goal. The analysis in Section 2.2 is somewhat laborious, although the material is standard, and can be found in many other books, as well as in innumerable PhD theses. The field definitions of Eqs (2.4)–(2.5) are used repeatedly throughout the book, and are worth studying carefully.

In Section 2.3, the coupled-wave equations are solved for SHG in the simplest approximation. The results are readily extended to the slightly more complicated cases of sum and difference frequency generation, and optical parametric amplification, which we move on to in Section 2.4. The important case of Gaussian beams is treated in Section 2.5, where the effect of the Gouy phase shift in the waist region of a focused beam is highlighted.

2.2 Electromagnetic theory

We start from Maxwell's equations for a non-magnetic medium with no free charges namely

$$\text{div } \mathbf{D} = 0; \quad \text{div } \mathbf{B} = 0; \quad \text{curl } \mathbf{E} = -\partial \mathbf{B}/\partial t; \quad \text{curl } \mathbf{B} = \mu_0 \partial \mathbf{D}/\partial t. \qquad (2.1\text{a–d})$$

By taking the curl of Eq. (2.1c) and the time derivative of Eq. (2.1d), the $\mathbf{B}$ vector can be eliminated. This leads to[1]

$$-\nabla^2 \mathbf{E} + \frac{1}{c^2} \frac{\partial^2 \mathbf{E}}{\partial t^2} = -\mu_0 \frac{\partial^2 \mathbf{P}}{\partial t^2} \qquad (2.2)$$

---

[1] In writing this equation, a term $-\varepsilon_0^{-1}$ grad div $\mathbf{P}$ has been dropped from the right-hand side. While div $\mathbf{D}$ is always zero in a charge-free medium, it cannot necessarily be assumed that div $\mathbf{E}$ and div $\mathbf{P}$ are zero for waves propagating in a medium that is not optically isotropic. However, in practice one is usually justified in assuming that div $\mathbf{E}$ and div $\mathbf{P}$ are small.

where $c = (\mu_0 \varepsilon_0)^{-1/2}$ is the velocity of light in vacuum. Ignoring the vectorial nature of the fields for the moment, and specialising Eq. (2.2) to the case of plane waves travelling in the $+z$-direction, the equation reduces to

$$-\frac{\partial^2 E}{\partial z^2} + \frac{1}{c^2}\frac{\partial^2 E}{\partial t^2} = -\mu_0 \frac{\partial^2 (P^{\mathrm{L}} + P^{\mathrm{NL}})}{\partial t^2} \qquad (2.3)$$

where P^{L} and P^{NL} are the respective linear and nonlinear contributions to the polarisation, and the electric field and the polarisation are so far both real quantities.

We now express the field and polarisation waves in complex form, writing them (as is conventional in nonlinear optics) as sums of discrete frequency components with angular frequencies ω_n namely[2]

$$E(z,t) = \tfrac{1}{2}\sum_n \hat{E}_n \exp\{i\omega_n t\} + \text{c.c.} = \tfrac{1}{2}\sum_n \tilde{E}_n \exp\{i(\omega_n t - k_n z)\} + \text{c.c.} \qquad (2.4)$$

$$P(z,t) = \tfrac{1}{2}\sum_n \hat{P}_n \exp\{i\omega_n t\} + \text{c.c.} = \tfrac{1}{2}\sum_n \tilde{P}_n \exp\{i(\omega_n t - k_n z)\} + \text{c.c.} \qquad (2.5)$$

where c.c. is short for 'complex conjugate'. As we will discover shortly, the natural choice of angular wave number[3] is $k_n = n_n \omega_n / c$, where n_n is the refractive index at ω_n.

Although the time dependence of the optical carrier waves has been separated out, the complex amplitudes $\hat{E}_n, \hat{P}_n, \tilde{E}_n$ and $\tilde{P}_n$ in Eqs (2.4) and (2.5) may themselves vary in time in the case of pulsed excitation. However, we will assume for the moment that the waves are strictly monochromatic, so the complex amplitudes are time independent. The two parameters carrying tildes are distinguished from those carrying carets by the fact that the rapidly varying spatial factors $\exp\{-ik_n z\}$ have been removed.[4]

In terms of $\tilde{E}_n$ and $\tilde{P}_n$, the linear polarisation is $\tilde{P}_n^{\mathrm{L}} = \varepsilon_0 \chi^{(1)} \tilde{E}_n$ where $\chi^{(1)}$ is the linear susceptibility. Substituting Eqs (2.4)–(2.5) into Eq. (2.3), and taking the linear part of the polarisation over to the left-hand side, we obtain

$$2ik_n \frac{\partial \tilde{E}_n}{\partial z} + \left(k_n^2 - \frac{\omega_n^2 (1 + \chi^{(1)})}{c^2}\right)\tilde{E}_n = \frac{\omega_n^2 \tilde{P}_n^{\mathrm{NL}}}{c^2 \varepsilon_0} \qquad (2.6)$$

where the second z-derivative of $\tilde{E}_n$ has been neglected. The natural choice of angular wave number is clearly the one that makes $\partial \tilde{E}_n / \partial z = 0$ in the linear approximation (where $\tilde{P}_n^{\mathrm{NL}} = 0$), and we duly set[5]

$$k_n = \frac{\omega_n}{c}\sqrt{1 + \chi^{(1)}} = \frac{n_n \omega_n}{c} \qquad (2.7)$$

---

[2] Some authors omit the factors of $\frac{1}{2}$ in Eqs (2.4) and (2.5), and some use $\exp\{i(k_n z - \omega_n t)\}$ instead of $\exp\{i(\omega_n t - k_n z)\}$. These choices affect many subsequent equations, so you should always check what conventions are being used when looking for definitive results in books or journal articles. See Appendix A for a review of different conventions.

[3] The term *angular wave number* is used throughout this book for $k = 2\pi/\lambda$.

[4] Note that the complex amplitudes are not uniquely defined in terms of the real field and polarisation. See Appendix E for further discussion of this issue.

[5] Of course, we could have made a different choice for k_n at the price of making $\tilde{E}_n$ a function of z.

where $n_n = \sqrt{1 + \chi^{(1)}}$ is the refractive index at frequency ω_n. The susceptibility $\chi^{(1)}$ will turn out to be frequency dependent, which is a manifestation of dispersion.

Equation (2.6) reduces, with the help of Eq. (2.7), to

$$\frac{\partial \tilde{E}_n}{\partial z} = -i \frac{\omega_n}{2c n_n \varepsilon_0} \tilde{P}_n^{\mathrm{NL}}. \tag{2.8}$$

This simple differential equation is the basis for all the coupled-wave equations of nonlinear optics. It contains critical information about the effect of different phase relationships between the nonlinear polarisation and the field. If the polarisation wave has a phase lead of $\pi/2$ with respect to the field (i.e. the arguments of $\tilde{P}_n^{\mathrm{NL}}$ and $i\tilde{E}_n$ are the same), Eq. (2.8) indicates that $\tilde{E}_n$ will grow, whereas, if the sign of the phase difference is reversed (a phase *lag* of $\pi/2$), the field will decay. If, on the other hand, the polarisation and field are in phase, the amplitude of the field will be unchanged, but the phase rather than the amplitude of the field will change under propagation. In this case, it is the angular wave number and hence the refractive index that are modified.

It must be stressed that any time dependence of the complex amplitudes has been ignored in the derivation of Eq. (2.8), so a more sophisticated equation will have to be developed before we can study the nonlinear optics of short pulses. This is done in Section 7.2.

2.3 Second harmonic generation

2.3.1 Coupled-wave equations

While discussing second harmonic generation, we will use the labels $n = 1, 2$ for the fundamental and harmonic waves, so $\omega_1 = \omega$ and $\omega_2 = 2\omega$. The first task is to find $\tilde{P}_1^{\mathrm{NL}}$ and $\tilde{P}_2^{\mathrm{NL}}$ for use in Eq. (2.8), which can be done by evaluating $\varepsilon_0 \chi^{(2)} E^2$ for a field of the form

$$E(z, t) = \tfrac{1}{2}[\hat{E}_1 \exp\{i\omega_1 t\} + \hat{E}_2 \exp\{i\omega_2 t\}] + \text{c.c.} \tag{2.9}$$

The relevant contributions to the second harmonic polarisation (see Eq. 2.5) are $\hat{P}_2^{\mathrm{NL}} = \tfrac{1}{2}\varepsilon_0 \chi^{(2)} \hat{E}_1^2$ and $\hat{P}_1^{\mathrm{NL}} = \varepsilon_0 \chi^{(2)} \hat{E}_2 \hat{E}_1^*$ or, equivalently $\tilde{P}_2^{\mathrm{NL}} = \tfrac{1}{2}\varepsilon_0 \chi^{(2)} \tilde{E}_1^2 \exp\{i\Delta kz\}$ and $\tilde{P}_1^{\mathrm{NL}} = \varepsilon_0 \chi^{(2)} \tilde{E}_2 \tilde{E}_1^* \exp\{-i\Delta kz\}$ in terms of parameters that are slowly varying in space. Notice in the second two equations the appearance of the phase-mismatch parameter Δk defined by

$$\Delta k = k_2 - 2k_1 = 2\omega(n_2 - n_1)/c. \tag{2.10}$$

This originates from the slightly different spatial dependencies of the nonlinear polarisations and the fields they are driving.

Substituting the expressions for the nonlinear polarisation into Eq. (2.8), and relabelling $\chi^{(2)}$ as χ^{SHG} yields

$$\frac{\partial \tilde{E}_2}{\partial z} = -i\frac{\omega_2}{4cn_2}\chi^{\text{SHG}}\tilde{E}_1^2 e^{i\Delta kz} \tag{2.11}$$

$$\frac{\partial \tilde{E}_1}{\partial z} = -i\frac{\omega_1}{2cn_1}\chi^{\text{SHG}}\tilde{E}_2\tilde{E}_1^* e^{-i\Delta kz}. \tag{2.12}$$

These are the coupled-wave equations governing second harmonic generation. Notice the slight asymmetry in the factors on the right-hand side.[6] This can be traced back to the factor of 2 in the cross-term of $(a+b)^2 = a^2 + 2ab + b^2$; Eq. (2.12) contains a cross-term in the electric fields on its right-hand side, whereas Eq. (2.11) does not.

2.3.2 The low-depletion approximation

The simplest solution to the coupled-wave equations is obtained when the energy conversion to the second harmonic is so small that the fundamental field remains essentially undepleted. In this 'low-depletion' approximation, Eq. (2.12) can be ignored, and $\tilde{E}_1$ treated as a constant in Eq. (2.11), which is now readily integrated. Setting $\tilde{E}_2 = 0$ at $z = 0$, one obtains[7]

$$\tilde{E}_2(z) = -i\frac{\omega_2\chi^{\text{SHG}}\tilde{E}_1^2}{4cn_2}\frac{e^{i\Delta kz}-1}{i\,\Delta k} = -ie^{i\Delta kz/2}\frac{\omega_2\chi^{\text{SHG}}\tilde{E}_1^2}{4cn_2}z\,\text{sinc}\left\{\frac{\Delta kz}{2}\right\} \tag{2.13}$$

where $\text{sinc}\,x = (\sin x)/x$. Equation (2.13) is easily written in terms of intensities (in $\text{W}\,\text{m}^{-2}$) rather than fields (in $\text{V}\,\text{m}^{-1}$) with the help of the formula

$$I_n(z) = \tfrac{1}{2}c\varepsilon_0 n_n \left|\tilde{E}_n(z)\right|^2 \qquad (n = 1, 2). \tag{2.14}$$

The second harmonic intensity is readily shown to be

$$I_2(z) = \frac{(\omega_2\,\chi^{\text{SHG}}I_1)^2}{8\varepsilon_0 c^3 n_2 n_1^2}\,\text{sinc}^2\left\{\frac{\Delta kz}{2}\right\}. \tag{2.15}$$

This is a crucial result. Notice that I_2 varies as the square of both the nonlinear coefficient, and the fundamental intensity I_1. Moreover, if $\Delta k = 0$, the sinc function goes to unity, and I_2 grows as the square of the distance too. On the other hand, when $\Delta k \neq 0$, Eq. (2.15) becomes

$$I_2(z) = \frac{(\omega_2\chi^{\text{SHG}}I_1)^2}{2\varepsilon_0 c^3 n_2 n_1^2}\frac{\sin^2\{\Delta kz/2\}}{\Delta k^2}. \tag{2.16}$$

In this case, the intensity oscillates between zero and a maximum of

$$I_2^{\max} = \frac{1}{2\varepsilon_0 c^3 n_2 n_1^2}\left(\frac{\omega_2\chi^{\text{SHG}}I_1}{\Delta k}\right)^2. \tag{2.17}$$

---

[6] Had we written $2\omega_1$ instead of ω_2 in Eq. (2.11), it would have been even easier to overlook the asymmetry.

[7] It is a common notational convention to replace χ^{SHG} with $2d_{\text{eff}}$; see Chapter 4.

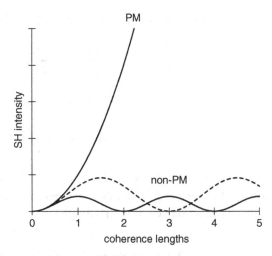

Fig. 2.1 Second harmonic intensity as a function of propagation distance under phase-matched (PM) and two non-PM
conditions. The distance scale applies to the non-PM solid line. The dotted line is for a coherence length that is 1.5
times greater.

The intensity reaches this value for the first time at $z = L_{\mathrm{coh}}$ where

$$L_{\mathrm{coh}} = \frac{\pi}{|\Delta k|} \tag{2.18}$$

which is known as the *coherence length*.[8]

Within the present plane-wave analysis, the conversion efficiency is given by $I_2(z)/I_1(0)$.
However, in the more realistic case where the fundamental and harmonic are in the form of
beams (see Section 2.5), a more sophisticated definition that takes the beam profiles into
account is needed.

Figure 2.1 shows the behaviour of $I_2(z)$ under phase-matched (PM), and two non-phase-
matched (non-PM) conditions. The distance axis is in units of coherence length for the
non-phase-matched case plotted as a solid line; the dotted line is for a coherence length 1.5
times larger. Notice that I_2^{max} increases as Δk gets smaller, in line with Eq. (2.17).

2.3.3 Manley–Rowe relations for second harmonic generation

Naturally, when significant energy is transferred to the second harmonic, the fundamental
field will be depleted, and the approximation on which Eqs (2.15)–(2.17) are based will break
down. It is then necessary to solve Eqs (2.11) and (2.12) as a coupled pair. In the present
context, we simply note that these equations ensure that the total energy (and therefore
the intensity) of the field is conserved in the energy exchange between fundamental and

---

[8] Sadly, there is no generally agreed definition of 'coherence length' in nonlinear optics. Definitions with $1, 2, \pi$,
and 2π in the numerator of Eq (2.18) can be found in different textbooks. The definition in Eq (2.18) is the one
that was used almost universally in the 1960s, and I see no reason to change it.

harmonic. For it is easy to show (see Problem 2.1) that Eqs (2.11) and (2.12) ensure that

$$\frac{dI_1}{dz} = -\frac{dI_2}{dz}.$$ (2.19)

In terms of the number of photons Φ_n in each beam, it is also easy to show that

$$\frac{d\Phi_1}{dz} = -2\frac{d\Phi_2}{dz}$$ (2.20)

which is a statement that for every photon gained (or lost) at 2ω, two photons are lost (or gained) at ω. Equation (2.20), and its counterpart in more general situations (see Section 2.4.1), are known as Manley–Rowe relations.

2.3.4 Interlude: third harmonic generation

We interrupt the discussion of second harmonic generation in order to highlight the close analogy with third harmonic generation (THG), which is its third-order counterpart. The equations governing the two processes are in fact so similar that one can almost write them down by guesswork. The key difference is of course that SHG requires a non-centrosymmetric medium, whereas THG can be observed in centrosymmetric media including amorphous solids, liquids and gases.

Third harmonic generation arises from the frequency combination $\omega + \omega + \omega = 3\omega$ and, for the purposes of this section, we write $\omega_1 = \omega$ and $\omega_3 = 3\omega$. The analogue of Eq. (2.13) is

$$\tilde{E}_3(z) = -ie^{i\Delta kz/2}\frac{3\omega\frac{1}{4}\chi^{\mathrm{THG}}\tilde{E}_1^3}{2cn_3}z\,\mathrm{sinc}\left\{\frac{\Delta kz}{2}\right\}$$ (2.21)

where the mismatch parameter is now defined as $\Delta k = k_3 - 3k_1$. The corresponding expression for the third harmonic intensity is

$$I_3(z) = \frac{(3\omega\,\chi^{\mathrm{THG}})^2}{4\varepsilon_0^2 c^4 n_3 n_1^3}I_1^3\left(\frac{\sin\{\Delta kz/2\}}{\Delta k}\right)^2$$ (2.22)

which is analogous to Eq. (2.16) for SHG. Notice the cubic relationship between the fundamental and the third harmonic intensities, a feature that is clearly evident in the experimental results of Fig. 2.2.

When $\Delta k = 0$, Eq. (2.22) indicates that $I_3(z)$ (like $I_2(z)$) increases as the square of the distance of propagation. However, it is much more difficult to achieve phase matching for THG than for SHG. Under *non*-phase-matched conditions, it is clear from Eq. (2.22) that $I_3(z)$ varies as $(\chi^{\mathrm{THG}}/\Delta k)^2$. Since the nonlinear coefficient and the mismatch parameter are both proportional to the number density of the nonlinear medium, this implies that the harmonic intensity is *independent* of the number density. It follows that the harmonic signals generated by gases and solids may be of comparable magnitude.

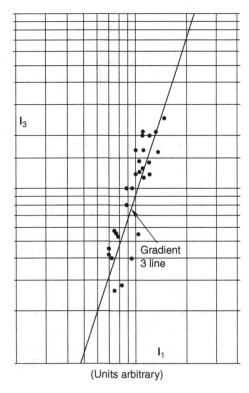

(Units arbitrary)

Fig. 2.2   Third harmonic intensity as a function of fundamental intensity plotted on a log-log scale and showing wide experimental scatter.

2.3.5 The antenna picture

After this brief excursion into the third harmonic case, we return to second harmonic generation. A useful way of understanding the SHG process is to write the second harmonic field at the exit surface of the nonlinear medium as the sum of contributions from all upstream points in the nonlinear medium, shown as slices of width dz' in Fig. 2.3. Each slice acts as an optical frequency antenna, radiating its contribution to the total harmonic field in the forward direction.[9] In this spirit, we write

$$\hat{E}_2(z) \equiv \tilde{E}_2(z)\exp\{-ik_2 z\} = -\frac{i\omega}{\varepsilon_0 c n_2}\int_0^z dz'\,\hat{P}_2(z')\underbrace{\exp\{-ik_2(z-z')\}}_{\text{harmonic propagation}} \qquad (2.23)$$

where $\hat{P}_2(z')dz'$ represents the strength of the antenna, and the final term is the phase change of the harmonic field generated at z' in propagating from z' to z. Substituting

---

[9] The antennae actually radiate in all directions, but the phasing is correct for constructive interference only in the forward direction.

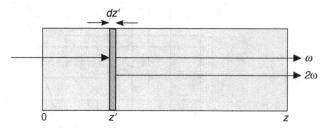

Fig. 2.3   The antenna model. The harmonic signal at the exit of the crystal is viewed as the sum of contributions from upstream sources.

$\hat{P}_2(z') = \frac{1}{2}\varepsilon_0 \chi^{\text{SHG}} \tilde{E}_1^2 \exp\{-i2k_1 z'\}$ into the equation yields

$$\tilde{E}_2(z) = -\frac{i\omega\chi^{\text{SHG}}\tilde{E}_1^2}{2cn_2} \int_0^z dz' \exp\{i\,\Delta k z'\} \tag{2.24}$$

which leads immediately to Eq. (2.13).

Within the present plane-wave approach, the antenna picture of Eq. (2.23) merely offers an alternative interpretation of Eq. (2.13). It will, however, be used in the next section to provide important insight into the SHG process, and will prove to be an indispensable tool for the treatment of focused beams in Section 2.5.

2.3.6 Phase relationships and quasi-phase matching

We will now focus our attention on the way in which the phase of the harmonic field develops in second harmonic generation, particularly when the process is not phase matched. To display the results, we will use a graphical representation of the complex line integral of Eq. (2.24), known as a *phasor diagram*. This is constructed by treating the integral as a sum of infinitesimal vectors in the complex plane, as shown in Fig. 2.4.

If the fundamental field $\tilde{E}_1$ is real and positive, Eq. (2.24) indicates that $\tilde{E}_2$ is negative imaginary under perfect phase-matching conditions ($\Delta k = 0$). In this case, the phasor diagram consists of a straight line running in the negative imaginary direction from O towards A in Fig. 2.4. The length of the line (representing the second harmonic field amplitude) increases linearly with propagation distance, so the intensity goes as the square of the distance, as in Fig. 2.1.

In the presence of dispersion, the phasor plot becomes a curve. For normal dispersion ($n_2 > n_1$), Eq. (2.10) indicates that $\Delta k > 0$, and the effect (through Eq. 2.24) is to cause the phasor line to veer away from OA, and to trace out the circular arc OB.[10] The change in the phase-matching conditions does not affect the length of the phasor path, only its shape. At point B in the figure, the net field $\tilde{E}_2(z_B)$ is given by the straight line OB. It is easy to see from Eq. (2.24) that the (obtuse) angle between OA and the tangent to the

---

[10] The distinction between normal dispersion ($n_2 > n_1$) and anomalous dispersion ($n_2 < n_1$) is discussed in Section 6.2 of Chapter 6.

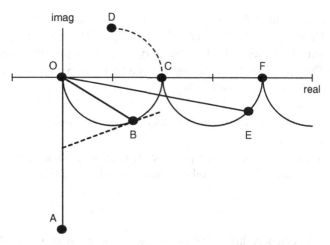

Fig. 2.4 Phasor representation of second harmonic generation in the complex plane. OA represents the phase-matched harmonic field, OBC the phasor path for the first coherence length under non-phase-matched conditions, and CD part of the second coherence length. For quasi-phase-matching, the path is OBCEF.

circle at B (dotted in Fig. 2.4) is $\Delta k z$, and it can also be shown (see Problem 2.3) that $OB = OC \sin\{\Delta k z/2\}$.

After one coherence length, the phasor plot reaches C, where $z = L_{coh} = \pi/|\Delta k|$. The net harmonic field is real at this point, so it is tempting to conclude that the fundamental and harmonic fields are in some sense 'in phase'. However, the slowly varying envelopes only provide phase information in combination with the appropriate exponential factors from Eq. (2.4) and, in any case, one should always treat statements like 'in phase' or 'in quadrature' with caution for two waves of different frequency. (After all, if the fundamental and second harmonic fields interfere constructively at one point, they will interfere destructively half a cycle later.) What counts in SHG is the relative phases of the harmonic field and the associated nonlinear polarisation, which is basically the 2ω component of the fundamental field squared. After one coherence length, this has changed by π, which is why the phasor path at C in Fig. 2.4 is running in the positive imaginary direction, in exact opposition to its initial direction at O. The amplitude of the second harmonic field $\tilde{E}_2(z = L_{coh})$ has clearly reached a maximum at C and, thereafter, the phasor path will continue via D, returning to O (zero harmonic field) after two coherence lengths.

But Fig. 2.4 invites us to consider another possibility. Suppose that at C, the sign of the nonlinear coefficient could be changed. The phasor path would then reverse direction, proceeding via E to F, where the amplitude of the harmonic field would be double what it was at C. And if, at F, the sign could be changed again, and the reversal repeated after every coherence length, the harmonic field could be increased without limit.

Fortunately, changing the sign of the nonlinear coefficient at regular intervals can be realised experimentally by growing a non-centrosymmetric crystal as a series of slabs, each slab one coherence length thick, and with the orientation of alternate slabs reversed. Imagine a loaf of sliced bread with every other slice turned upside down. This method of overcoming refractive index mismatch was in fact proposed by Armstrong, Bloembergen, Ducuing and Pershan (see Fig. 10 of [9]). However, it was many years before it was reliably achieved in the laboratory by using a process known as *periodic poling* [11]. This approach to overcoming the phase-mismatch problem is called *quasi-phase matching* (QPM) to distinguish it from *birefringent phase matching* (BPM) [5]. Notice from Fig. 2.4 that the second harmonic field generated in QPM is phased in a direction close to the real axis, roughly in quadrature to the harmonic field in BPM under phase-matched conditions.

The poling period of a periodically poled material is defined as the thickness of one pair of inverted slabs or, in other words, two coherence lengths. Hence

$$L_{\text{pole}} = 2L_{\text{coh}} = \frac{2\pi}{|\Delta k|}. \tag{2.25}$$

In practice, this is typically a few tens of microns, so the slices of bread are very thin! The intensity growth of a second harmonic under QPM conditions is shown in Fig. 2.5; notice that the curve follows the non-phase-matched curve for the first coherence length, but the paths separate as soon as the polarity is reversed. The step-like nature of the QPM line follows directly from the geometry of Fig. 2.4, although that figure represented the field whereas Fig. 2.5 shows the intensity. Had the field rather than the intensity been plotted, the steps would have been of equal height.

It is clear from Figs 2.4 and 2.5 that the second harmonic field under QPM is smaller than for perfect phase matching. After integral numbers of coherence lengths, the relevant factor is $2/\pi = 0.637$, which is the ratio of the diameter of a circle to half its circumference and a result that follows immediately from Fig. 2.4. Since intensity goes as the square of the field, the QPM intensity after two coherence lengths (at F in Fig. 2.4) is four times that at C, and the intensity ratio is $(2/\pi)^2 = 0.405$.

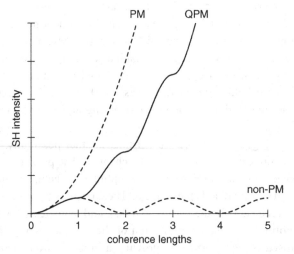

Fig. 2.5   Second harmonic intensity as a function of propagation distance for quasi-phase-matching (QPM). The dotted lines are reproduced from Fig. 2.1 for comparison.

2.3.7 Additional features

A number of important features have been left out of the simple treatment given in this section. In particular:

(A) The plane wave is an idealisation that is never realised in practice. SHG experiments are normally conducted using beams of finite lateral extent that may pass through a focus in or near the nonlinear medium.
(B) Most nonlinear materials used for SHG are birefringent, and at least one of the coupled fields is usually an extraordinary wave. As we shall see in the next chapter, this means that the energy will angle off to the side, reducing the efficiency of the nonlinear interaction.
(C) The nonlinear coefficient that couples the fields is a tensor, and the fundamental and harmonic fields are often orthogonally polarised.

Item (A) is discussed in Section 2.5 below, and items (B) and (C) in Chapters 3 and 4. Notwithstanding its limitations, the simplified treatment we have presented has revealed the crucial importance of phase matching for SHG, and indeed for many other second-order frequency-mixing processes, some of which we consider next.

2.4 General three-wave processes

2.4.1 Coupled-wave equations and the Manley–Rowe relations

We now move on to the general case where three waves with frequencies linked by the equation $\omega_1 + \omega_2 = \omega_3$ interact through the second-order nonlinearity.[11] As for SHG, we need to find a set of coupled-wave equations based on Eq. (2.8), but this time we must use a field of the form

$$E(z,t) = \tfrac{1}{2}[\hat{E}_1 \exp\{i\omega_1 t\} + \hat{E}_2 \exp\{i\omega_2 t\} + \hat{E}_3 \exp\{i\omega_3 t\}] + \text{c.c.} \tag{2.26}$$

We work out $\varepsilon_0 \chi^{(2)} E^2$ as before and, of the terms that then appear, those of special interest are $\tilde{P}_3^{\text{NL}} = \varepsilon_0 \chi^{(2)} \tilde{E}_1 \tilde{E}_2 \exp\{i\Delta kz\}$, $\tilde{P}_2^{\text{NL}} = \varepsilon_0 \chi^{(2)} \tilde{E}_3 \tilde{E}_1^* \exp\{-i\Delta kz\}$ and $\tilde{P}_1^{\text{NL}} = \varepsilon_0 \chi^{(2)} \tilde{E}_3 \tilde{E}_2^* \exp\{-i\Delta kz\}$. Substitution of these expressions into Eq. (2.8) yields the three coupled differential equations.

$$\frac{\partial \tilde{E}_3}{\partial z} = -i\frac{\omega_3}{2cn_3}\chi^{\text{TWM}}\tilde{E}_1\tilde{E}_2 e^{i\Delta kz} \tag{2.27}$$

---

[11] In this section, the only numerical significance of the subscripts is that ω_3 is the highest frequency of the three. Frequencies ω_2 and ω_3 are no longer the second and third harmonics as in earlier sections.

$$\frac{\partial \tilde{E}_2}{\partial z} = -i \frac{\omega_2}{2cn_2} \chi^{\text{TWM}} \tilde{E}_3 \tilde{E}_1^* e^{-i\Delta kz} \tag{2.28}$$

$$\frac{\partial \tilde{E}_1}{\partial z} = -i \frac{\omega_1}{2cn_1} \chi^{\text{TWM}} \tilde{E}_3 \tilde{E}_2^* e^{-i\Delta kz} \tag{2.29}$$

where the label TWM stands for three-wave mixing, and $\Delta k = k_3 - k_2 - k_1$.[12] These equations are evidently close relatives of Eqs (2.11)–(2.12). The equation paralleling Eq. (2.19) is

$$\frac{dI_1}{dz} + \frac{dI_2}{dz} = -\frac{dI_3}{dz} \tag{2.30}$$

which is a statement of conservation of energy among the three interacting fields. For the photon numbers, Eq. (2.20) generalises to

$$\frac{d\Phi_1}{dz} = \frac{d\Phi_2}{dz} = -\frac{d\Phi_3}{dz} \tag{2.31}$$

where the set of identities within Eq. (2.31) are the Manley–Rowe relations for the general three-wave mixing process. In physical terms, they state that, for every photon gained (or lost) at ω_3, one is lost (or gained) at both ω_1 and ω_2.

Notice that the derivation of Eqs (2.27)–(2.29) was based on selecting 'terms of special interest', and discarding all the others. But what allowed us to throw other terms away? What about terms at $2\omega_1$ for example? Or at $\omega_1 + \omega_3$? The answer is that we are assuming that the $\omega_1 + \omega_2 = \omega_3$ process is the only one close to phase matching, while other frequency combinations are not phase matched and can therefore be ignored. The assumption is normally justified in practice, which highlights the importance of phase matching in nonlinear frequency mixing.

While Eqs (2.27)–(2.29) govern all nonlinear processes of this type, there are a number of different cases depending on which waves are present initially and their relative strengths and phases. In brief these are:

- Sum frequency generation (SFG), where two waves at ω_1 and ω_2 add to give $\omega_3 = \omega_1 + \omega_2$. As the wave at ω_3 is enhanced, the waves at ω_1 and ω_2 are depleted (or vice versa if and when the process goes into reverse). When one of the incident frequencies is much stronger than the other, SFG may be termed 'parametric up-conversion', the thought being that the strong wave converts the weak wave to the sum frequency.
- Second harmonic generation is a special case of SFG. In quantitative work, it makes a difference whether there is a single fundamental beam, or two distinct beams at the same frequency that might for instance be non-collinear. The analysis of Section 2.3 applies to the former case. In the latter case, the formulae for SFG should be used with $\omega_1 = \omega_2$; see Section C2.3 of Appendix C for further discussion.

---

[12] This is our standard definition of the mismatch parameter in which the high-frequency value comes first; see Appendix A.

- Difference frequency generation (DFG), where ω_3 and one of the two lower frequencies (say ω_2) subtract to give $\omega_3 - \omega_2 = \omega_1$. Once ω_1 is present, it also subtracts from ω_3 to interact with ω_2. DFG is just SFG in reverse, so ω_3 is depleted, and both ω_1 and ω_2 grow. It may perhaps come as a surprise that the wave at ω_2 increases in strength, but this becomes obvious as soon as one writes the DFG process in terms of photons, i.e. $\hbar\omega_3 \to \hbar\omega_1 + \hbar\omega_2$; see also Eq. (1.23). Every time a photon at ω_1 is created, so is a photon at ω_2. Again, when one of the incident frequencies is much stronger than the other, DFG may be termed 'parametric down-conversion'.
- Optical parametric amplification, which is based on exactly the same equations as DFG. The only difference is that the two lower frequencies ω_1 and ω_2 are normally very weak initially, and one or both may even be generated from noise. As shown below, this affects the nature of the solution to Eqs (2.27)–(2.29).

Although general analytical solutions to Eqs (2.27)–(2.29) can be obtained, approximate solutions are usually more than adequate, and more manageable too. As for SHG, approximations are possible if at least one of the incident fields is sufficiently strong that the change in its magnitude resulting from the nonlinear interaction can be neglected. However, the results exhibit some quite subtle features, and care is necessary.

2.4.2 Modified field variables

For the discussion of specific three-wave processes such as sum and difference frequency generation, we will write the coupled-wave equations in terms of the modified field variables [20]

$$\tilde{A}_i = \sqrt{\frac{n_i}{\omega_i}}\,\tilde{E}_i. \tag{2.32}$$

Equations (2.27)–(2.29) then simplify to

$$\frac{\partial \tilde{A}_3}{\partial z} = -i\eta \tilde{A}_1 \tilde{A}_2 e^{i\,\Delta kz}; \quad \frac{\partial \tilde{A}_2}{\partial z} = -i\eta \tilde{A}_3 \tilde{A}_1^* e^{-i\,\Delta kz}; \quad \frac{\partial \tilde{A}_1}{\partial z} = -i\eta \tilde{A}_3 \tilde{A}_2^* e^{-i\,\Delta kz} \tag{2.33}$$

where the coupling coefficient common to all three equations is

$$\eta = \sqrt{\frac{\omega_3 \omega_2 \omega_1}{n_3 n_2 n_1}}\,\frac{\chi^{\text{TWM}}}{2c}. \tag{2.34}$$

The simplification occurs because $\left|\tilde{A}_i\right|^2$ is directly proportional to the photon flux F_i (photons per area per time) through the relation

$$F_i = \frac{I_i}{\hbar\omega_i} = \frac{\frac{1}{2}c\varepsilon_0 n_i \left|\tilde{E}_i\right|^2}{\hbar\omega_i} = \frac{c\varepsilon_0}{2\hbar}\left|\tilde{A}_i\right|^2. \tag{2.35}$$

Since the coupled-wave equations govern processes where gain (or loss) of a photon at ω_3 is accompanied by the corresponding loss (or gain) of photons at ω_1 and ω_2, the nearer one can get to photon language, the simpler the equations are bound to become.

2.4.3 Sum and difference frequency generation

In the case of SFG, if ω_1 and ω_2 are roughly equal in strength and the energy conversion to ω_3 is small, depletion of both incident waves can be ignored, which leads to

$$\tilde{A}_3(z) = -i\,\mathrm{e}^{i\Delta kz/2}\eta\tilde{A}_1\tilde{A}_2\, z\,\mathrm{sinc}\left\{\frac{\Delta kz}{2}\right\}. \tag{2.36}$$

This equation is clearly a close analogue of Eq. (2.13), and it is worth comparing the formulae carefully to appreciate the close relationships between then.

A parallel result can be obtained for DFG ($\omega_3 - \omega_2 = \omega_1$) if the energy conversion to ω_1 is sufficiently small that both $\tilde{A}_3$ and $\tilde{A}_2$ can be treated as constants. In this case, one obtains

$$\tilde{A}_1(z) = -i\,\mathrm{e}^{-i\Delta kz/2}\eta\tilde{A}_3\tilde{A}_2^*\, z\,\mathrm{sinc}\left\{\frac{\Delta kz}{2}\right\}. \tag{2.37}$$

However, care must be exercised with this equation, as it is only valid under these very restricted conditions. The more general case where $\tilde{A}_2$ is allowed to vary is treated in the following section.

2.4.4 Optical parametric amplification

In an optical parametric amplifier (OPA), the high frequency ω_3 is called the *pump*, the lower frequency of primary interest is called the *signal*, and the remaining frequency is called the *idler*. In practice, the signal usually has a higher frequency than the idler, and we will identify the signal with the wave at ω_2 for the purpose of the present discussion.

We now examine the behaviour of Eqs (2.33) in the case where depletion of the pump beam at ω_3 can be neglected, so that $\tilde{A}_3$ is a constant, and the first of the three coupled equations can be ignored. Both $\tilde{A}_1$ and $\tilde{A}_2$ are allowed to vary, and combining the remaining two differential equations yields

$$\frac{\partial^2 \tilde{A}_2}{\partial z^2} = g^2 \tilde{A}_2 - g\Delta k\, \tilde{A}_1^* e^{-i\Delta kz} \tag{2.38}$$

$$\frac{\partial^2 \tilde{A}_1}{\partial z^2} = g^2 \tilde{A}_1 - g\Delta k\, \tilde{A}_2^* e^{-i\Delta kz} \tag{2.39}$$

where g is the *parametric gain coefficient* given by

$$g = \eta\tilde{A}_3 = \frac{\chi^{\mathrm{TWM}}\tilde{A}_3}{2c}\sqrt{\frac{\omega_3\omega_2\omega_1}{n_3\, n_2\, n_1}}. \tag{2.40}$$

For convenience, we will assume that χ^{TWM} and $\tilde{A}_3$ (and hence also g) are real quantities.[13]

---

[13] The nonlinear coefficient will be real provided the fields are well away from material resonances. The pump field can be taken to be real without any loss of generality.

Remarkably, Eqs (2.38)–(2.39) possess the analytic solution[14]

$$\tilde{A}_2(z)e^{i\Delta kz/2} = \tilde{A}_2(0)\cosh g'z + i\left(\tfrac{1}{2}\Delta k\tilde{A}_2(0) - g\tilde{A}_1^*(0)\right)\left(\frac{\sinh g'z}{g'}\right) \tag{2.41}$$

$$\tilde{A}_1(z)e^{i\Delta kz/2} = \tilde{A}_1(0)\cosh g'z + i\left(\tfrac{1}{2}\Delta k\tilde{A}_1(0) - g\tilde{A}_2^*(0)\right)\left(\frac{\sinh g'z}{g'}\right) \tag{2.42}$$

where $g' = \sqrt{g^2 - (\Delta k/2)^2}$. The important general conclusion to draw from Eqs (2.41)–(2.42) is that, provided $g > \Delta k/2$, the equations support the exponential growth of both signal and idler with an effective gain coefficient of g'. On the other hand, when $g < \Delta k/2$, g' becomes imaginary, the hyperbolic functions become trigonometric functions, and the solutions are oscillatory.

One can gain valuable insight (and have some fun too) by plotting graphs of Eqs (2.41)–(2.42). A good starting point is the case where $\tilde{A}_1(0) = 0$ (zero initial idler), for which the signal and idler fields are

$$\tilde{A}_2(z) = \tilde{A}_2(0)e^{-i\Delta kz/2}\left(\cosh g'z + i\frac{\Delta k\sinh g'z}{2g'}\right) \tag{2.43}$$

$$\tilde{A}_1(z) = \tilde{A}_2^*(0)e^{-i\Delta kz/2}\left(-i\frac{g\sinh g'z}{g'}\right). \tag{2.44}$$

Figure 2.6 traces the growth of $\left|\tilde{A}_1(z)\right|^2$ and $\left|\tilde{A}_2(z)\right|^2$ in the case where $\tilde{A}_2(0) = 1$, $g = 3$ and $\Delta k = 4$. The distance scale is in units of the nominal coherence length ($= \pi/\Delta k$),

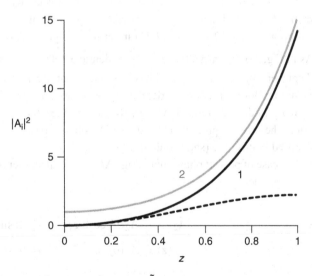

Fig. 2.6 Solutions of Eqs (2.41)–(2.42) for $g = 3$, $\Delta k = 4$, and $\tilde{A}_2(0) = 1$. The distance scale is in units of the nominal coherence length ($= \pi/\Delta k$).

[14] This result is quoted by Yariv in Chapter 17 of the 3rd edition of *Quantum Electronics* [20], albeit in a slightly different notation, and with a different definition of Δk.

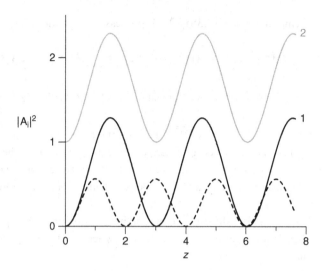

Fig. 2.7   Solutions of Eqs (2.41)–(2.42) for $g = 1.5$, and other parameters as Fig. 2.6.

although that is of little significance for these parameter values. The two curves are identical in shape, and are separated by 1 unit in the vertical direction (see Problem 2.5). After 5 units of distance, they rise to roughly 2×10^8, and are virtually indistinguishable. At this huge gain factor, the assumption that the pump is undepleted will probably have been infringed.

The dramatic effect of reducing g from 3 to 1.5 is shown in Fig. 2.7. Since g is now less than $\Delta k/2$, g' is imaginary, and it makes sense to define $\Delta k'/2 = ig' = \sqrt{(\Delta k/2)^2 - g^2}$ and to write Eqs (2.43) and (2.44) in terms of trigonometric functions (see Problem 2.6). As in Fig. 2.6, the curve for $\left|\tilde{A}_1(z)\right|^2$ is identical to the one for $\left|\tilde{A}_2(z)\right|^2$ apart from a vertical displacement of 1 unit, but this is the only significant feature the two figures have in common. Both curves are evidently now oscillatory and restricted to low values. Notice the increase in the coherence length by the factor of $\Delta k/\Delta k' = 4/\sqrt{7} \approx 1.5$. The dotted line traces the solution predicted by Eq. (2.37), which is based on conditions that are clearly not fulfilled in this case (see Problem 2.7).

The case of perfect phase matching ($\Delta k = 0$) is also interesting. Equations (2.41)–(2.42) then become

$$\tilde{A}_2(z) = \tilde{A}_2(0) \cosh gz - i\tilde{A}_1^*(0) \sinh gz \qquad (2.45)$$

$$\tilde{A}_1(z) = \tilde{A}_1(0) \cosh gz - i\tilde{A}_2^*(0) \sinh gz \qquad (2.46)$$

which highlight the fact that, when $\tilde{A}_1(0) \neq 0$, the initial phase relationship of the three waves has a crucial effect on the evolution, particularly when $\tilde{A}_2$ and $\tilde{A}_1$ are of comparable magnitude. This is a standard feature of multi-wave nonlinear interactions whenever all the interacting fields are present at the outset.

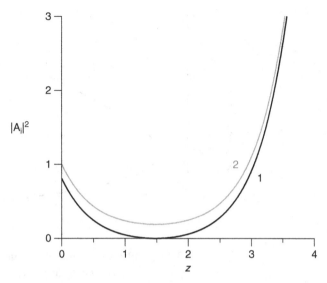

Solutions of Eqs (2.41)–(2.42) for $g = 1$, $\Delta k = 0$, $\tilde{A}_2(0) = 1$ and $\left|\tilde{A}_1(0)\right| = 0.9$. It is left to the reader to discover the argument of $\tilde{A}_1(0)$; see Problem 2.8.

An example is shown in Fig. 2.8 in which $g = 1$, $\tilde{A}_2(0) = 1$ (and real), and $\left|\tilde{A}_1(0)\right| = 0.9$. It is left to the reader to deduce the phase of $\tilde{A}_1(0)$, and to work out what would happen if $\left|\tilde{A}_1(0)\right|$ were increased to 1 (see Problem 2.8).

2.4.5 Tuning characteristics

In a typical practical application, the idler alone might be absent initially (as in Eqs (2.43)–(2.44) above), or the signal and idler might both grow from noise. In the former case, the idler phase adjusts automatically to optimise the interaction. In the latter case, the division of the pump photon energy between signal and idler is determined by the phase-matching conditions, as explained in Chapter 1. It follows that varying the phase-matching conditions enables the optical parametric amplification process to be tuned.

A thorough discussion of *angle tuning* using birefringent phase matching (BPM) must wait until we have studied the properties of birefringent media in Chapter 3, and the tensor nature of the nonlinear coefficients in Chapter 4. However, the principle of tunability can be illustrated in the case of quasi-phase matching (QPM), which was discussed in Section 2.3.6. As we saw there, if the mismatch parameter of a nonlinear interaction is Δk, *effective* phase matching can be achieved by using a periodically poled medium with a poling period of $L_{\text{pole}} = 2\pi / |\Delta k|$. Figure 2.9 illustrates how the signal and idler wavelengths in an optical parametric amplifier depend on the poling period in lithium niobate. For a given period (on the abscissa), the two wavelengths can be read off the ordinate. The upper and lower branches of the curve meet at 1.568 μm, which is twice the pump wavelength of 784 nm.

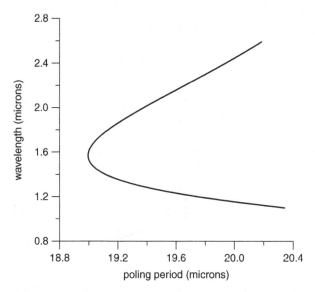

Fig. 2.9   Tuning curve for optical parametric amplification as a function of poling period. The lower and upper branches of the curve trace the signal and idler wavelengths.

Unfortunately, once a poled sample has been fabricated, the period is fixed, so continuous tuning is not possible using this technique. However, samples are frequently made with several different periodicities on the same wafer, so a limited degree of tunability can be achieved by switching from one to the next.

2.4.6 Optical parametric oscillators

Figure 2.6 showed that an optical parametric amplifier delivers gain at the signal and idler frequencies in the presence of a suitably strong pump, while it was shown in the previous section how the process can be tuned. Taken together, these ingredients provide a solution to one of the key problems of the laser era: how to create *tunable* coherent radiation. Of course, broadband lasers can sometimes be tuned to a limited degree, but the tuning range is invariably restricted, and the freedom to generate any desired wavelength at will is lacking. But now, all that is necessary to achieve this long-sought-after goal is to put an optical parametric amplifier within an optical cavity, where the nonlinear crystal replaces the amplifying medium of a traditional laser. Such a device is called an optical parametric oscillator (or OPO).

As noted already, the first OPO was demonstrated in 1965 [10], although 20 years or so were to elapse before efficient and reliable OPOs could be realised. In designing an OPO, one has to decide which of the three participating frequencies (pump, signal and idler) the cavity mirrors should be designed to reflect and which they should transmit. Obviously the pump beam has to reach the nonlinear medium somehow or other. Various different OPO configurations can be employed but, for longitudinal pumping, it is clearly necessary for at least one of the mirrors to transmit the pump beam. As far as the signal and idler are

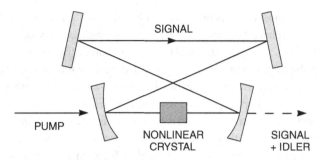

Fig. 2.10 Schematic diagram of an optical parametric oscillator.

concerned, there are two options: in a 'singly-resonant' OPO, the mirrors reflect the signal and transmit the idler, whereas in a 'doubly-resonant' device, both beams are strongly reflected.

Broadly speaking, singly-resonant OPOs tend to be better behaved, because problems with conflicting mode conditions that afflict doubly-resonant systems are avoided. The point at issue here is that an optical cavity of length L exhibits Fabry–Perot resonances at angular frequencies given by $\omega_q = q\pi c/L$ where q is a (normally very large) integer. Even if the signal is the only frequency of the three constrained by this feature, the smooth tunability of an OPO may be somewhat impaired. But if both the signal and idler must simultaneously satisfy a mode condition of this kind, the problems become much more severe.

A schematic diagram of a longitudinally pumped singly-resonant OPO is shown in Fig. 2.10. If birefringent phase matching is employed, tuning can be achieved by rotating the crystal in conjunction with other tuning elements. Short-pulse OPOs can be realised by synchronous pumping with a mode-locked laser, which involves accurate matching of the length of the OPO cavity to the periodicity of the mode-locked pulse train of the pump. In this case, fine tuning can be achieved by harnessing group velocity dispersion (GVD). Because of GVD (see Chapter 6), the cavity round-trip period of the signal pulse will be wavelength dependent. But the period is tied to that of the pump so, if the physical length of the cavity is changed, the signal wavelength changes too to maintain synchronism. In other words, the signal shifts to a wavelength where the group time delay in the nonlinear medium cancels the change in the mirror spacing.

Further material on OPOs can be found in [8] and [21].

2.5 Focused beams

2.5.1 Introduction

Everything said so far has been based on a plane-wave analysis in which the transverse variation of the interacting fields has been ignored. But in the real world, one is invariably working with beams, and this affects nonlinear interactions in several different ways.

The most obvious property of a beam is that it is more intense in the centre line than on the periphery. So, since the efficiency of nonlinear frequency mixing depends on intensity (see, e.g. Eqs 2.16 and 2.22), the conversion efficiency will be higher in the centre than at the edges, and we should expect output beams in nonlinear frequency conversion to be narrower than input beams. In a rough-and-ready approach to SHG with finite beams, one could simply apply the formulae of Sections 2.3 and 2.4 at arbitrary points across the fundamental wavefront, and work out the associated profile of the harmonic. Beams expand and contract as they propagate, of course, and that would need to be taken into account. However, if a beam passes through a focus in the course of a nonlinear interaction, the crucial phase relationships between the various frequency components will be upset by a feature known as the *Gouy phase shift*, and a more sophisticated analysis is needed. The origin and effects of the Gouy phase on nonlinear frequency mixing will be studied in the following sections.

2.5.2 Gaussian beams and the Gouy phase

The simplest beam-like profile is the Gaussian beam. This is an exact solution of the paraxial wave equation (PWE), which is derived in Appendix G, and reads

$$\frac{\partial \tilde{E}}{\partial z} = -\frac{i}{2k} \nabla_T^2 \tilde{E} \tag{2.47}$$

where $\nabla_T^2 = \partial^2/\partial x^2 + \partial^2/\partial y^2$ is the transverse Laplacian operator. In most practical situations, the PWE is an excellent approximation to Maxwell's equations, which means in turn that a Gaussian beam is very close to an exact solution of Maxwell's equations.

The spatial structure of a Gaussian beam is given by

$$\tilde{E}(r, z) = \left(\frac{E_0}{1 - i\zeta} \right) \exp \left\{ \frac{-r^2}{w_0^2(1 - i\zeta)} \right\} \tag{2.48}$$

where $r = \sqrt{x^2 + y^2}$ is the transverse displacement from the axis, w_0 is a measure of the beam width in the focal plane ($z = 0$), and $\zeta = z/z_R$ is the axial distance[15] normalised to the *Rayleigh length* defined by[16]

$$z_R = \frac{\pi w_0^2}{\lambda} = \tfrac{1}{2} k w_0^2. \tag{2.49}$$

Note that λ and k in this equation are the wavelength and angular wave number in the medium of refractive index n.

The basic geometry of a Gaussian beam is shown in Fig. 2.11. The beam is symmetrical about the origin (or focal point) at $x = y = z = 0$ where the transverse profile is narrowest.

---

[15] The symbol ζ is lower-case zeta, the sixth letter of the Greek alphabet.
[16] The confocal parameter $b = 2z_R$ is sometimes defined instead of the Rayleigh length.

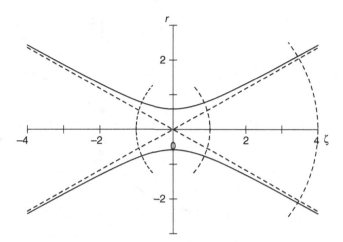

Fig. 2.11 Geometry of a Gaussian beam. The distance scale is in units of the Rayleigh length. The solid lines track the half-maximum intensity contour.

The beam converges towards the focal point from the upstream side, and expands away from it in an entirely symmetrical manner on the downstream side. In the vicinity of the focus, the beam exhibits a waist-like structure that extends between roughly $z = \pm z_R$ (or $\zeta = \pm 1$). The field strength on-axis ($r = 0$) is given by the first term in Eq. (2.46), which can be rewritten

$$\tilde{E}(0, z) = \left(\frac{E_0}{\sqrt{1 + \zeta^2}} \right) \exp\{i\phi_{\text{Gouy}}\}. \tag{2.50}$$

Squaring the modulus of $\tilde{E}(0, z)$ yields the on-axis intensity, namely

$$\left| \tilde{E}(0, z) \right|^2 = \frac{E_0^2}{1 + \zeta^2}. \tag{2.51}$$

The phase term in Eq. (2.50) is the crucial Gouy phase mentioned at the start of this section and given by

$$\phi_{\text{Gouy}} = \tan^{-1} \zeta. \tag{2.52}$$

As shown in Fig. 2.12, ϕ_{Gouy} rises from $-\pi$ at $\zeta = -\infty$ to $+\pi$ at $\zeta = +\infty$. In combination with the fast-varying spatial term $\exp\{-ikz\}$, its effect is to reduce the effective value of k in the waist region, and correspondingly to increase the local wavelength. As we will discover shortly, the Gouy phase has a profound effect on nonlinear frequency-mixing processes in focused beams.

The beam profile transverse to the axis is governed by the second term in Eq. (2.48) which can be recast in the form

$$\exp \left\{ \frac{-r^2}{w_0^2(1 - i\zeta)} \right\} = \exp \left\{ -\frac{r^2}{w^2} - i\frac{\zeta r^2}{w^2} \right\}. \tag{2.53}$$

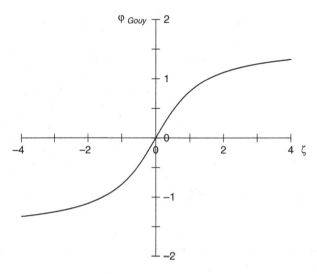

The Gouy phase as a function of normalised distance.

The parameter $w(\zeta)$, which governs the variation of the beam width along the axis, is given by

$$w(\zeta) = w_0\sqrt{1 + \zeta^2}. \tag{2.54}$$

It is easy to show that the beam diameter measured at half-maximum intensity is given by

$$d(\zeta) = w(\zeta)\sqrt{2\log 2} = 1.18w_0\sqrt{1 + \zeta^2}. \tag{2.55}$$

Clearly w_0 is the value of w at the focal point where the diameter is $d(0) = 1.18w_0$. The solid lines in Fig. 2.11 trace the half-maximum intensity contours.

The complex term in Eq. (2.53) shows how the phase of the beam varies as one moves away from the axis, and this enables the *radius of curvature R* of the beam wavefront to be found. By using the well-known intersecting chords theorem, it can be shown (see Problem 2.9) that the link between the phase $\psi(r, \zeta) = -\zeta r^2/w^2$ and R is

$$r^2 \cong (2R)\left(\frac{\lambda\,|\psi(r,\zeta))|}{2\pi}\right). \tag{2.56}$$

and it follows that

$$R \cong z_R\left(\zeta + \frac{1}{\zeta}\right) = z + \frac{z_R^2}{z}. \tag{2.57}$$

This result indicates firstly that $R \to \infty$ as $z \to 0$ (the wavefront is plane at $z = 0$), and secondly that $R \to z$ as $z \to \infty$, so when $z \gg z_R$ the centre of curvature tends to the focal point. The curved dotted lines in Fig. 2.11 indicate the curvature of the wavefront at $\zeta = \pm 1$ and $\zeta = 4$.

2.5.3 Harmonic generation with Gaussian beams

The theory of harmonic generation with Gaussian beams is quite involved [22, 23], so we will merely present the most important results for second and third harmonic generation (SHG and THG), and argue that they are reasonable.

The first important result to emerge from a detailed study of the problem is that the Rayleigh length of the harmonic beam (at $q\omega$) is the same as that of the fundamental beam that created it. This means that the dimensionless distance measure $\zeta = z/z_R$ is the same for both beams. Furthermore, since $z_R = \frac{1}{2}kw_0^2$ with $k = n(q\omega)/c$ (see Eq. 2.49), it follows that the harmonic beam is narrower than the fundamental. For SHG, the factor is roughly[17] $\sqrt{2}$, a result that is entirely reasonable given that $I_2 \sim I_1^2$ (see Eq. 2.15). In the general case of qth harmonic generation, $I_q \sim I_1^q$ and the narrowing factor is roughly $\sqrt{q}$.

The most insightful way of presenting the expression for the harmonic field is through the antenna picture of Eq. (2.23). Assuming the conversion efficiency is small so that depletion of the fundamental field can be ignored, the formulae for the harmonic field in the SHG and THG cases are[18]

$$\tilde{E}_2(z) = -i\left(\frac{2\omega\chi^{SHG}}{4cn_2}\right)\left(\frac{E_0^2}{(1-i\zeta)}\exp\left\{-\frac{2k_1r^2}{2z_R(1-i\zeta)}\right\}\right)z_R\int_{-\infty}^{\infty}\frac{d\zeta'\exp\{i\Delta_2\zeta'\}}{(1-i\zeta')}$$

(2.58)

$$\tilde{E}_3(z) = -i\left(\frac{3\omega\chi^{THG}}{8cn_3}\right)\left(\frac{E_0^3}{(1-i\zeta)}\exp\left\{-\frac{3k_1r^2}{2z_R(1-i\zeta)}\right\}\right)z_R\int_{-\infty}^{\infty}\frac{d\zeta'\exp\{i\Delta_3\zeta'\}}{(1-i\zeta')^2}$$

(2.59)

where $\Delta_q = \Delta k_q z_R = (k_q - qk_1)z_R$ and $\zeta' = z'/z_R$.

These equations contain critical information about the effect of the Gouy phase on optical harmonic generation. An important point to appreciate is that, in each case, the overall Gaussian beam structure of the harmonic beam resides in the central bracket, while the final integral governs the phase-matching characteristics including the role played by the Gouy phase. Think carefully about the $(1-i\zeta)^{q-1}$ factor in the denominators of these integrals. Why is this factor raised to the $(q-1)$th power? The reason is that it represents the *difference* between the strength and phase of the polarisation antenna driving the harmonic field, and the harmonic field itself. Notice how this factor introduces a π phase difference between the extremes of the integral for the SHG case, and a 2π phase difference for THG.

The key properties of the phase-matching integrals are summarised in Table 2.1, which also includes the general case of qth harmonic generation.[19] The most important feature to emerge is that all the integrals are zero when $\Delta k_q > 0$. This means that when a fundamental beam is focused through a medium that exhibits normal dispersion *and is of infinite extent*,

---

[17] The factor will only be exact under phase-matched conditions when the refractive indices at the fundamental and harmonic frequencies are the same.
[18] Throughout this section, the indices 2 and 3 refer once more to the second and third harmonic waves.
[19] The leading factor of $-i$ from Eqs (2.58)–(2.59) has been included with the integrals in Table 2.1.

Table 2.1 Properties of the integral of Eqs (2.53)–(2.54).

		$\Delta k_q > 0$	$\Delta k_q = 0$	$\Delta k_q < 0$
SHG	$-i \int\limits_{-\infty}^{\infty} \dfrac{d\zeta'\, \exp\{i\,\Delta k_2 z'\}}{(1 - i\zeta')}$	0	$-i\pi$	$-2\pi i\, \exp\{\Delta k_2 z_R\}$
THG	$-i \int\limits_{-\infty}^{\infty} \dfrac{d\zeta'\, \exp\{i\,\Delta k_3 z'\}}{(1 - i\zeta')^2}$	0	0	$-2\pi i\, \lvert\Delta k_3\rvert\, z_R\, \exp\{\Delta k_3 z_R\}$
qth HG	$-i \int\limits_{-\infty}^{\infty} \dfrac{d\zeta'\, \exp\{i\,\Delta k_q z'\}}{(1 - i\zeta')^{q-1}}$	0	0	$-2\pi i\, (\lvert\Delta k_q\rvert\, z_R)^{q-2}\, \exp\{\Delta k_3 z_R\}$

no second or third harmonic will be generated, or in fact, any higher harmonic either. In practice, 'infinite' means that the focus lies many Rayleigh lengths from either the front or back surface.

Further insight into these properties can be gained by displaying the integrals as phasor diagrams, as explained in Section 2.3.5 above. Some examples are shown in Fig. 2.13. Finite limits were used to create some of the plots in this diagram to prevent them from running off the page. For example, curve a shows the phasor plot of $-i \int\limits_{-\zeta}^{\zeta} d\zeta'(1 - i\zeta')^{-1}$ with $\zeta = 16$, which applies to SHG under perfect phase matching. For infinite limits, the length of the curve is itself infinite, so the two arms of the curve would disappear off the right-hand side of the page. The grey arrow connecting the end points at $\zeta' = \pm 16$ represents the value of the integral. This is clearly negative imaginary, and inspection shows that its magnitude is roughly 3, which is consistent with its value of $-i\pi$ for infinite limits (see Table 2.1). The locations of $\zeta' = \pm 1$ on the phasor plot are also indicated. The π phase difference between the extremes of the integral (associated with the factor $(1 - i\zeta)^{-1}$) is clearly evident in the 180 degree swing in the direction of the curve in the figure.

Curve b in Fig. 2.13 represents SHG for $\Delta k_2 z_R = +0.02$ between integration limits of $\zeta' = \pm 1000$. The two end-points converge to the same point on the real axis in this case, so the value of the integral is tending to zero, as indeed it does for any positive value of Δk_2.

Curves c and d are for THG with $\Delta k_3 z_R = 0$ and -0.5, respectively, and corresponding limits of $\zeta' = \pm 10$ and ± 1000. In the former case, the end-points converge towards $+1$ on the real axis; the curve is in fact a perfect circle. Because the factor in the denominator of the integral is $(1 - i\zeta)^{-2}$, the phase difference between the ends of the curve is now 2π. For curve d, Table 2.1 gives the value of the integral for infinite limits as $-2\pi i\, \lvert\Delta k_3\rvert\, z_R\, \exp\{\Delta k_3 z_R\}$, which is $-1.905i$ in this case.

Finally, it is reassuring to see that Eqs (2.58) and (2.59) are consistent with earlier results in the plane-wave limit. For instance, if we consider Eq. (2.58) on-axis ($r = 0$), close to the beam waist ($\zeta \simeq 0$), and for a thin nonlinear medium extending from $z' = 0$ to $z' = z$ ($\ll z_R$), the second harmonic field reduces to

$$\tilde{E}_2(z) = -\frac{i(2\omega)\chi^{\mathrm{SHG}} E_0^2}{4cn_2} \int\limits_0^z dz'\, \exp\{i\,\Delta k_2 z'\} \tag{2.60}$$

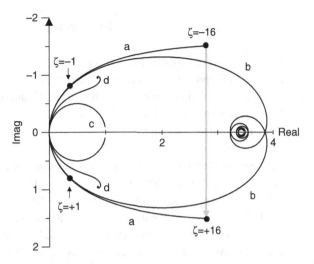

Fig. 2.13 Phasor diagrams for harmonic generation in a Gaussian beam: (a) SHG at $\Delta k_2 = 0$ for a symmetric focus in a medium extending from $\zeta = -16$ to $+16$, (b) $\Delta k_2 z_R = +0.02$, (c) THG for $\Delta k_3 = 0$ in the range $\zeta = -16$ to $+16$, (d) $\Delta k_3 z_R = -0.5$. See text for details.

which is equivalent to Eq. (2.23). If $\Delta k_2 = 0$, the field is clearly negative imaginary, just as in the plane-wave case; see the line OA in Fig. 2.4.

2.5.4 Focused beams: summary

The key conclusion of Section 2.5 is that if a beam is focused into an infinite nonlinear medium exhibiting normal dispersion ($\Delta k > 0$), no net harmonic radiation will be generated. In practical terms, this mathematical statement means that *negligible* harmonic generation will result if the focal point lies within the medium and *many Rayleigh lengths from either the front or back surface*. The physical interpretation of this remarkable result is that the Gouy phase conspires with normal dispersion to ensure that the net contributions to the harmonic from sources upstream and downstream of the focus interfere destructively.

If there is no way of making Δk negative (and there rarely is), the best strategy for circumventing the problem is to restrict the extent of the nonlinear medium within the focal region. This ensures that at least one of the integration limits in Table 2.1 is no longer $\pm\infty$, which means that the cancellation of the contributions to the harmonic signal from the two sides of the focus is no longer complete. For a relatively thick nonlinear medium, the focus can be moved towards the front or the back surface, or even located outside the medium altogether, which may be a sensible measure anyway if one wants to avoid destroying expensive nonlinear crystals. The narrow gas jets used in experiments on high harmonic generation are good examples of thin nonlinear media; see Chapter 10.

Problems

2.1 Use Eq. (2.14) to show that Eqs (2.11) and (2.12) ensure that the total energy in the two waves is conserved in the SHG interaction.

2.2 A square law relationship between the harmonic and fundamental intensities applies for the second harmonic (Eq. 2.16), and a cubic law for the third harmonic (Eq. 2.22). Convince yourself that, in the case of pulsed excitation, the same relationships apply to the energy.

2.3 Assuming perfect phase matching in a second harmonic generation experiment, find the second harmonic intensity when $\chi^{SHG} = 0.5 \, \text{pm V}^{-1}$, the sample thickness is 3 mm, the fundamental vacuum wavelength is 600 nm, and the fundamental intensity is $1 \, \text{GW cm}^{-2}$. Assume that the fundamental and harmonic indices are 1.52. Assume the fundamental beam is effectively undepleted, and comment on the validity of this approximation under these circumstances. (Hint: Watch out for mixed units in this question.)

2.4 Prove that the angle $\widehat{AOB}$ in Fig. 2.4 is equal to $\Delta kz/2$, and hence that $OB = OC \, \sin\{\Delta kz/2\}$.

2.5 Show that, under quasi-phase-matched conditions, the field amplitude during the second coherence length is $E(z) = E_{CL}\sqrt{\frac{1}{2}(5 - 3\cos\{\Delta k(z - L_{coh})/2\})}$ where E_{CL} is the field after one coherence length.

2.6 Convince yourself that $\left|\tilde{A}_2(z)\right|^2 - \left|\tilde{A}_1(z)\right|^2$ in Eqs (2.41)–(2.42) is a constant. What fundamental physical principle does this fact reflect?

2.7 Rewrite Eqs (2.41)–(2.42) in terms of trigonometric functions in the case where $g < \Delta k/2$.

2.8 Under what conditions is Eq. (2.37) a good approximation to Eq. (2.42)?

2.9 Deduce the modulus and argument of the initial idler field in Fig. 2.8, and discover how the figure would change if the modulus were increased from 0.9 to 1. Could this outcome be achieved experimentally?

2.10 Verify that the Gaussian beam solution of Eq. (2.48) is a solution of the paraxial wave equation of Eq. (2.47).

2.11 Prove Eq. (2.56) and sketch a graph of $R(\zeta)$ in Eq. (2.57).

2.12 A beam at 1 μm and its second harmonic (at 500 nm) are phased in such a way that zeros of the fundamental are aligned with positive peaks of the harmonic. After what distance of propagation in air do zeros of the fundamental harmonic coincide with zeros of the harmonic? ($n_{air} - 1 = 2.7412 \times 10^{-4}$ at 1 μm; $n_{air} - 1 = 2.7897 \times 10^{-4}$ at 500 nm) (As noted in Section 2.3.6, phase relationships in SHG are best handled by comparing the harmonic field with the *square* of the fundamental field.)

2.13 Since the cosh and sinh terms in Eqs (2.41)–(2.42) separately satisfy the original differential equations, why are both necessary in the solution? Equally, why does the initial idler field occur in the solution for the signal, and vice versa? After all, Eqs (2.38)–(2.39) seems to suggest that signal and idler are completely decoupled when $\Delta k = 0$.

2.14 Show that curve c in Fig. 2.13 traces a perfect circle.

Crystal optics

3.1 Preview

The central feature of this chapter is the phenomenon of *birefringence*, also known as *double refraction*, which occurs in crystals that are *optically anisotropic*. Given that birefringence is a linear optical effect, why is the whole of Chapter 3 being devoted to it? Firstly, most nonlinear crystals are birefringent, and so one naturally needs to know how light propagates in these media. Secondly, several important nonlinear optical techniques (the most obvious being phase matching) exploit birefringence to achieve their goal. Lastly, the material in this chapter provides essential background for the following chapter on the *nonlinear* optics of crystals.

Section 3.2 is a brief tutorial on crystal symmetry. Crystallography is something of a world on it own, and many people find it a complete mystery. Although the summary offered here is very basic, it should provide everything needed for what comes later.

Section 3.3 discusses the propagation of EM waves in optically anisotropic media, and contains a fairly detailed analysis of birefringence (double refraction), ordinary and extraordinary waves, and associated topics. The treatment is mainly centred on uniaxial media because of their relative simplicity and the fact that most nonlinear crystals are of this type.

Section 3.4 describes how birefringence can be exploited in the construction of wave plates, while Section 3.5 is reserved for a brief mention of biaxial media in which the propagation characteristics are considerably more complicated.

3.2 Crystal symmetry

Two questions about a crystalline medium are of critical importance in nonlinear optics:

(1) Is the crystal centrosymmetric or non-centrosymmetric? In other words, does it or does it not possess inversion symmetry?
(2) Is the crystal optically isotropic or anisotropic? Does it, in other words, exhibit birefringence (double refraction)?

If the crystal is non-centrosymmetric, second-order nonlinear optical effects can of course be observed. Because the second-order tensor governing the nonlinear electromechanical phenomenon of piezoelectricity has identical symmetry properties to $\chi_{ijk}^{(2)}$ in Eq. (3.1), all second-order nonlinear crystals are also piezoelectric.[1] If, on the other hand, the medium is centrosymmetric (i.e. it possesses inversion symmetry), all even-order nonlinear processes are forbidden, and so one is restricted to observing odd-order effects. A simple explanation of this principle was given in Chapter 1. A slightly more general version is based on the tensor/matrix form of Eq. (1.29), in which the second-order polarisation reads

$$P_i = \varepsilon_0 \sum_{jk} \chi_{ijk}^{(2)} E_j E_k \quad (i, j, k = x, y, z). \tag{3.1}$$

If the medium possesses inversion symmetry, the polarisation must change sign if all components of the field change sign. But if both fields change sign in Eq. (3.1), the right-hand side is unchanged. So the polarisation must be zero, which can only be true if all 27 elements of $\chi_{ijk}^{(2)}$ are zero. The way this conclusion arises in a quantum mechanical calculation of the coefficients is discussed in Chapter 8. The argument obviously applies equally to any even-order term in the polarisation of order higher than 2.

We turn now to the second question at the start of this section. As we shall discover shortly, most common nonlinear optical crystals are both non-centrosymmetric and optically anisotropic, but it is important to understand that there is no fundamental connection between the two properties. Some crystals have one property without the other, some have both, and some neither. An optically isotropic crystal will not of course exhibit birefringence and so, even if it is non-centrosymmetric and second-order nonlinear processes are allowed, it will not be possible to employ birefringent phase matching (BPM). But there are of course other methods of phase matching (e.g. quasi-phase matching).

Every crystalline medium belongs to one of 32 point symmetry classes, which are grouped into seven symmetry systems as shown in Table 3.1 [24,25]. Materials in six of the seven systems (containing 27 of the 32 classes) are optically anisotropic, while those in the seventh (the cubic system, containing five classes) are optically isotropic.[2] The seven crystal systems are referred to by name, and the 32 classes by point group.[3] The classes are sometimes referred to by class number, but there is more than one numbering scheme, so caution is necessary. The numbers in Table 3.1 are those used by Cady [25].

The data in column 3 of the table indicate whether or not a crystal possesses inversion symmetry. It can be seen immediately that the presence of birefringence says nothing about whether or not a crystal is non-centrosymmetric. For example, copper sulphate (class 2) is biaxial (the more complicated kind of anisotropy), yet it possesses a centre of symmetry. On the other hand, gallium arsenide (class 31) is optically isotropic, but lacks inversion symmetry.

---

[1] Indeed, the word 'piezoelectric' is sometimes used as a synonym for 'non-centrosymmetric'.

[2] The word 'isotropic' meaning 'the same in all directions', carries two distinct meanings in optics and crystallography depending on whether the reference is to *optical* isotropy or to *structural* isotropy. Gases, liquids and amorphous solids are structurally isotropic, but crystals are by definition structurally anisotropic. On the other hand, media that do not exhibit birefringence are said to be *optically* isotropic, and these include all crystals in classes 28–32 in Table 3.1. But since they are crystals, none is *structurally* isotropic. In the absence of external perturbations, all structurally isotropic media are optically isotropic too.

[3] Each point group possesses particular symmetry properties with respect to a point. Fortunately, it is not necessary to have a detailed understanding of the point group codes listed in Hermann–Mauguin (international) notation in the second column of Table 3.1. Readers are referred to [6] or [7], or to other specialist texts on crystallography.

System/Class No.	Symmetry code	Inversion sym.	Examples
Table 3.1	The 32 crystal classes.		

System/Class No.	Symmetry code	Inversion sym.	Examples
Biaxial crystals			
Triclinic system			
1	1	no	
2	$\bar{1}$	yes	Copper sulphate
Monoclinic system			
3	2	no	
4	m	no	
5	2/m	yes	
Orthorhombic system			
6	2 2 2	no	
7	m m 2	no	LBO, KTP, KTA
8	2/m 2/m 2/m	yes	
Uniaxial crystals			
Tetragonal system			
9	$\bar{4}$	no	
10	4	no	
11	$\bar{4}$ 2 m	no	KDP, ADP, CDA
12	4 2 2	no	Nickel sulphate
13	4/m	yes	
14	4 m m	no	
15	4/m 2/m 2/m	yes	
Trigonal system			
16	3	no	Sodium periodate
17	$\bar{3}$	yes	
18	3 2	no	α-quartz
19	3 m	no	BBO, Lithium niobate
20	$\bar{3}$ 2/m	yes	Calcite
Hexagonal system			
21	$\bar{6}$	no	
22	$\bar{6}$ 2 m	no	Gallium selenide
23	6	no	Lithium iodate
24	6 2 2	no	β-quartz
25	6/m	yes	
26	6 m m	no	Cadmium selenide
27	6/m 2/m 2/m	yes	
Optically isotropic crystals			
Cubic system			
28	2 3	no	Sodium chlorate
29	4 3 2	no	
30	3m = 2/m $\bar{3}$	yes	Pyrite
31	$\bar{4}$ 3 m	no	Gallium arsenide, zinc blende
32	4/m $\bar{3}$ 2/m = m3m	yes	Sodium chloride, diamond

	Anisotropic	Isotropic[4]	All
Table 3.2 Crystal classes with key properties summarised.			
Non-centrosymmetric	18	3	21
Centrosymmetric	9	2	11
All	27	5	32

The statistics summarised in Table 3.2 indicate that 18 of the 32 crystal classes are both non-centrosymmetric and optically anisotropic. Second-order nonlinear processes (such as second harmonic generation) are allowed, and birefringent phase matching can be used to enhance them. As noted earlier, most common nonlinear materials fall into this category, and a number of these are listed against their class in the right-hand column of Table 3.1. Most common nonlinear crystals happen to belong to classes 7, 11 or 19.

The shape and size of a crystal's *unit cell* is defined by three vectors **a**, **b** and **c**, and the associated *abc*-axes are known as the *crystallographic axes*. The *abc*-axes are orthogonal in the orthorhombic, tetragonal, and cubic systems, but are non-orthogonal in the others. A right-handed set of orthogonal *xyz*-axes, known as the *principal dielectric axes* (or sometimes the *physical axes*), is invariably used in addition, but unfortunately the conventions for relating these to the *abc*-axes are not as consistent as one might wish. Even when the crystallographic axes are themselves orthogonal, it cannot necessarily be assumed that *abc* maps to *xyz*. In the orthorhombic system, for instance, the translation might be $bac \rightarrow xyz$. Reference works such as Dmitriev *et al.* [26] should be consulted for detailed information of this kind. We shall return to this confusing feature in Chapter 4.

Interestingly, the geometric structure of a crystal (i.e. the location of the atoms within the lattice) does not in itself determine its symmetry properties. For example, the geometry of gallium arsenide (class 31; see Fig. 3.1) is identical to that of diamond (class 32), but diamond possesses inversion symmetry while gallium arsenide does not. The reason is that the centre of symmetry in diamond lies mid-way between two nearest neighbours, both of which are of course carbon atoms in that case. But nearest neighbours in GaAs are always gallium and arsenic, and since these are different atoms, the inversion symmetry of diamond is lost [27]. Conversely, crystals of the same point symmetry group may have a wide variety of different internal geometries. For example, diamond and sodium chloride are both members of class 32, but the arrangement of atoms in the unit cell is different.

3.3 Light propagation in anisotropic media

3.3.1 Maxwell's equations

We now discuss how birefringence originates, and how light propagates in birefringent media (see also [28,29]). As the first step, we undertake a simple exercise in electromagnetic

---

4 In this context 'isotropic' refers to optical isotropy. See footnote 2 on page 46.

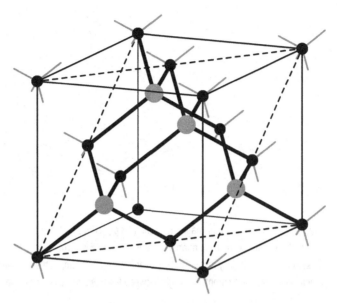

Fig. 3.1 The face-centred cubic structure of GaAs. Ga and As have equivalent status in the lattice so, if each black dot represents Ga, each grey dot represents As (or vice versa). Every atom lies at the centre of a tetrahedron with atoms of the other species at the vertices. If the cube in the figure is divided into eight cubes of half the side length, then four of the eight contain tetrahedra. The stubs indicate bonds to atoms in adjacent cubes. Diamond has the same geometrical structure, but all atoms are carbon.

(EM) wave propagation to determine the constraints imposed by Maxwell's equations on the directions of the four vectors **E**, **D**, **B** and **H**. In most of what follows, we consider plane EM waves of angular frequency ω propagating in the direction of the unit vector $\hat{\mathbf{s}}$. The space-time dependence is then of the form $\exp i(\omega t - \mathbf{k}.\mathbf{r})$ where the wave vector $\mathbf{k}$ ($= k\hat{\mathbf{s}}$) is perpendicular to the wavefront, which is the plane of constant phase. To make the argument simpler (but with no loss of generality), we will temporarily assume that $\mathbf{k}$ points in the $+z$-direction. In this case, the four vectors depend only on z through their space-time dependence $\exp i(\omega t - kz)$. In a non-magnetic medium (in which $\mathbf{B} = \mu_0 \mathbf{H}$), containing no free charges, Maxwell's equations read

$$\text{div } \mathbf{D} = 0; \quad \text{div } \mathbf{B} = 0; \quad \text{curl } \mathbf{E} = -\partial \mathbf{B}/\partial t; \quad \text{curl } \mathbf{B} = -\partial \mathbf{D}/\partial t. \qquad (3.2a\text{–}d)$$

Three simple deductions can now be made:

- Since **B** (like the other three field vectors) is independent of x and y, Eq. (3.2b) reduces to $\partial B_z/\partial z = 0$, which can only be true if $B_z = 0$; hence **B** and **H** lie in the x-y plane, which is the plane of the wavefront.
- Applying the same argument to Eq. (3.2a) leads to a similar conclusion for **D**, which therefore lies in the plane of the wavefront too.
- If for convenience we choose the y-axis to lie along **B** and **H**, the only non-zero term on the left-hand side of Eq. (3.2d) is $-\hat{\mathbf{x}}\partial B_y/\partial z$, and it follows that **D** lies along the x-axis; hence **B**, **H** and **D** all lie in the plane of the wavefront, with **D** perpendicular to **B** and **H**.

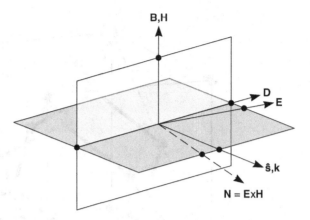

Fig. 3.2 Field vector directions in a non-magnetic medium as prescribed by Maxwell's equations. **D**, **B** and **H** lie in the plane of the wavefront, while **ŝ** and **k** are normal to the wavefront. **E** is perpendicular to **B** and **H**, and so is coplanar with **D**, **ŝ** and **k**. But **E** is not normally parallel to **D** in an anisotropic medium, so the Poynting vector **N** = **E** × **H** is not usually perpendicular to the wavefront. The angle between **N** and **k** is the walk-off angle.

- It remains to consider the direction of **E**, which is restricted by Eq. (3.2c). Since the only non-zero terms in curl **E** are $-\hat{\mathbf{x}}\partial E_y/\partial z + \hat{\mathbf{y}}\partial E_x/\partial z$, and since **B** only has a y-component, it follows that $E_y = 0$. Hence **E** is perpendicular to **B**, but Maxwell's equations impose no other constraint.

The directions of the four field vectors are therefore as shown in Fig. 3.2. The vectors **D**, **B** (and **H**), and **k** are mutually perpendicular, while **E** is coplanar with **D** and **k**, but its direction is otherwise not determined. These conclusions are in no way affected by the assumption that **k** points along the z-axis, which was merely a convenient choice that led quickly to the general conclusion. From here on, the assumption is withdrawn and, in subsequent figures, the direction of **k** is arbitrary once again.

The direction in which light is polarised is related to the directions of the vectors in Fig. 3.2. As mentioned in Chapter 1, the word *polarisation* is used in optics in two senses: in relation to the *polarisation of light,* on the one hand, and to the *polarisation of matter,* on the other. The vector **P** representing the macroscopic electric dipole moment per unit volume of a medium is of course the polarisation in its second sense. Since **D** = ε_0**E** + **P**, the three vectors are coplanar.

Most modern authors associate the *polarisation of light* (i.e. the direction in which it is polarised) with the direction of the electric field vector **E**. However, Born and Wolf [29] comment that it has been more common in the history of optics to associate the polarisation of light with the direction of the *magnetic* field! To avoid ambiguity, they speak of the 'direction of vibration' or the 'plane of vibration', taking care to specify the vector concerned in all cases.

In this book, we define the plane of polarisation of light as the plane perpendicular to **B** and **H**, and containing the four vectors **D**, **E**, **P** and **k**, which is drawn as a horizontal plane in Fig. 3.2.

3.3.2 The constitutive relation

If Maxwell's equations do not define a unique direction for **E**, what does? The answer is the so-called *constitutive relation* of the medium, which determines the magnitude and direction of **E** for a given **D**, and vice versa. In its simplest form, this relation is simply

$$\mathbf{D} = \varepsilon_0 \varepsilon \mathbf{E} \tag{3.3}$$

where the scalar $\varepsilon = 1 + \chi^{(1)}$ is called the *dielectric constant,* and $\chi^{(1)}$ is the linear susceptibility. However, this only applies in an optically isotropic medium. In an *anisotropic* (*non*-isotropic) material, the constitutive relation reads

$$\mathbf{D} = \varepsilon_0 \boldsymbol{\varepsilon} \mathbf{E} \tag{3.4}$$

where $\boldsymbol{\varepsilon}$ is now the *dielectric constant tensor*. Written in matrix form, Eq. (3.4) reads

$$\begin{pmatrix} D_X \\ D_Y \\ D_Z \end{pmatrix} = \varepsilon_0 \begin{pmatrix} \varepsilon_{XX} & \varepsilon_{XY} & \varepsilon_{XZ} \\ \varepsilon_{YX} & \varepsilon_{YY} & \varepsilon_{YZ} \\ \varepsilon_{ZX} & \varepsilon_{ZY} & \varepsilon_{ZZ} \end{pmatrix} \begin{pmatrix} E_X \\ E_Y \\ E_Z \end{pmatrix} \tag{3.5}$$

in which the vectors have become column matrices and the (second-rank) tensor has become the square *dielectric constant matrix*. Capital letters have been used to define an arbitrary set of orthogonal axes; lower case letters will be reserved for the *principal axes* to be introduced shortly.

The key message of Eqs. (3.4) and (3.5) is that **D** and **E** are not necessarily parallel in an anisotropic material, and the tensor/matrix machinery is needed to specify the precise relation between them. The matrix in Eq. (3.5) makes it look as if nine numbers are needed for this purpose but, fortunately, only three are sufficient even for a medium of the most general symmetry. For a start, it is easy to show that the dielectric constant matrix is symmetric, and consequently that a special set of orthogonal axes known as the *principal dielectric axes* always exists with respect to which the matrix is diagonal. We shall write these axes *xyz*. A mathematician would say that 'the dielectric constant matrix can be diagonalised in a similarity transformation'. With respect to the principal axes, Eq. (3.5) becomes

$$\begin{pmatrix} D_x \\ D_y \\ D_z \end{pmatrix} = \varepsilon_0 \begin{pmatrix} \varepsilon_{xx} & 0 & 0 \\ 0 & \varepsilon_{yy} & 0 \\ 0 & 0 & \varepsilon_{zz} \end{pmatrix} \begin{pmatrix} E_x \\ E_y \\ E_z \end{pmatrix} \tag{3.6}$$

where the three diagonal elements are the three *principal dielectric constants* of the medium. Note that the principal axes are not the same as the crystallographic axes (*abc*) which, as noted earlier, are not necessarily even orthogonal. More information about the relationship between the two sets of axes in a given crystal can be found in the next chapter.

From Eq. (3.6), there are clearly three different types of material. Media in which all three dielectric constants are different are called *biaxial*, those in which two of the three are the same (by convention $\varepsilon_{xx} = \varepsilon_{yy}$) are called *uniaxial*, and when all three are the same, Eq. (3.6) reverts to Eq. (3.3), which is the isotropic case. Optically anisotropic media therefore divide into biaxial and uniaxial materials, represented respectively, by classes 1–8 and 9–27 in Table 3.1 Members of classes 28–32 are optically isotropic.

We can now begin to understand some of the characteristic features of EM wave propagation in anisotropic media. The conditions imposed by Maxwell's equations are shown in Fig. 3.2, while the missing condition determining the direction of **E** for a given **D** is provided by Eq. (3.6). A little thought suggests however that these conditions will normally conflict. What happens, for example, if the constitutive relation specifies a direction of **E** that is *not* perpendicular to **B** and **H**, as it almost invariably does? The answer is that such a wave will not propagate. Indeed, it turns out that, for an arbitrary direction of **k**, Maxwell's equations and the constitutive relation are simultaneously satisfied for *only two* directions of **D**. These directions are mutually orthogonal, and define the planes of polarisation of the two allowed EM wave solutions. A wave entering the medium polarised in any other direction will divide into two orthogonally polarised components, which will propagate according to different rules.

3.3.3 The index ellipsoid: ordinary and extraordinary waves

The characteristics of anisotropic media can be displayed pictorially in several different ways. The first of the two representations used in this chapter is called the *index ellipsoid* (or the *wave normal ellipsoid*), and is defined by the formula

$$\frac{x^2}{\varepsilon_{xx}} + \frac{y^2}{\varepsilon_{yy}} + \frac{z^2}{\varepsilon_{zz}} = 1 \tag{3.7}$$

or equivalently

$$\frac{x^2}{n_x^2} + \frac{y^2}{n_y^2} + \frac{z^2}{n_z^2} = 1. \tag{3.8}$$

In Eq. (3.8), the three principal dielectric constants have been replaced by the three associated refractive indices, the significance of which will emerge shortly.

In the general (biaxial) case, the three semi-axes of the ellipsoid n_x, n_y and n_z are all different. However, to keep things simple, we will confine ourselves mostly to *uniaxial media* in which $n_x = n_y = n_o$ (the *ordinary refractive index*) and $n_z = n_e \neq n_o$ (the *extraordinary refractive index*). In this case, Eq. (3.8) reduces to

$$\frac{x^2 + y^2}{n_o^2} + \frac{z^2}{n_e^2} = 1. \tag{3.9}$$

The x and y semi-axes are now of equal length, the x-y section is circular (so the ellipsoid technically becomes a spheroid), and the z-axis is known as the *optic axis*. There are now two distinct cases depending on which index is the larger. If $n_e > n_o$, the spheroid is 'prolate' (like a rugby ball) and the medium is termed 'positive uniaxial'; on the other hand, if $n_e < n_o$, the spheroid is 'oblate' (more like a pumpkin) and the medium is 'negative uniaxial'. Most crystals used in nonlinear optics happen to be negative uniaxial, and the form of the spheroid in this case is shown in Fig. 3.3.

Detailed analysis [28, 29] reveals that sections of the spheroid through the origin O have a particular significance in determining the properties of EM wave propagation. Consider a wave for which the vector **k** is directed at an arbitrary angle θ to the z- (or optic) axis.

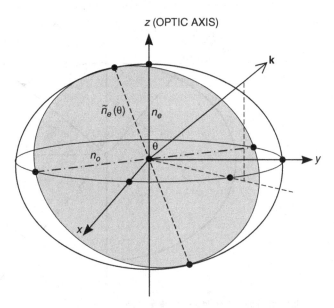

Fig. 3.3   The index ellipsoid for a negative uniaxial medium. The section in the x-y plane is a circle, so the object is technically a spheroid. The shaded ellipse is perpendicular to **k**.

The plane perpendicular to **k** and parallel to the wavefront (shaded in the figure), generally cuts the spheroid in an ellipse, except in the special case when $\theta = 0$ when the section is circular. This ellipse has several important properties:

(a) The axes of the ellipse define the two orthogonal directions of **D** that simultaneously satisfy Maxwell's equations and the constitutive relation. One of the two axes (the major axis in the negative uniaxial case, shown dash-dotted in the figure) always lies in the x-y plane. It corresponds to the direction of polarisation of the *ordinary wave*, and has the same length irrespective of the direction of **k**. The other axis (shown dotted) relates to the *extraordinary wave*, and its length depends on the angle θ between **k** and the z-axis.

(b) The lengths of the semi-axes of the ellipse are the refractive indices of the associated waves, namely n_o for the *ordinary wave*, and $\tilde{n}_e(\theta)$ for the *extraordinary wave*. The value of $\tilde{n}_e(\theta)$ is easily deduced from Fig. 3.4. This shows the projection of the index ellipsoid onto the **k**–$\hat{z}$ plane or, equivalently, the view along the dash-dotted line in Fig. 3.3. The length of the bold line perpendicular to **k** in Fig. 3.4 is the value of $\tilde{n}_e(\theta)$ and, since the section of a spheroid is necessarily an ellipse, it is easy to show that

$$\tilde{n}_e(\theta) = \left\{ \frac{\cos^2\theta}{n_o^2} + \frac{\sin^2\theta}{n_e^2} \right\}^{-1/2}. \tag{3.10}$$

Notice that $\tilde{n}_e(90°) = n_e$ and, for propagation along the optic axis, $\tilde{n}_e(0°) = n_o$, in which case the refractive index for all polarisations is the same.

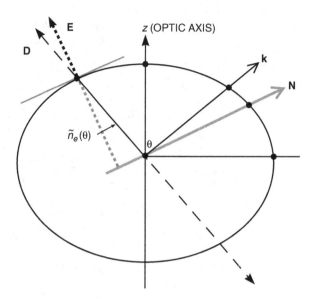

Fig. 3.4   Projection of the index ellipsoid onto the **k-ẑ** plane.

(c) Where the axes of the ellipse (representing the directions of **D**) intersect the spheroid, the normals to the surface define the corresponding directions of **E**. For *ordinary waves*, **D** and **E** are parallel and lie in the x-y plane, and this wave propagates in the usual way. However, the two vectors are generally not parallel for the *extraordinary wave* (see Fig. 3.4), which leads to peculiar propagation properties discussed in Section 3.3.5 below.

The fact that the *extraordinary refractive index* n_e is *not* the same as the refractive index of the extraordinary wave given by $\tilde{n}_e(\theta)$ in Eq. (3.10) is a perennial source of confusion. It is vital to distinguish between the two. The parameter $\tilde{n}_e(\theta)$ is angle-dependent and varies between n_o at $\theta = 0°$ and n_e at $\theta = 90°$; on the other hand, n_e is the maximum value of $\tilde{n}_e(\theta)$ for positive uniaxial media ($n_e > n_o$), and the minimum value of $\tilde{n}_e(\theta)$ for negative uniaxial media ($n_e < n_o$). Values of n_o and n_e as a function of wavelength are quoted for different materials in the literature.

3.3.4 Index surfaces

An alternative method of displaying the properties of an anisotropic medium now presents itself. For every direction of the propagation vector **k**, we measure out the values of n_o and $\tilde{n}_e(\theta)$, creating a double surface of rotation comprising a sphere of radius n_o for the ordinary wave, and a spheroid formed by rotating the ellipse of Eq. (3.10) about the optic axis.[5] For a positive uniaxial medium, the spheroid lies outside

---

[5] Instead of creating the double surface from the two effective refractive indices for each direction of propagation, Yariv and Yeh [28] plot the components of **k**; their diagram, which they call the *normal surface*, is therefore a scaled version of ours. On the other hand, Born and Wolf [29] construct what they also call the *normal surface* by plotting the two *phase velocities* in each direction. Their diagram is therefore based on the inverse refractive indices.

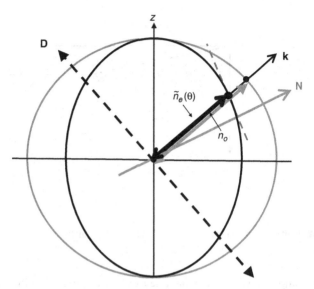

Fig. 3.5 Index surfaces for a negative uniaxial medium.

the sphere; for a negative medium, it lies inside. The two surfaces touch along the optic axis. A section through the index surfaces for a negative uniaxial medium is shown in Fig. 3.5. Careful thought shows that the inner extraordinary surface is simply the ellipse of Fig. 3.4 rotated by 90°. Notice that the bold line representing $\tilde{n}_e(\theta)$, which was perpendicular to **k** in Fig. 3.4, lies along **k** in Fig. 3.5. The radius of the circle outside the ellipse corresponds to the ordinary refractive index n_0.

The fact that Figs 3.3 and 3.5 both involve spheroids, circles and ellipses is another potential source of confusion. It is worth repeating that the index ellipsoid of Fig. 3.3 is a single surface object from which the refractive indices of the two orthogonally polarised waves in a given direction of propagation can be obtained by a geometrical construction. The double-layered surface of Fig. 3.5, on the other hand, traces those two refractive indices for any given direction of propagation. The plane of polarisation of the waves can also be deduced from the figures. You should convince yourself that the ordinary wave is polarised perpendicular to the plane of the paper in Figs 3.4 and 3.5, while the extraordinary wave is polarised in the plane of the paper.

3.3.5 Walk-off

The fact that **E** and **D** are generally not parallel for an extraordinary wave means that the wave exhibits peculiar propagation properties. According to EM theory, the direction of energy flow in an EM wave is given by the Poynting vector $\mathbf{N} = \mathbf{E} \times \mathbf{H}$. Hence, if **E** is not parallel to **D**, the energy does not flow in the same direction as the normal to the wavefront defined by **k**, but walks off to the side as indicated in both Figs 3.4 and 3.5.

As mentioned in Section 3.3.3, **E** lies along the normal to the ellipse of Fig. 3.4. The *walk-off angle, ρ* (the angle between **D** and **E** and equivalently between **N** and **k**), is therefore

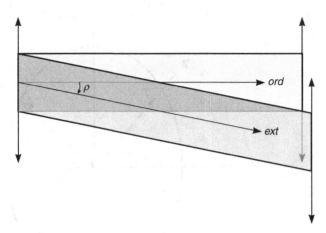

Fig. 3.6   Walk-off of an extraordinary wave, showing the walk-off angle ρ.

the angle between the radial line from the origin to the ellipse, and the normal to the ellipse at that point. Some tricky trigonometry leads to the formula

$$\rho(\theta) = \pm \left[\tan^{-1}\{(n_\mathrm{o}/n_\mathrm{e})^2 \tan\theta\} - \theta \right] \qquad (3.11)$$

where the upper and lower signs apply to negative and positive crystals, respectively. Positive ρ means that the Poynting vector points away from the optic axis (as in Fig. 3.5). The walk-off angle is typically a few degrees; see Problem 3.5.

Figure 3.6 contrasts the propagation of an ordinary and an extraordinary wave with parallel wavefronts. It highlights the way in which the energy in an extraordinary wave slips sideways, so that the extraordinary beam ultimately separates completely from the ordinary beam. The effect is similar to the behaviour of an aircraft attempting to fly (say) north in a strong easterly cross-wind. Though the plane is flying on the correct northerly bearing, the sideways air flow pushes it to the right of its intended flight path. Some supermarket trolleys exhibit a similar property, although the problem there is wear and tear!

3.4 Wave plates

The principle behind quarter- and half-wave plates can now be understood. Consider the propagation of ordinary and extraordinary waves in the plane normal to the optic (z-) axis, where $\theta = 90°$. To simplify the discussion, we assume that the direction of propagation is along the y-axis which points into the paper in Fig. 3.7. Suppose that a wave that is linearly polarised at 45° to the x- and z-axes (along the line PP) is incident on the entry surface of the sample at $y = 0$. The wave must be resolved into two equal components, an ordinary component (refractive index n_o) polarised in the x-direction, and an extraordinary component (refractive index n_e) polarised in the z-direction. The respective electric fields are

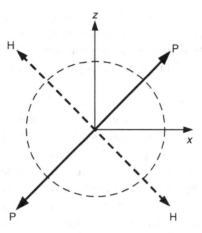

Fig. 3.7   Change in polarisation direction from PP to HH in a half-wave plate.

$$E_x = \frac{A}{\sqrt{2}} \cos\{\omega(t - n_o y/c)\}$$
$$E_z = \frac{A}{\sqrt{2}} \cos\{\omega(t - n_e y/c)\}$$

(3.12)

where A is the amplitude of the original field. Although the two components are in phase at $y = 0$, a phase difference develops during propagation given by $\phi(y) = 2\pi y \Delta n / \lambda_0$, where λ_0 is the vacuum wavelength and $\Delta n = n_o - n_e$. After propagating a distance $y = \lambda_0/4|\Delta n|$, the components are $\pi/2$ out of phase (technically they are said to be 'in quadrature'), and careful inspection will convince you that the total field is now circularly polarised (see Problem 3.3). An element that creates circularly polarised light in this way is called a *quarter-wave plate*.

The phase difference continues to increase under further propagation, and reaches π at $y = \lambda_0/2 |\Delta n|$. The components are now in anti-phase, so that $E_x = -E_z$. This implies that the total field is once again linearly polarised, but at 90° to the original orientation, as indicated by the dotted line HH in Fig. 3.7. A device with these properties is called a *half-wave plate*.

Quarter- and half-wave plates are used routinely for manipulating the polarisation of optical beams. Bear in mind that, because the refractive indices are wavelength dependent, wave plates have to be specified for a particular wavelength.

3.5 Biaxial media

3.5.1 General features

Professor Arthur Schawlow, one of the inventors of the laser and a great humourist too, liked to define a diatomic molecule as a molecule with one atom too many. In the same spirit, one might define a biaxial crystal as a crystal with one *axis* too many, because the optical properties of biaxial media are much more complicated than those of their uniaxial counterparts. Happily, most common nonlinear materials are uniaxial, but several members of biaxial class 7 (e.g. LBO and KTP) are in widespread use, so one cannot escape the topic of biaxial media altogether.

In a biaxial medium, the semi-axes of the index ellipsoid, which represent the values of n_x, n_y and n_z in Eq. (3.8), are all different, and the principal axes are always chosen so that n_y is intermediate between n_x and n_z.[6] Most of the ideas set out in Section 3.3.3 for finding the polarisation of waves propagating in a particular direction are unchanged in biaxial crystals, but the details are more complicated than before. Two waves with orthogonal directions of **D** still propagate for every value of **k** but, except in special cases, both solutions have extraordinary wave characteristics, with the Poynting vector angled to **k** and associated walk-off effects. Moreover, these media are called *bi*-axial rather than *uni*-axial because they have two optic axes rather than one.

3.5.2 Propagation in the principal planes: the optic axes

Fortunately, biaxial crystals are almost invariably used in nonlinear optics with all the interacting waves in one of the principal planes. The propagation characteristics are then relatively straightforward, and similar to those in uniaxial media.

The index surfaces of a biaxial crystal with $n_z > n_y > n_x$ are represented in Fig. 3.8, where sections of the double surface in the principal planes are shown. The three solid lines are quadrants of circles; the three dotted lines are quadrants of ellipses. Diagrams of

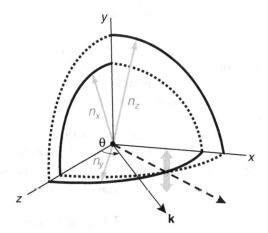

Fig. 3.8　Index surfaces for a biaxial medium showing sections in the principal planes. Solid lines indicate circular quadrants and dotted lines elliptical quadrants.

---

[6] Some highly respected sources (e.g. Born and Wolf [29] and Yariv and Yeh [28]) choose principal axes to ensure that $n_z > n_y > n_x$. While this is certainly a possible convention in principle, leading reference works such as Dmitriev *et al.* [26] quote the indices of some biaxial crystals (e.g. KNbO$_3$ and KB5) with $n_x > n_y > n_z$.

this kind can be quite confusing, so it is worth spending a little time understanding how to interpret this one. Propagation along one of the principal axes is the simplest case. The figure indicates that for propagation along (for example) the z-axis, the two indices are n_x and n_y ($> n_x$); these are the radii of the two solid circles touching the z-axis. The waves are, of course, polarised transversely to the z-direction, in the x and y directions respectively. The **D** and **E** vectors are parallel in this special case, so walk-off effects do not occur, and both solutions can be regarded as 'quasi-ordinary' waves.

When **k** is not aligned with one of the principal axes, but still lies in one of the three principal planes, the situation is straightforward too, because there is a close analogy with uniaxial media. Under these circumstances, one of the two solutions is always a quasi-ordinary wave; for any direction in the x-z plane, for instance, one wave is always polarised in the y-direction and has index n_y. A similar situation applies in the x-y and y-z planes. However, the choice of n_x and n_z (in either order) as the highest and lowest of the three refractive indices means that the x-z plane is different from the other two. This is evident from Fig 3.8, which shows that the two index surfaces intersect in the x-z plane. The line through the intersection (drawn dashed in Fig. 3.8) defines one of the two optic axes of a biaxial medium, and the other one lies symmetrically on the other side of the z-axis in the x-z plane, as shown in Fig. 3.9.

The directions of the optic axes are easily found. Consider a wave propagating in the x-z plane with wave vector **k** at an angle θ to the z-axis. We already know that the y-polarised wave has index n_y for all θ, and it is easy to show that the index of the other wave rises from n_x at $\theta = 0°$ to n_z at $\theta = 90°$, varying as it does so according to the formula

$$\tilde{n}(\theta) = \left\{ \frac{\cos^2 \theta}{n_x^2} + \frac{\sin^2 \theta}{n_z^2} \right\}^{-1/2} \tag{3.13}$$

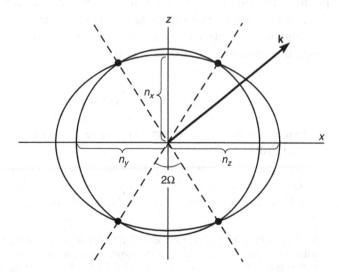

Fig. 3.9   The x-z sections of the index surfaces of a biaxial medium, showing the two optic axes angled at Ω to the z-axis.

which is analogous to Eq. (3.10). The directions of the optic axes can now be found from Eq. (3.13) by setting $\tilde{n}(\theta) = n_y$. Both waves then have the same refractive index and the same phase velocity, which is the defining characteristic of an *optic axis*. The angles of the optic axes to the z-axis, conventionally written Ω (or V_z), are readily found from Eq. (3.13) to be

$$\Omega = \pm \cos^{-1} \left\{ \frac{n_x}{n_y} \sqrt{\frac{n_z^2 - n_y^2}{n_z^2 - n_x^2}} \right\}. \tag{3.14}$$

Notice from this equation that when $n_x = n_y$ (the case of a uniaxial medium), the two axes merge into the single axis of a uniaxial medium along the z-axis, i.e. $\Omega = 0$.

It is a surprising fact that the two optic axes of a biaxial crystal are the *only* directions for which the two effective refractive indices are the same [29]. This is strange because, given that the optic axis in Fig. 3.8 lies at the intersection of the two index surfaces, one would expect there to be other directions close to the optic axis (but out of the x-z plane, indicated by the grey double arrow) where the surfaces would also intersect and the indices were also identical. After all, two planes normally intersect in a line, not at a point. The answer to this conundrum is that the index surfaces do not behave as simple surfaces, and it turns out that the surface intersection is 'avoided' (in the mathematical sense of an avoided crossing) for all directions other than the two optic axes.[7]

3.5.3 Positive and negative crystals

As in the uniaxial case, biaxial crystals are classified into positive and negative types, although the criterion is unfortunately not immediately obvious. The rule is that, if 2Ω is acute, a biaxial medium is positive if $n_z > n_x$ and negative if $n_x > n_z$. The designations are reversed if 2Ω is obtuse. There is a logic to this apparently arbitrary system insofar as positive biaxial media are those in which the optic axes are closer (an acute angle between them) to the axis of the index ellipsoid exhibiting the highest refractive index. This matches the designation in uniaxial media (with a single optic axis along the z-axis), where the crystal is positive uniaxial if $n_z \equiv n_e > n_o$.

Problems

3.1 Express the elliptical x-z section of Eq. (3.9) in polar form. How does it differ from Eq. (3.10)?

3.2 Consider a wave propagating along the x-axis of a uniaxial medium. In what directions are the two allowed wave solutions polarised? What are the respective refractive indices? Do any walk-off effects occur in this case?

---

[7] The fact that there really are just two directions for which the indices are the same follows also from a geometrical property of an ellipsoid. It can be shown that an ellipsoid with unequal axes has only two circular sections through its centre, which are perpendicular to the plane containing the largest and smallest semi-axes. The two optic axes are normals to these circular sections.

3.3 A wave propagates for a distance $\lambda/4\,|\Delta n|$ in the $+y$-direction of a positive uniaxial crystal ($n_e > n_o$); see Fig. 3.7. Is the resulting circularly polarised wave polarised in a clockwise or an anticlockwise direction (looking in the direction of wave propagation)?

3.4 Prove Eq. (3.14).

3.5 Beta-barium borate (BBO) is used in a second harmonic generation experiment. The fundamental wavelength is $\lambda = 1.064\ \mu$m, and the angle between the direction of propagation and the optic axis is $22.8°$. Find the walk-off angle.

(BBO refractive indices at $0.532\ \mu$m: $n_o = 1.67421$, $n_e = 1.55490$.)

4 Nonlinear optics in crystals

4.1 Introduction and preview

As noted in Section 1.8, the treatment presented in Chapter 1 was greatly over-simplified. The fact that the frequency dependence of the coefficients in the polarisation expansion was neglected gave the false impression that the coefficients governing all processes of a given order are the same, apart from simple factors. The tensor nature of the coefficients was completely ignored too.

Unfortunately, if one wants to understand nonlinear optical interactions in crystalline media, one cannot avoid getting to grips with the tensor nature of the nonlinear coefficients, a topic that is intricate and hard to simplify. At the very least, one needs to be able to interpret the numerical notation used to label the coefficients, and to know how to apply the data supplied in standard reference works in a given crystal geometry.

There is no disguising the fact that Section 4.3 in the present chapter is rather complicated. Readers who want to avoid the worst of the difficulties should skim it, always bearing in mind that the situation turns out to be far less alarming at the end of the journey than it seemed it might be at the beginning. Early on in that section, it looks as if there could be literally hundreds of separate nonlinear coefficients to deal with. But it soon emerges that the number is almost certainly no larger than 18, and perhaps only 10. And even this worst-case scenario applies only in crystals of the triclinic class, of which there are no members in regular use in nonlinear optics. Indeed, in most common crystals, the number of independent coefficients is typically only two or three and, for a given geometry, there may be only one.

4.2 The linear susceptibility

We start by writing the relationship between the polarisation and the field in linear optics in its general form. The defining equation is[1]

$$\hat{P}_i(\omega) = \varepsilon_0 \sum_j \chi_{ij}^{(1)}(\omega)\hat{E}_j(\omega) \quad (i, j = x, y, z) \tag{4.1}$$

---

[1] The summation sign may be omitted because, under the *repeated suffix convention*, summation is implied over any repeated suffix on the right-hand side.

where $\chi_{ij}^{(1)}(\omega)$ is the frequency-dependent matrix of the linear susceptibility (the component form of the associated tensor), and $\hat{P}_i(\omega)$ and $\hat{E}_i(\omega)$ are the (complex) amplitudes of quasi-monochromatic terms in the summations of Eqs (2.4) and (2.5). Since the indices i and j can each be x, y or z, Eq. (4.1) is shorthand for three equations.

When combined with $\mathbf{D} = \varepsilon_0 \mathbf{E} + \mathbf{P}$, Eq. (4.1) becomes

$$\hat{D}_i(\omega) = \varepsilon_0 \hat{E}_i(\omega) + \varepsilon_0 \sum_j \chi_{ij}^{(1)}(\omega) \hat{E}_j(\omega) = \varepsilon_0 \sum_j \varepsilon_{ij}(\omega) \hat{E}_j(\omega) \qquad (4.2)$$

where $\varepsilon_{ij}(\omega)$ is the dielectric constant matrix from Section 3.3.2. From Eq. (3.6), we know that, with respect to the principal dielectric axes, only the three diagonal elements of $\varepsilon_{ij}(\omega)$ are non-zero and, from the subsequent discussion, that these elements are the squares of the associated refractive indices. The frequency dependence of Eq. (4.2) is therefore simply the well-known phenomenon of dispersion.

A more detailed discussion of Eq. (4.1) can be found in Appendix B. The connection between the frequency dependence of the susceptibility and the non-instantaneous response of the polarisation to the field is established, and it is shown that the equation is in fact a hybrid form, with some time-domain and some frequency-domain characteristics.

4.3 Structure of the nonlinear coefficients

4.3.1 Formal definition

We now turn our attention to the structure of the second-order nonlinear coefficients, which is the central topic of this chapter. For the second-order nonlinear polarisation governing the three-wave interaction $\omega_1 + \omega_2 = \omega_3$ of sum frequency generation (SFG), we write by analogy with Eq. (4.1) the expression[2]

$$\hat{P}_i(\omega_3) = \tfrac{1}{2}\varepsilon_0 \sum_p \sum_{jk} \chi_{ijk}^{\mathrm{SFG}}(\omega_3;\ \omega_1, \omega_2) \hat{E}_j(\omega_1) \hat{E}_k(\omega_2)$$

$$= \tfrac{1}{2}\varepsilon_0 \sum_{jk} \Big[\chi_{ijk}^{\mathrm{SFG}}(\omega_3;\ \omega_1, \omega_2)\, \hat{E}_j(\omega_1) \hat{E}_k(\omega_2)$$

$$+ \chi_{ikj}^{\mathrm{SFG}}(\omega_3; \omega_2, \omega_1)\, \hat{E}_j(\omega_2) \hat{E}_k(\omega_1) \Big] \qquad (4.3)$$

where $\omega_1 \neq \omega_2$, and the suffices i, j and k can each be x, y, or z. The symbol $\sum_p$ in the first line of the equation implies that the terms with ω_1 and ω_2 permuted must be included, as shown explicitly in the second line. The pre-factor of $\tfrac{1}{2}$ is a consequence of the field definitions of Eqs (2.4) and (2.5).

Equation (4.3) merits detailed discussion. Notice first of all that the nonlinear coefficients have three frequency arguments representing the frequency of the polarisation to the left of

---

[2] A detailed discussion of Eq. (4.3) can be found in Appendix C.

the semicolon, and those of the two fields in the order that they appear to the right. Since $\omega_3 = \omega_1 + \omega_2$, only two frequency arguments are strictly necessary, but it is conventional to quote all three. As regards the coordinate suffices ijk, each of which can be x, y or z, the component of the polarisation (i) comes first, followed by those of the fields (jk) in the appropriate order. Under the repeated suffix convention, the summation $\sum_{jk} \{rhs\}$ is redundant, and could have been omitted. Negative frequencies are associated with complex conjugate fields and polarisations, and must be treated as distinct from their positive counterparts.

4.3.2 Intrinsic permutation symmetry

Equation (4.3) can be simplified once it is recognised that some pairs of terms on the right-hand side are indistinguishable in their effect, for example $\chi_{xyz}^{\text{SFG}}(\omega_3; \omega_1, \omega_2)$ $\hat{E}_y(\omega_1)\hat{E}_z(\omega_2)$ and $\chi_{xzy}^{\text{SFG}}(\omega_3; \omega_2, \omega_1)\hat{E}_z(\omega_2)\hat{E}_y(\omega_1)$. As explained in Appendix C, it is common practice in this situation to equate the two contributions, in other words to assume that a coefficient is unchanged if the two field frequencies are reversed, together with the associated coordinate indices, i.e. $\chi_{ijk}^{(\text{SFG})}(\omega_3; \omega_1, \omega_2) = \chi_{ikj}^{(\text{SFG})}(\omega_3; \omega_2, \omega_1)$. This assumption of *intrinsic permutation symmetry* (IPS) enables Eq. (4.3) to be abbreviated to

$$P_i(\omega_3) = \varepsilon_0 \sum_{jk} \chi_{ijk}^{\text{SFG}}(\omega_3; \omega_1, \omega_2) E_j(\omega_1) E_k(\omega_2) \tag{4.4}$$

where $\chi_{ijk}^{(2)}(\omega_3; \omega_1, \omega_2)$ and $\chi_{ikj}^{(2)}(\omega_3; \omega_2, \omega_1)$ are both included in this expression. It must be stressed that IPS is a convenience, not a fact that can be proved. The point is that only the sum of the two coefficients can be measured, and so it is sensible to assume that the individual components are the same.

Equation (4.4) needs to be adjusted when $\omega_1 = \omega_2$, or when any of the frequencies is zero. Further discussion of the issues involved, and a general formula covering all situations, can be found in Appendix C. Here we simply quote results for the most important cases:

Second harmonic generation ($\omega_3 = 2\omega$; $\omega_1 = \omega_2 = \omega$)

$$\hat{P}_i(2\omega) = \tfrac{1}{2}\varepsilon_0 \sum_{jk} \chi_{ijk}^{\text{SHG}}(2\omega; \omega, \omega) \hat{E}_j(\omega) \hat{E}_k(\omega). \tag{4.5}$$

The Pockels effect ($\omega_3 = \omega$; $\omega_1 = \omega$, $\omega_2 = 0$)

$$\hat{P}_l(\omega) = 2\varepsilon_0 \sum_{jk} \chi_{ijk}^{\text{PE}}(\omega; \omega, 0) \hat{E}_j(\omega) E_k(0). \tag{4.6}$$

Optical rectification ($\omega_3 = 0$; $\omega_1 = \omega$, $\omega_2 = -\omega$)

$$P_i(0) = \tfrac{1}{2}\varepsilon_0 \sum_{jk} \chi_{ijk}^{\text{OR}}(0; \omega, -\omega) \hat{E}_j(\omega) \hat{E}_k^*(\omega) \tag{4.7}$$

where $\hat{E}_j^*(\omega) = \hat{E}_j(-\omega)$. As explained in Appendix C, in all these equations the order of the two frequencies after the semicolon is fixed by definition, and terms with the frequencies in reverse order do not appear. Notice that the pre-factors are identical to those that arose naturally in the simple treatment of Chapter 1.

4.3.3 Full permutation symmetry and the Manley–Rowe relations

At this point, one is bound to feel daunted by the accelerating complexity of the analysis. Consider, for example, the case of sum frequency generation. It is already clear from Eq. (4.4) that there is the potential for 27 separate nonlinear coefficients (since ijk can each be x, y or z), and that is bad enough. But what about negative frequencies? Do we have to deal with coefficients like $\chi_{xyz}^{(2)}(-\omega_3; -\omega_1, -\omega_2)$ as well, which would double the number of coefficients to 54? And there is another problem too. In Chapter 2, we discovered that a proper treatment of SFG involves three coupled-wave equations, but so far we have only considered the coefficient driving the equation for the field at ω_3. If the coefficients involved in the other two equations have to be included as well, that could multiply the number of coefficients by a further factor of 3, raising the total to 162.

Happily, the worry about the large number of coefficients turns out to be far less serious that it appears at first sight. Firstly, reversing the signs of all participating frequencies turns a coefficient into its complex conjugate. But in a lossless medium where the frequencies of the fields are far from atomic and molecular resonances, the coefficients are real to an excellent approximation, so reversing all the signs makes no difference.

Secondly, under loss-free conditions, a principle known as *full permutation symmetry* can be invoked under which the nonlinear coefficients are unchanged if the three Cartesian indices ijk are permuted *together with their associated frequency arguments*. This enables one to write, for example,

$$\chi_{xyz}^{(2)}(\omega_3; \omega_1, \omega_2) = \chi_{yzx}^{(2)}(\omega_1; -\omega_2, \omega_3) = \chi_{zxy}^{(2)}(\omega_2; \omega_3, -\omega_1) \tag{4.8}$$

where the signs have been adjusted to maintain the convention that the first frequency is the sum of the other two. The three coefficients in Eq. (4.8) are those needed in the coupled-wave equations for SFG. We recall that the Manley–Rowe relations of Eqs (2.20) and (2.31) rely on the three coefficients being the same, and it is entirely consistent that this is ensured by the principle of full permutation symmetry, which only applies under loss-free conditions.

That brings us back to the 27 independent coefficients with which we started this section. But we will shortly discover that, for second harmonic generation at least, the number is actually only 18. And, better still, even that worst-case scenario never arises in practice because all common nonlinear materials belong to symmetry classes in which most of the coefficients are zero in any case.

4.3.4 Contracted suffix notation

We now introduce the commonly used contracted suffix notation for the nonlinear coefficients. We start from Eq. (4.5) for SHG which reads (for $j \neq k$)

$$\hat{P}_i(2\omega) = \tfrac{1}{2}\varepsilon_0 \left(\chi_{ijk}^{\text{SHG}}(2\omega; \omega, \omega) + \chi_{ikj}^{\text{SHG}}(2\omega; \omega, \omega) \right) \hat{E}_j(\omega)\hat{E}_k(\omega) \quad (j \neq k). \tag{4.9}$$

	i	x	y	z			
	n	1	2	3			
	jk	xx	yy	zz	yz, zy	xz, zx	xy, yx
	p	1	2	3	4	5	6

Table 4.1 Contracted suffix notation for uniaxial media.

The two coefficients in this equation are clearly indistinguishable, so we take them as equal (under IPS) and define $\chi^{\text{SHG}}_{ijk} = \chi^{\text{SHG}}_{ikj} = 2d_{i(jk)}$, where the brackets around jk serve as a reminder that the order is irrelevant.[3] The first suffix i and the suffix pair jk are now translated into numbers n and p $(d_{i(jk)} \rightarrow d_{np})$ according to the prescription shown in Table 4.1.[4] Thus, for example $\chi^{\text{SHG}}_{xyz} = \chi^{\text{SHG}}_{xzy} = 2d_{14}$ and for these coefficients Eq. (4.9) yields the contribution

$$\hat{P}_x(2\omega) = 2\varepsilon_0 d_{14}\hat{E}_y(\omega)\hat{E}_z(\omega). \tag{4.10}$$

Similarly, when the second two suffices are the same, we have for example $\chi^{(2)}_{yzz} \rightarrow 2d_{23}$, which leads to

$$\hat{P}_y(2\omega) = \varepsilon_0 d_{23}\hat{E}_y^2(\omega). \tag{4.11}$$

Combining everything into a single equation we have

$$\begin{pmatrix} \hat{P}_x(2\omega) \\ \hat{P}_y(2\omega) \\ \hat{P}_z(2\omega) \end{pmatrix} = \varepsilon_0 \begin{pmatrix} d_{11} & d_{12} & d_{13} & d_{14} & d_{15} & d_{16} \\ d_{21} & d_{22} & d_{23} & d_{24} & d_{25} & d_{26} \\ d_{31} & d_{32} & d_{33} & d_{34} & d_{35} & d_{36} \end{pmatrix} \begin{pmatrix} \hat{E}_x^2(\omega) \\ \hat{E}_y^2(\omega) \\ \hat{E}_z^2(\omega) \\ 2\hat{E}_y(\omega)\hat{E}_z(\omega) \\ 2\hat{E}_z(\omega)\hat{E}_x(\omega) \\ 2\hat{E}_x(\omega)\hat{E}_y(\omega) \end{pmatrix}. \tag{4.12}$$

From the form of the d-matrix, it is clear that the number of independent SHG coefficients is now no more than 18, even for a crystal of the most general symmetry. In practice, the number is normally far less, as we will shortly discover.

4.3.5 The Kleinmann symmetry condition

The notational contraction of Table 4.1 can be applied more generally if all the frequencies involved in the interaction are small compared to the resonance frequencies of the medium.

[3] The origin of the factor 2 in the definition of d is lost in the history of the subject. Note that when d appears in the literature with three suffices, the contracted form is equal to the original separate coefficients, not to their sum. So, for instance, $d_{14} = d_{xyz}(= d_{123}) = d_{xzy}(= d_{132})$. We could incidentally have defined $2d_{i(jk)} = \frac{1}{2}\left(\chi^{(2)}_{ijk} + \chi^{(2)}_{ikj}\right)$ in which case we would not have needed to equate $\chi^{(2)}_{ijk}$ and $\chi^{(2)}_{ikj}$; see Appendix C.
[4] The mapping in Table 4.2 is not necessarily the same in biaxial media; see Table 4.4.

Dispersion can then be neglected, making the nonlinear susceptibilities independent of frequency, in which case all coefficients with the same set of *ijk* indices become equal, e.g. $\chi_{xyy} = \chi_{xyx} = \chi_{yyx}$, etc. In fact, this identity, which is called the *Kleinmann symmetry condition* (KSC), is often approximately true even when significant dispersion is present.

Under the KSC, χ_{ijk} is the same if j and k are interchanged, so the contracted notation of Eq. (4.10) can now be used in the general three-wave case, not just for SHG. But the KSC has the further consequence that the number of independent coefficients drops from 18 to 10, one for each of the sets *xxx, xxy, xxz, xyy, xyz, xzz, yyy, yyz, yzz* and *zzz*. Hence, some of the elements in the 3×6 d matrix of Eq. (4.12) are now interrelated, namely

$$
\begin{pmatrix}
d_{11} & d_{12} & d_{13} & d_{14} & d_{15} & d_{16} \\
d_{21} & d_{22} & d_{23} & d_{24} & d_{25} & d_{26} \\
d_{31} & d_{32} & d_{33} & d_{34} & d_{35} & d_{36}
\end{pmatrix}
\rightarrow
\begin{pmatrix}
d_{11} & d_{12} & d_{13} & d_{14} & d_{15} & d_{16} \\
\mathbf{d_{16}} & d_{22} & d_{23} & d_{24} & \mathbf{d_{14}} & \mathbf{d_{12}} \\
\mathbf{d_{15}} & \mathbf{d_{24}} & d_{33} & \mathbf{d_{23}} & \mathbf{d_{13}} & \mathbf{d_{14}}
\end{pmatrix}
$$

$$(4.13)$$

where the eight elements now linked to others are printed in bold. The equation for sum frequency generation analogous to Eq. (4.12) now reads

$$
\begin{pmatrix}
\hat{P}_x(\omega_3) \\
\hat{P}_y(\omega_3) \\
\hat{P}_z(\omega_3)
\end{pmatrix}
= 2\varepsilon_0
\begin{pmatrix}
d_{11} & d_{12} & d_{13} & d_{14} & d_{15} & d_{16} \\
\mathbf{d_{16}} & d_{22} & d_{23} & d_{24} & \mathbf{d_{14}} & \mathbf{d_{12}} \\
\mathbf{d_{15}} & \mathbf{d_{24}} & d_{33} & \mathbf{d_{23}} & \mathbf{d_{13}} & \mathbf{d_{14}}
\end{pmatrix}
$$

$$
\times
\begin{pmatrix}
\hat{E}_x(\omega_1)\hat{E}_x(\omega_2) \\
\hat{E}_y(\omega_1)\hat{E}_y(\omega_2) \\
\hat{E}_z(\omega_1)\hat{E}_z(\omega_2) \\
\hat{E}_y(\omega_1)\hat{E}_z(\omega_2) + \hat{E}_z(\omega_1)\hat{E}_y(\omega_2) \\
\hat{E}_z(\omega_1)\hat{E}_x(\omega_2) + \hat{E}_x(\omega_1)\hat{E}_z(\omega_2) \\
\hat{E}_x(\omega_1)\hat{E}_y(\omega_2) + \hat{E}_y(\omega_1)\hat{E}_x(\omega_2)
\end{pmatrix}.
$$

$$(4.14)$$

4.4 Crystal symmetry

Invariably, crystal symmetry drastically reduces the number of non-zero nonlinear coefficients still further. The non-zero SHG coefficients for each of the 21 non-centrosymmetric crystal classes, and the relationships between them, are detailed in Appendix D. The most important conclusion is that the worst-case scenario of 18 independent coefficients only occurs in triclinic class 1 crystals, no member of which is in regular use in nonlinear optics. In fact, most popular nonlinear materials are in uniaxial classes 11 and 19, and biaxial class 7; common members of these three categories are listed in Tables 4.2 and 4.3, and the non-zero coefficients of the two uniaxial classes are shown in Table 4.4. The situation is more complicated in biaxial class 7 because of the awkward relationship between the physical and the crystallographic axes in these materials shown in column 6 of Table 4.3. The implications of this latter point will be discussed later.

Table 4.2 Common uniaxial crystals.

	Abbreviation	Formula	Sign
Class 11 ($\bar{4}$ 2 m)			
Potassium dihydrogen phosphate	KDP	KH_2PO_4	negative
Ammonium dihydrogen phosphate	ADP	$NH_3H_2PO_4$	negative
Cesium dihydrogen arsenate	CDA	CsH_2AsO_4	negative
Class 19 (3 m)			
Lithium niobate		$LiNbO_3$	negative
β-barium borate	BBO	β-BaB_2O_4	negative
Proustite		Ag_3AsS_3	negative

Table 4.3 Common biaxial crystals.

	Abbrev.	Formula	Sign	Indices	$xyz =$
Class 7 (m m 2)					
Lithium triborate	LBO	LiB_3O_5	negative	$n_x < n_y < n_z$	acb
Potassium titanyl phosphate	KTP	$KTiOPO_4$	negative	$n_x < n_y < n_z$	abc
Potassium niobate		$KNbO_3$	negative	$n_x > n_y > n_z$	bac
Potassium pentaborate tetrahydrate	KB5	$KB_5O_8.4H_2O$	positive	$n_x > n_y > n_z$	abc
Potassium titanyl arsenate	KTA	$KTiOAsO_4$	positive	$n_x < n_y < n_z$	abc

Table 4.4 Non-zero d-elements in class 11 and 19 crystals.

	1	2	3	4	5	6
Class 11 (3 m): $d_{14} = d_{25}$; d_{36}						
1				d_{14}		
2					$= d_{14}$	
3						d_{36}
Class 19 ($\bar{4}$ 2 m): $d_{15} = d_{24}$; $d_{16} = d_{21} = -d_{22}$; $d_{31} = d_{32}$; d_{33}						
1					d_{15}	d_{16}
2	$= d_{16}$	$= -d_{16}$		$= d_{15}$		
3	d_{31}	$= d_{31}$	d_{33}			

4.5 Second harmonic generation in KDP

Let us consider second harmonic generation by birefringent phase matching in a uniaxial class 11 crystal such as KDP, which is a fairly straightforward case. The simplest scenario, envisaged in Chapter 2, and shown in Fig. 4.1, is where two ordinary waves at ω combine to create an extraordinary wave at 2ω, a process that is described by the shorthand o + o → e

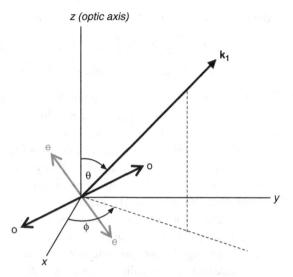

Fig. 4.1 Type 1 phase-matching geometry in a class 11 crystal. $\mathbf{k_1}$ is the wave vector of the fundamental, which is polarised in the x-y plane (indicated by o-o) and is therefore an ordinary wave. The polarisation of the extraordinary second harmonic wave (shown as e-e) is perpendicular to o-o and to $\mathbf{k_1}$.

and called *Type-I phase-matching*. Phase matching requires that

$$n_{\text{o}}^{\omega} = \tilde{n}_{\text{e}}^{2\omega}(\theta) = \left\{ \frac{\cos^2 \theta}{(n_{\text{o}}^{2\omega})^2} + \frac{\sin^2 \theta}{(n_{\text{e}}^{2\omega})^2} \right\}^{-1/2} \tag{4.15}$$

where the second step follows from Eq. (3.10), and θ is the angle between the direction of propagation and the z-axis, as shown in the figure. Rearranging Eq. (4.15) yields

$$\cos \theta = \frac{n_{\text{o}}^{2\omega}}{n_{\text{o}}^{\omega}} \sqrt{\frac{(n_{\text{e}}^{2\omega})^2 - (n_{\text{o}}^{\omega})^2}{(n_{\text{e}}^{2\omega})^2 - (n_{\text{o}}^{2\omega})^2}}. \tag{4.16}$$

For a fundamental wavelength of 1.064 μm, $n_{\text{o}}^{\omega} = 1.4938$, $n_{\text{o}}^{2\omega} = 1.5124$ and $n_{\text{e}}^{2\omega} = 1.4705$, and if these values are inserted, Eq. (4.16) gives $\theta = 41.21°$.

But this is not the end of the story, because it is still necessary to find the optimum value of the azimuthal angle ϕ. From Table 4.4, the three elements d_{14}, d_{25} and d_{36} are non-zero in KDP and, since the ordinary fundamental wave will be polarised in the x-y plane (see the double arrow o-o in the figure), it is clear from Eq. (4.12) that the operative term in the polarisation is

$$\hat{P}_z(2\omega) = 2\varepsilon_0 d_{36} \hat{E}_x(\omega) \hat{E}_y(\omega). \tag{4.17}$$

From Fig. 4.1, it can be seen that $\hat{E}_x(\omega) = \hat{E}_{\text{o}}(\omega) \sin \phi$ and $\hat{E}_y(\omega) = \hat{E}_{\text{o}}(\omega) \cos \phi$, where $\hat{E}_{\text{o}}(\omega)$ is the field amplitude of the incident ordinary wave, while the second harmonic polarisation transverse to the direction of propagation (represented by the double arrow e-e in the figure) is $\hat{P}_z(2\omega)\sin\theta$. The conclusion is that

$$\hat{P}_e(2\omega) = 2\varepsilon_0 d_{36}\hat{E}_o(\omega)^2 \sin\theta\sin\phi\cos\phi$$
$$= \varepsilon_0\left[d_{36}\sin\theta\sin 2\phi\right]\hat{E}_o(\omega)^2 = \varepsilon_0 d_{\text{eff}}\hat{E}_o(\omega)^2 \qquad (4.18)$$

where the final step defines the effective nonlinear coefficient d_{eff} for the ooe interaction as $d_{\text{eff}} = d_{36}\sin\theta\sin 2\phi$. To ensure phase matching, θ must be close to the value set by Eq. (4.16); the optimum value of ϕ is clearly $45°$.

Notice that the role of the effective nonlinear coefficient is to link the polarisation of the extraordinary wave to the field of the ordinary wave, in contrast to the individual d coefficients which relate the Cartesian components of P and E.

Phase-matched SHG in KDP is also possible by combining an ordinary and an extraordinary fundamental wave to create an extraordinary second harmonic. The condition for this oee interaction (*Type II phase matching*) is that $n_o(\omega) + n_e(\omega) = 2n_e(2\omega)$. This is satisfied at $\theta = 59.13°$, although it is not possible to write down an exact analytical formula in this case. An intricate piece of three-dimensional geometry (see Problem 4.3) shows that [30]

$$\hat{P}_e(2\omega) = \varepsilon_0\left[\tfrac{1}{2}(d_{36}+d_{14})\sin 2\theta\sin 2\phi\right]\hat{E}_o(\omega)\hat{E}_e(\omega) = \varepsilon_0 d_{\text{eff}}\hat{E}_o(\omega)\hat{E}_e(\omega) \qquad (4.19)$$

where the effective nonlinear coefficient in this case is[5] $d_{\text{eff}} = \tfrac{1}{2}(d_{36}+d_{14})\sin 2\theta\cos 2\phi$.

4.6 Second harmonic generation in LBO

It is quite easy to treat the case of second harmonic generation in one of the principal planes of a biaxial crystal such as lithium triborate (LBO), once a potentially confusing feature is understood. For class 7 crystals, the crystallographic axes *abc* on which the fundamental symmetry properties are based do not always translate in the obvious way to the *xyz* axes used to define the index ellipsoid and the direction of the optic axes. Of course, problems of this kind are to be expected given that the crystallographic axes are not even orthogonal in four of the seven crystal symmetry systems. But even in the orthorhombic system (that includes class 7) where the crystallographic axes *are* orthogonal, one cannot necessarily assume that $a \to x$, $b \to y$, $c \to z$. The different mappings for the five class 7 crystals listed in Table 4.3 are shown in column 6.

Why is this a problem? Because it is the *abc* axes, not the *xyz* axes, that determine the structure of the d-matrices, and the numerical suffix notation is based on the former, not the latter. For LBO, the axes mapping is $a \to x$, $b \to z$, $c \to y$, and the non-zero d-elements are $d_{31} = d_{caa} = d_{yxx}$, $d_{32} = d_{cbb} = d_{yzz}$, $d_{33} = d_{ccc} = d_{yyy}$ as shown in Table 4.5. Notice that 3 now means y, and 2 means z!

Apart from this, it is straightforward to analyse SHG in LBO. The refractive indices at 1.06 μm and 0.53 μm are listed in Table 4.6, and the x-y section of the index surfaces is

---

[5] The formula for the effective coefficient quoted by Dmitriev *et al.* [26] assumes that $d_{14} = d_{36}$ which is a reasonable approximation in KDP.

Table 4.5 Non-zero d-elements in LBO.

	$1 = aa$ $= xx$	$2 = bb$ $= zz$	$3 = cc$ $= yy$	$4 = bc, cb$ $= zy, yz$	$5 = ac, ca$ $= xy, yx$	$6 = ab, ba$ $= xz, zx$
$1 = a = x$					d_{15}	
$2 = b = z$				d_{24}		
$3 = c = y$	d_{31}	d_{32}	d_{33}			

Table 4.6 Refractive indices of LBO.

LBO	n_x	n_y	n_z
1.06 µm	1.5648	1.5904	1.6053
0.53 µm	1.5785	1.6065	1.6216

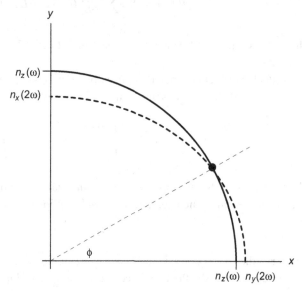

Fig. 4.2   The x-y section of the LBO index surfaces showing the phase-matching direction for SHG. The index variations are exaggerated.

shown in Fig. 4.2. Two waves can propagate at an angle ϕ to the x-axis: the one polarised in the z-direction is a quasi-ordinary wave[6] with refractive index n_z, while the refractive index of the other wave falls from n_y at $\phi = 0°$ to n_x at $\phi = 90°$. Since $n_y(2\omega) > n_z(\omega) > n_x(2\omega)$, phase matching is possible at a value of ϕ that can be found from

$$\cos \phi = \frac{n_x^{2\omega}}{n_z^{\omega}} \sqrt{\frac{(n_y^{2\omega})^2 - (n_z^{\omega})^2}{(n_y^{2\omega})^2 - (n_x^{2\omega})^2}} \tag{4.20}$$

---

[6] A 'quasi-ordinary' wave has the propagation characteristics of an ordinary wave in a uniaxial medium, although there are strictly speaking no ordinary waves in biaxial media; see Section 3.5.

which has the same structure as Eq. (4.16). Using the data of Table 4.6 yields a value of ϕ around 11.6°.

The d coefficient mediating the generation of the second harmonic wave is $d_{yzz} = d_{cbb} = d_{32}$; another possibility would be $d_{xzz} = d_{abb} = d_{12}$, but this coefficient is zero in class 7 crystals. Since the angle between the y-axis and the plane of the wavefront is ϕ, the effective nonlinear coefficient is $d_{\text{eff}} = d_{32} \cos \phi_{\text{pm}}$ which is close to d_{32} because ϕ_{pm} is small.

4.7 The Pockels effect and the Pockels cell

As described in Chapter 1, the Pockels effect relates to the change in refractive index of a non-centrosymmetric medium induced by an applied DC field. From Eq. (4.6) we have

$$\hat{P}_i(\omega) = 2\varepsilon_0 \sum_{jk} \chi^{\text{PE}}_{ijk}(\omega; \, \omega, 0) \hat{E}_j(\omega) E_k^{\text{DC}} \tag{4.21}$$

where, as explained in Section 4.3.3, the order of the fields on the right-hand side is fixed by definition.

The Pockels effect is in fact more usually described in terms of the effect of the DC field on the inverse dielectric constant matrix, which reads

$$\boldsymbol{\varepsilon}^{-1} = \begin{pmatrix} n_{\text{o}}^{-2} & 0 & 0 \\ 0 & n_{\text{o}}^{-2} & 0 \\ 0 & 0 & n_{\text{e}}^{-2} \end{pmatrix} \tag{4.22}$$

when referred to the principal axes of a uniaxial medium. Use of $\boldsymbol{\varepsilon}^{-1}$ turns out to be convenient insofar as the equation of the index ellipsoid (Eq. 3.6) can be written

$$\bar{\mathbf{r}} \boldsymbol{\varepsilon}^{-1} \mathbf{r} = \mathbf{I} \tag{4.23}$$

where $\bar{\mathbf{r}}$ is the row (xyz), $\mathbf{r}$ is the associated column, and $\mathbf{I}$ is the identity matrix. In terms of its effect on $\boldsymbol{\varepsilon}^{-1}$, the Pockels effect can be represented by

$$\Delta(\boldsymbol{\varepsilon}^{-1})_{ij} = \sum_k r_{ijk} E_k^{\text{DC}}. \tag{4.24}$$

Since the symmetry of $\boldsymbol{\varepsilon}^{-1}$ means that the order of the indices ij is irrelevant, a similar contracted notation to that of Table 4.1 can be introduced. Equation (4.24) then reads

$$\Delta(\boldsymbol{\varepsilon}^{-1})_p = \Delta\left(\frac{1}{n^2}\right)_p = \sum_n r_{pn} E_n^{\text{DC}} \quad (n = 1 - 3, p = 1 - 6) \tag{4.25}$$

where the index p represents a pair of subscripts as before. The relationship between the r and χ coefficients can be shown to be

$$r_{ijk} = r_{jik} = r_{pn} = -\frac{\chi^{\text{PE}}_{ijk}(\omega; \, \omega, 0) + \chi^{\text{PE}}_{jik}(\omega; \, \omega, 0)}{\varepsilon_{ii}\varepsilon_{jj}} = -\frac{2\chi^{\text{PE}}_{ijk}(\omega; \, \omega, 0)}{\varepsilon_{ii}\varepsilon_{jj}}. \tag{4.26}$$

Consider the simple case of a class 11 crystal such as KDP. In close analogy with the d coefficients in the upper section of Table 4.4, the only non-zero r coefficients are $r_{41} = r_{52} \neq r_{63}$. For a wave polarised in the x-y plane propagating in the $+z$-direction through a sample of thickness l with a DC voltage V between the exit and entrance surfaces, Eq. (4.25) reads

$$\Delta\left(\frac{1}{n^2}\right)_{xy} = \Delta\left(\frac{1}{n^2}\right)_{yx} = \frac{r_{63}V}{l}. \tag{4.27}$$

Equations (4.22) and (4.23) now yield for the equation of the index ellipsoid

$$\frac{x^2 + y^2}{n_o^2} + \frac{z^2}{n_e^2} + \frac{2xy\,r_{63}V}{l} = 1. \tag{4.28}$$

It is readily shown that the $z = 0$ section of the ellipsoid is an ellipse whose axes are inclined at 45° to the x- and y-axes, as shown in Fig. 4.3. The lengths of the semi-axes, which represent the refractive indices of the two orthogonally polarised fields that can propagate in the z-direction are

$$n_{\pm} \cong n_0(1 \pm \tfrac{1}{2}n_0^2 r_{63}V/l). \tag{4.29}$$

The nonlinear medium therefore behaves as a wave plate whose strength can be controlled by varying the applied voltage; this is the principle on which the Pockels cell is based. Section 3.4 gives the quarter-wave distance as $\lambda/4\,|\Delta n|$ and this, with $\Delta n = n_0^3 r_{63}V/l$ from Eq. (4.29), yields for the quarter-wave voltage the formula

$$V_{\lambda/4} = \frac{\lambda}{4n_0^3 r_{63}}. \tag{4.30}$$

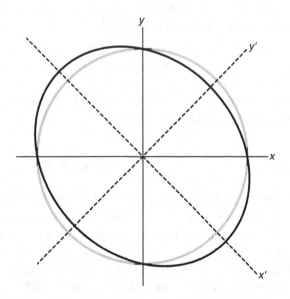

Fig. 4.3   The Pockels effect causes the x-y section of the index ellipsoid in a uniaxial medium to become elliptical. The circular (zero DC field) profile is shown in grey.

Notice that the thickness of the crystal has cancelled out. This is because doubling the crystal length doubles the interaction length but halves the longitudinal electric field for a given applied voltage.

Unfortunately, the voltages required to operate Pockels cells on this basis are very high (see Problem 4.5), but it is possible in principle to construct wave plates, modulators, and beam deflectors using these principles. An alternative configuration in which the DC potential is transverse to the beam has the advantage that the interaction length can be increased without lowering the DC field.

4.8 Optical rectification

In Chapter 1, we saw how a beam propagating in a non-centrosymmetric medium can create a DC polarisation proportional to the intensity. This process, known as optical rectification, was studied extensively in the early days of nonlinear optics, and was then largely forgotten for 20 years until it was rediscovered as a means to create terahertz radiation. The process is described by the formula

$$P_i^{DC} = \tfrac{1}{2}\varepsilon_0 \sum_{jk} \chi_{ijk}^{OR}(0;\ \omega, -\omega)\hat{E}_j(\omega)\hat{E}_k^*(\omega) \tag{4.31}$$

from Eq. (4.7). At this point, the principle of full permutation symmetry enables us to write

$$\chi_{kji}^{OR}(0;\ \omega, -\omega) = \chi_{ijk}^{PE}(\omega; \omega, 0). \tag{4.32}$$

This exact relationship between the Pockels effect and optical rectification coefficients can be incorporated into Eq. (4.26) to yield

$$r_{ijk} = r_{jik} = r_{pn} = -\frac{2\chi_{ijk}^{PE}(\omega;\ \omega, 0)}{\varepsilon_{ii}\varepsilon_{jj}} = -\frac{2\chi_{kji}^{OR}(0;\ \omega, -\omega)}{\varepsilon_{ii}\varepsilon_{jj}}. \tag{4.33}$$

We have therefore discovered two nonlinear optical effects with a direct numerical relationship between their coefficients.

Problems

4.1 What are the units of the second-order nonlinear coefficients defined in this chapter?

4.2 Consider a possible set-up for implementing the situation envisaged in Problem 1.5. What beam directions could be used and what nonlinear coefficient would mediate the interaction?

4.3 Try to prove Eq. (4.19). A good start is to write down the components of unit vectors in the o and e directions. It does not matter which direction of the double arrows you count as positive, provided you are consistent throughout the calculation. This is a tricky question.

4.4 Prove Eq. (4.29). This formula, which involves an approximation, was used to plot Fig. 4.3.

4.5 The electro-optic coefficient r_{63} for ammonium dihydrogen phosphate (ADP) is 8.5 pm V^{-1}. Find the quarter-wave voltage at the wavelength of a HeNe laser. ($n_0(632.8 \text{ nm})$ in ADP $= 1.522$.)

5 Third-order nonlinear processes

5.1 Introduction

Third-order nonlinear processes are based on the term $\varepsilon_0 \chi^{(3)} E^3$ in the polarisation expansion of Eq. (1.24). Just as the second-order processes of the previous chapter coupled three waves together, so third-order processes couple four waves together, and hence are sometimes called *four-wave processes*.

Third-order processes are in some ways simpler and in other ways more complicated than their second-order counterparts. Perhaps the most important difference is that third-order interactions can occur in centrosymmetric media. This means that, while crystals can be used if desired, one is often dealing with optically isotropic materials in which the complexity of crystal optics is absent. However, the tensor nature of the third-order coefficients is still not entirely straightforward, even in isotropic media.

A second issue is that phase matching is automatic for many third-order processes and, even when it is not, the existence of *four* waves, potentially travelling in four different directions, makes phase matching easier to achieve. Of course, phase matching was also automatic at second order for the Pockels effect and optical rectification, but those were somewhat special cases involving DC fields.

At first sight, automatic phase matching sounds like a good thing, but the benefit (if it is such) comes at a serious price. When phase matching was critical, and one had to work to achieve it for a particular combination of frequencies, it did at least provide a way of promoting one particular nonlinear process, while discriminating against all the others. The problem with the third-order processes tends to be that, when one of them occurs, so do several others, and disentangling them can then be very difficult.

In principle, we could follow the same routine at third order as we did for the second-order processes in Chapter 4. Although all the ideas are the same, the higher order makes things more complicated than before, and then it was bad enough! Here, we shall make things simpler by confining attention almost exclusively to structurally isotropic media, i.e. non-crystalline media such as gases, liquids and amorphous solids.

In Section 5.2, we will discuss several basic third-order interactions including third harmonic generation (THG), the DC and optical Kerr effects, and intensity-dependent refractive index (IDRI). IDRI in particular has widespread applications in optical pulse technology, some of which are covered in Chapter 7. Stimulated scattering processes are introduced in Section 5.3 by extending the Kerr effect analysis to include a material resonance. The stimulated Raman effect is discussed from two different perspectives, one of which highlights the travelling wave nature of the Raman excitation. The insight gained from this exercise carries over into Section 5.4, which is concerned with the interaction of optical and acoustic waves. We consider two cases involving the diffraction of light by ultrasound, before moving on to stimulated Brillouin scattering in Section 5.5. Section 5.6 deals with optical phase conjugation, and the chapter ends with a brief comment on supercontinuum generation in Section 5.7.

5.2 Basic third-order processes

5.2.1 Definitions

The most general third-order nonlinear process involves the interaction of waves at four different frequencies linked by $\omega_1 + \omega_2 + \omega_3 = \omega_4$. Fortunately, in all common cases, some of the frequencies are the same, and some may also be zero, or the negatives of others. Important examples including third harmonic generation and three different variants of the Kerr effect were mentioned in Chapter 1 in connection with Eqs (1.25) and (1.27). Those equations were of course based on a simple formalism where the nonlinear coefficients were assumed to be scalar quantities. However, to appreciate all the features of these effects, and especially their response to different polarisation states of light, one cannot avoid treating the coefficients as tensors, even in the case of structurally isotropic media. Since the associated mathematics is intricate, most of the detail has been relegated to Appendix C.

According to the most general formula (see Eq. C3.1), and in close analogy with Eq. (4.3), the polarisation at ω_4 is given by

$$\hat{P}_i(\omega_4) = \tfrac{1}{4}\varepsilon_0 \sum_p \sum_{jkl} \chi^{(3)}_{ijkl}(\omega_4;\ \omega_1,\omega_2,\omega_3) \hat{E}_j(\omega_1)\hat{E}_k(\omega_2)\hat{E}_l(\omega_3) \qquad (5.1)$$

where $ijkl$ can each be x, y or z, and $\sum_p$ indicates that the right-hand side is to be summed over all distinct permutations of ω_1, ω_2 and ω_3. This means that the form of the polarisation depends on the frequency set, and so is process-specific. Some particular examples are:

Third harmonic generation $(\omega + \omega + \omega = 3\omega)$

$$\hat{P}_i(3\omega) = \tfrac{1}{4}\varepsilon_0 \sum_{jkl} \chi^{\text{THG}}_{ijkl}(3\omega;\ \omega,\omega,\omega) \hat{E}_j(\omega)\hat{E}_k(\omega)\hat{E}_l(\omega). \qquad (5.2)$$

Third harmonic generation is closely analogous to second harmonic generation at second order, and is discussed in Section 5.2.3 below.

The DC Kerr effect $(\omega = 0 + 0 + \omega)$

$$\hat{P}_i(\omega) = 3\varepsilon_0 \sum_{jkl} \chi^{\text{K}}_{ijkl}(\omega;\ 0,0,\omega) E_j(0) E_k(0)\hat{E}_l(\omega). \qquad (5.3)$$

The first nonlinear effect to be observed (see Chapter 1), this concerns refractive index changes caused by an applied DC field; see Section 5.2.4.

The optical Kerr effect $(\omega_1 = \omega_2 - \omega_2 + \omega_1)$

$$\hat{P}_i(\omega_1) = \tfrac{3}{2}\varepsilon_0 \sum_{jkl} \chi_{ijkl}^{OK}(\omega_1; \omega_2, -\omega_2, \omega_1)\hat{E}_j(\omega_2)\hat{E}_k^*(\omega_2)\hat{E}_l(\omega_1). \tag{5.4}$$

In this case, the refractive index of an optical wave at ω_1 is modified in the presence of a wave at ω_2; more details are given in Section 5.2.5. If the difference between the frequencies of the two waves coincides with a resonance in the nonlinear medium, the same formula governs stimulated Raman scattering, which is discussed in Section 5.3.

Intensity-dependent refractive index $(\omega = \omega - \omega + \omega)$

$$\hat{P}_i(\omega) = \tfrac{3}{4}\varepsilon_0 \sum_{jkl} \chi_{ijkl}^{IDRI}(\omega; \omega, -\omega, \omega)\hat{E}_j(\omega)\hat{E}_k^*(\omega)\hat{E}_l(\omega). \tag{5.5}$$

This is a special case of the optical Kerr effect in which the two waves of Eq. (5.4) are one and the same, and a single wave modifies its own refractive index; see Section 5.2.6.

A full justification of Eqs (5.2)–(5.5) is given in Section C3 of Appendix C. It is worth pointing out that the different numerical pre-factors in Eqs (5.2)–(5.5) result from the permutation operation in Eq. (5.1). Once this has been performed, there is no more permutation to do and, thereafter, the order of the field frequencies in the coefficient arguments is fixed by definition. The pre-factors are incidentally identical to those in Eqs (1.25) and (1.27) of Chapter 1.

It should also be noted that in cases where two (or more) optical waves are involved, these need not necessarily travel in the same direction. There is, for example, no reason in principle why the two waves in Eq. (5.4) need be collinear, and in this case their frequencies could even be the same. An important special case known as 'degenerate four-wave mixing' (DFWM) involves four waves with the same frequency, but travelling in different directions; see Eq. (5.104).

5.2.2 Symmetry considerations

Now that the tensor character of the nonlinear coefficients is explicit, we need to consider the effect of symmetry. All the principles developed at second order in Chapter 4 extend to the third-order case. The key conclusion is that, in a lossless crystal of the most general triclinic symmetry, there are $3^4 = 81$ independent nonlinear coefficients. For other symmetry classes, the number is lower, and tables listing the independent coefficients in each class can be found in Boyd [1]. We are restricting ourselves to structurally isotropic media, in which there are no intrinsic axes, all directions are equivalent, and the orientation of the *xyz*-axes can be chosen to make calculations as simple as possible.[1] In this case, only 21 of the 81 coefficients are non-zero, and these are of four types: type 1 (three members) in which all indices are identical ($\chi_1 \equiv \chi_{iiii}$), and types 2, 3 and 4 (six members each) in which two pairs of indices are the same, namely $\chi_2 = \chi_{jjkk}$, $\chi_3 = \chi_{jkjk}$ and $\chi_4 = \chi_{jkkj}$ ($j \neq k$). The

---

[1] Once again, it is important to distinguish between the *structural* isotropy of gases, liquids and amorphous solids, and *optical* isotropy, which is exhibited by cubic crystals too (see Table 3.1). By its very nature, a crystal is not structurally isotropic although, in cubic crystals, all directions are optically equivalent.

numbering scheme ensures that indices 1 and 2 are the same in χ_2, indices 1 and 3 in χ_3, and 1 and 4 in χ_4. Within each type, all members are equal and, as we shall show in a moment, the symmetry of a structurally isotropic medium imposes the further constraint that

$$\chi_1 = \chi_2 + \chi_3 + \chi_4. \tag{5.6}$$

In terms of indices, the 21 non-zero coefficients can be listed as follows:

$$
\begin{aligned}
&\text{type 1}: \; xxxx = yyyy = zzzz \; (= xxyy + xyxy + xyyx)\\
&\text{type 2}: \; xxyy = yyzz = zzxx = yyxx = zzyy = xxzz\\
&\text{type 3}: \; xyxy = yzyz = zxzx = yxyx = zyzy = xzxz\\
&\text{type 4}: \; xyyx = yzzy = zxxz = yxxy = zyyz = xzzx
\end{aligned}
\tag{5.7}
$$

where the final step of the first line follows from Eq. (5.6). The key conclusion of Eqs (5.6) and (5.7) is that a structurally isotropic medium has just three independent third-order coefficients. As we shall discover in the following sections, further links between the different types of coefficient apply in certain cases. For collinear beams, it makes sense to set the z-axis along the direction of propagation, in which case all coefficients involving z in Eq. (5.7) can be ignored; this reduces the number of relevant non-zero coefficients from 21 to 8. Moreover, if all beams are plane polarised in the same direction, the x-axis can be chosen as the direction of polarisation, in which case the only relevant coefficient is $\chi_{xxxx} = \chi_1$.

In cubic crystals, which are optically (but not structurally) isotropic, the same 21 elements as in Eq. (5.7) are non-zero, but there are fewer links between them. For crystal classes 29, 31, and 32, the number of independent elements is four, the only change in this case being that Eq. (5.6) no longer applies. For classes 28 and 30, on the other hand, there are seven independent elements.[2]

5.2.3 Third harmonic generation

This is the simplest third-order effect to analyse. The generation process is a direct extension of the second harmonic case at second order, and has already been treated in Section 2.3.4 of Chapter 2. Since all field frequencies on the right-hand side of Eq. (5.2) are the same, permuting the last three coordinate indices (j, k and l) of the THG coefficients makes no difference, and so $\chi_2 = \chi_3 = \chi_4$. In the case of plane-polarised incident light, everything is especially straightforward since the only element of interest is χ_1. Of course, there is nothing to stop one choosing coordinate axes that are rotated (say by 45°) with respect to the incident field, as shown in Fig. 5.1. A simple exercise (see Problem 5.1) demonstrates that the effective coefficient with respect to the new axes is

$$\chi_{\text{eff}} = \tfrac{1}{2}(\chi_1 + \chi_2 + \chi_3 + \chi_4). \tag{5.8}$$

Clearly χ_{eff} must equal χ_1, since the coefficient cannot depend on the choice of axes, and this constitutes a proof of Eq. (5.6). It is also quite easy to show (see Problem 5.2) that the THG polarisation goes to zero in the case of circularly polarised light.

---

[2] The type 2, 3 and 4 coefficients divide into two sets of three at the mid-point in Eq. (5.7).

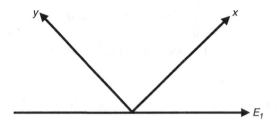

Fig. 5.1   Slanting axes used to prove Eq. (5.8); see Problem 5.1.

5.2.4 The DC Kerr effect and the Kerr cell

In the DC Kerr effect,[3] a strong DC field changes the refractive index of a medium. If the DC field is y-polarised, Eq. (5.3) indicates that the respective polarisations in the x- and y-directions are

$$\hat{P}_x(\omega) = 3\varepsilon_0 \chi^K_{xyyx}(\omega;\ 0,0,\omega) E_y^2(0)\hat{E}_x(\omega) = 3\varepsilon_0 \chi^K_4 E_y^2(0)\hat{E}_x(\omega)$$

$$\hat{P}_y(\omega) = 3\varepsilon_0 \chi^K_{yyyy}(\omega;\ 0,0,\omega) E_y^2(0)\hat{E}_y(\omega) = 3\varepsilon_0 \chi^K_1 E_y^2(0)\hat{E}_y(\omega).$$

$$(5.9)$$

It follows that the DC field creates a refractive index difference between the two polarisations given by

$$n_\| - n_\perp \cong \frac{3(\chi^K_1 - \chi^K_4)E_y^2(0)}{2n} = \frac{3\chi^K_2 E_y^2(0)}{n}$$

$$(5.10)$$

where $n_\|$ and $n_\perp$ are the respective indices for light polarised parallel and perpendicular to the DC field, and n is the zero field index. The second step follows from Eq. (5.6), and because χ^K_2 and χ^K_3 are indistinguishable from Eq. (5.3). The Kerr constant K of a medium is defined by the equation

$$\Delta n \equiv n_\| - n_\perp = \lambda_0 K E^2(0)$$

$$(5.11)$$

where λ_0 is the vacuum wavelength. This implies that $K = 3\chi^K_2/(\lambda_0 n)$ for consistency with Eq. (5.10). The sign of K is normally positive, although exceptions occur.

An optical switch based on the DC Kerr effect can be based on the same ideas as those used in a Pockels cell (see Section 4.7). A Kerr cell is placed between a pair of crossed polarisers set at 45° to the x- and y-axes. In the end-on view shown in Fig. 5.2, the electrodes are set to apply a DC field in the y-direction, while PP and AA indicate the respective alignments of the polariser and analyser. From Section 3.4, we know that to rotate the plane of polarisation of light by 90°, a phase change of π needs to be introduced between the x- and y-components of the field. From Eq. (5.10), this is clearly realised when the light propagates through the half-wave distance is given by

$$L_\pi = \frac{\lambda_0}{2\,|\Delta n|} \cong \frac{1}{2\,|K|\,E(0)^2}.$$

$$(5.12)$$

---

[3] Sometimes called simply the Kerr effect, this was the first nonlinear optical effect to be observed [2].

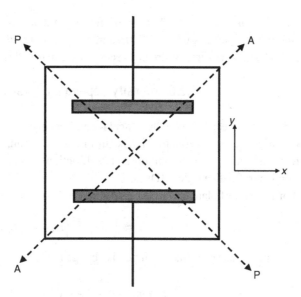

Schematic diagram of a Kerr cell looking along the line of the beam. The lines PP and AA show the alignment of polariser and analyser.

For liquids (such a nitrobenzene) commonly used in Kerr cells values of K are typically on the order of 10^{-12} m/V. This implies (see Problem 5.3) that Kerr cell switching voltages tend to be in the tens of kV range. Kerr liquids exhibit a multiplicity of relaxation times, mostly in the subpicosecond regime, but the response times of real Kerr cells are determined by other factors, and 100 ps is certainly achievable.

5.2.5 The optical Kerr effect

In the optical Kerr effect,[4] a strong wave at frequency ω_2 and intensity $I(\omega_2)$ changes the refractive index of a weak probe wave at ω_1, a process known as cross-phase modulation; the acronym XPM is often used. If the strong and the weak waves have the same polarisation, the operative term in the polarisation is

$$\hat{P}_x(\omega_1) = \tfrac{3}{2}\varepsilon_0 \chi^{OK}_{xxxx}(\omega_1;\ \omega_2, -\omega_2, \omega_1)\left|\hat{E}_x(\omega_2)\right|^2 \hat{E}_x(\omega_1) \tag{5.13}$$

which implies that the refractive index of the weak wave is changed by

$$\Delta n_x \cong \frac{3\chi^{OK}_{xxxx}I(\omega_2)}{2n(\omega_1)n(\omega_2)c\varepsilon_0}. \tag{5.14}$$

If, on the other hand, the weak and strong waves are, respectively, x- and y-polarised, Eq. (5.13) becomes

$$\hat{P}_x(\omega_1) = \tfrac{3}{2}\varepsilon_0 \chi^{OK}_{xyyx}(\omega_1;\ \omega_2, -\omega_2, \omega_1)\left|\hat{E}_y(\omega_2)\right|^2 \hat{E}_x(\omega_1). \tag{5.15}$$

---

[4] This is sometimes known as the AC Kerr effect.

This is the same as Eq. (5.13) apart from the fact that it contains a type 4 coefficient, and so the index change is weaker. One cannot say that it is *three times* as weak, because the type 2, 3 and 4 coefficients are not necessarily equal in this case.

5.2.6 Intensity-dependent refractive index

An important special case of the optical Kerr effect occurs when a single beam at $\omega = \omega_1 = \omega_2$ modifies its own refractive index. In the next section, we will discuss some physical effects that arise from this important 'self-action' effect, which is known as intensity-dependent refractive index (IDRI).

For the case of plane-polarised light, Eq. (5.5) can be written in the simple form

$$\hat{P}_x(\omega) = \tfrac{3}{4}\varepsilon_0 \chi_1^{\text{IDRI}} \left| \hat{E}_x(\omega) \right|^2 \hat{E}_x(\omega). \tag{5.16}$$

This implies that the refractive index is changed to[5]

$$n = n_0 + \left(\frac{3\chi_1^{\text{IDRI}}}{4n_0^2 c\varepsilon_0} \right) I = n_0 + n_2 I \tag{5.17}$$

where I is the intensity, n_0 is the low-intensity index, and the equation defines n_2 as the nonlinear refractive index. It is no surprise that the refractive index change implied by Eq. (5.17) is essentially the same as that of Eq. (5.14); the extra factor of 2 in the denominator of n_2 arises from the different pre-factors in Eqs (5.4) and (5.5).

Equations (5.16)–(5.17) include only the type 1 coefficient. However, IDRI exhibits some particularly interesting features in the case of circularly and elliptically polarised light, and so we should also consider these more complicated situations in which other coefficients are involved too. Because the second and fourth frequency indices of the IDRI coefficient are the same (see Eq. 5.5), it follows that $\chi_2^{\text{IDRI}} = \chi_4^{\text{IDRI}} (\neq \chi_3^{\text{IDRI}})$. Under these circumstances, it can be shown (see Eq. C3.8 of Appendix C, and Problem 5.4) that

$$\hat{P}_x = \tfrac{1}{4}\varepsilon_0 \left(A(\hat{\mathbf{E}}.\hat{\mathbf{E}}^*)\hat{E}_x + \tfrac{1}{2}B(\hat{\mathbf{E}}.\hat{\mathbf{E}})\hat{E}_x^* \right)$$
$$P_y = \tfrac{1}{4}\varepsilon_0 \left(A(\hat{\mathbf{E}}.\hat{\mathbf{E}}^*)\hat{E}_y + \tfrac{1}{2}B(\hat{\mathbf{E}}.\hat{\mathbf{E}})\hat{E}_y^* \right) \tag{5.18}$$

where we have adopted the traditional convention [31] that $A = 6\chi_2^{\text{IDRI}}$ and $B = 6\chi_3^{\text{IDRI}}$. Since we continue to focus on waves propagating in the z-direction, we are only considering the x- and y-components of the field and polarisation, and so $\hat{\mathbf{E}}.\hat{\mathbf{E}} = \hat{E}_x\hat{E}_x + \hat{E}_y\hat{E}_y$ etc.

A feature of special interest in Eq. (5.18) is the presence of E_x^* and E_y^* in the B terms. The complex conjugation indicates that, for circularly polarised light, the handedness of this contribution to the polarisation is opposite to that of the incident field. This becomes clearer if we write down the clockwise component of the polarisation namely[6]

$$\hat{P}_x - i\hat{P}_y = \tfrac{1}{4}\varepsilon_0 \left(A(\hat{\mathbf{E}}.\hat{\mathbf{E}}^*)(\hat{E}_x - i\hat{E}_y) + \tfrac{1}{2}B(\hat{\mathbf{E}}.\hat{\mathbf{E}})(\hat{E}_x + i\hat{E}_y) \right). \tag{5.19}$$

---

[5] See also Eq. (1.26).
[6] Clockwise looking in the $+z$-direction.

Incident light with counter-clockwise circular polarisation in E will therefore induce a clockwise contribution to the polarisation P of the medium.[7]

A detailed analysis of this feature and its ramifications can be found in Chapter 4 of Boyd [1]. The most important conclusion is that, for elliptically polarised light, the axes of the ellipse rotate under propagation. In contrast, for plane polarised and circularly polarised beams, the state of polarisation is unaffected by the intensity-dependent index change. In the former case, the change is given by Eq. (5.17). For circularly polarised light, the corresponding equation is

$$n = n_0 + \left(\frac{A}{4n_0^2 c\varepsilon_0} \right) I \tag{5.20}$$

so the B coefficient plays no part in this case.

5.2.7   Spatial solitons, self-focusing, and self-phase modulation

Intensity-dependent refractive index (IDRI) is probably the most important of the third-order processes in terms of its consequences. The spatial modulation of refractive index that it creates has a direct influence on optical beam propagation, while the corresponding temporal modulation affects the amplitude and phase structure of optical pulses, and the spectral structure of optical signals in general. In this section, we will concentrate on the spatial aspects of IDRI. Its contribution to supercontinuum generation is discussed in Section 5.7, and a detailed analysis of its effects on optical pulse propagation (and the many ways in which these can be exploited) is presented in Sections 7.3–7.6.

Consider first an intense optical beam with a bell-shaped transverse intensity profile propagating through a block of glass. The index change of Eq. (5.17) introduces an intensity-dependent spatial phase modulation, altering the shape of the wavefront, and hence its focusing and defocusing properties. Since the nonlinear index n_2 is almost invariably positive,[8] the refractive index will be raised at the centre of the beam where the intensity is greatest. This leads directly to a reduction in the wavelength, with the further consequence that the wavefronts bend in on themselves as shown in Fig. 5.3. If this effect is strong enough

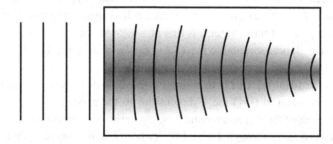

Fig. 5.3   Beam collapse due to self-focusing.

---

[7] Notice both meanings of the word 'polarisation' in this sentence. See footnote 1 in Chapter 1.
[8] In fused silica, n_2 is on the order of $3 \times 10^{-16} \text{cm}^2 \text{W}^{-1}$.

to overcome diffraction, which works in the opposite direction, the self-focusing tendency will predominate, and the beam will get narrower, causing the intensity to rise further and the process to accelerate. In the worst-case scenario, the end result is the catastrophic collapse of the beam, leading to severe and irreversible damage to the optical material.

Fortunately, the competition with diffraction means that 'whole beam self-focusing', which is what has been described, has a definite threshold. Unfortunately, a closely related phenomenon called 'small-scale self-focusing' also occurs, and leads to local hot-spots and filamentation. These effects were first considered by Kelley [32], and were observed by Alfano and Shapiro [33]. Precautions against all forms of self-focusing are routine in large laser systems. These are usually based on spatial filters in which the beam is focused through a small aperture to smooth out distortions in the wavefront. To prevent breakdown near the focus, the process takes place in a vacuum.

The theory of all types of self-focusing has had a long and chequered history. We will concentrate on the simplest case of whole-beam self-focusing in the case of one transverse dimension, i.e. a slit geometry. As we will discover, the competition in this case between diffraction and self-focusing can result in a stable balance, and the formation of spatial solitons. In two transverse dimensions, the analogous solution is normally unstable, which is unfortunate insofar as it would otherwise be possible to create optical beams that propagated without diffractive spreading.

We start by inserting the nonlinear polarisation term from the standard coupled-wave equation (2.8) into the paraxial wave equation (2.47). The resulting equation is

$$i\frac{\partial \tilde{E}^{\omega}}{\partial z} = \frac{1}{2k}\nabla_T^2 \tilde{E}^{\omega} + \frac{\omega n_2 I^{\omega} \tilde{E}^{\omega}}{c} \tag{5.21}$$

where the transverse Laplacian $\nabla_T^2 = \partial^2/\partial x^2 + \partial^2/\partial y^2$ takes care of diffraction, and a simplified version of Eq. (5.16) has been used to form the nonlinear term.

It is important to remember that, without the second term on the right-hand side, Eq. (5.21) is simply the paraxial wave equation, which has the Gaussian beam of Eq. (2.47) as an exact solution. However, when the second term is present, the potential exists for the equation to have a stationary solution, where the transverse profile does not vary with z. This is because, provided n_2 is positive (which it normally is), the phase of the second term is that of $+\tilde{E}^{\omega}$, whereas, for a beam with a bell-shaped transverse profile and a plane wavefront, the phase of the first term is that of $-\tilde{E}^{\omega}$. As we now show, in one transverse dimension, an analytical solution to Eq. (5.21) can be obtained in which the effects of diffraction and nonlinear refraction are in perfect balance.

To adapt Eq. (5.21) to a single transverse dimension, we assume that the electric field is constant in (say) the y-dimension, so the second derivative in y can be dropped. We introduce three characteristic length parameters, a beam width W, the diffraction length L_d linked to W through $L_d = kW^2$ (where k is the angular wave number), and a nonlinear propagation distance defined by

$$L_{nl} = \frac{c}{\omega |n_2| I_{pk}} \tag{5.22}$$

where I_{pk} is the peak intensity. Defined in this way, L_{nl} is the distance over which the nonlinear index change introduces a phase change of 1 radian, while $L_d = kW^2$ is related (though not identical) to the Rayleigh length in Section 2.5.2.[9]

We now use W and L_{nl} to define the dimensionless variables $\zeta' = z/L_d$ and $\xi = x/W$.[10] We also introduce a dimensionless variable for the field $\tilde{U} = \tilde{E}/\tilde{E}_{pk}$ and, for reasons that will emerge very shortly, we define

$$N = \sqrt{L_d/L_{nl}} = \sqrt{k_0^2 \omega n_0 |n_2| I_{pk} W^2}. \qquad (5.23)$$

In terms of these new variables, and assuming that $n_2 > 0$, Eq. (5.21) reduces to the simple form

$$i\frac{\partial \tilde{U}}{\partial \zeta'} = \left(\frac{1}{2}\frac{\partial^2}{\partial \xi^2} + N^2 \left|\tilde{U}\right|^2 \right) \tilde{U}. \qquad (5.24)$$

Equation (5.24) is in the form of the celebrated nonlinear Schrödinger equation, which possesses a hierarchy of remarkable analytical solutions known as solitons [34]. The parameter N defined in Eq. (5.23) is the soliton 'order'.

Solitons arise in a wide variety of physical systems when nonlinear and dispersive effects (diffraction in the present case) are in opposition. In the lowest order ($N = 1$) case, the two terms on the right-hand side of Eq. (5.24) are in perfect balance, and the soliton propagates as a beam of constant width. The $N = 1$ solution of Eq. (5.24) is

$$\tilde{U}(\xi, \zeta') = \text{sech}\{\xi\} \exp\{-\tfrac{1}{2}i\zeta'\}. \qquad (5.25)$$

When this is substituted into the right-hand side of Eq. (5.24), the result is zero, which is of course the mark of a stationary solution. Moreover, since $N = 1$ in this case, it follows from Eq. (5.23) that

$$\frac{c}{\omega |n_2| I_{pk}}(= L_{nl}) = kW^2(= L_{disp}) \qquad (5.26)$$

so the peak intensity of the soliton is related to the width parameter W through the equation

$$I_{pk}W^2 = \frac{c^2}{\omega^2 n_0 |n_2|}. \qquad (5.27)$$

The physical meaning of this equation is that the smaller the beam width W, the stronger the diffraction, and the higher the peak intensity needs to be to maintain the balance between the focusing effect of the nonlinear term, and the defocusing effect of diffraction. Rewriting Eq. (5.25) in terms of the original parameters yields

$$\tilde{E}(x, z) = \tilde{E}_{pk}\text{sech}\{x/W\} \exp\{-\tfrac{1}{2}iz/L_{nl}\}. \qquad (5.28)$$

---

[9] If W is identified with w_0 in Gaussian beam theory, the diffraction length $L_d = kW^2$ corresponds to twice the Rayleigh length, otherwise known as the confocal parameter $b = 2z_R$.

[10] The prime on ζ reflects the $\times 2$ difference between L_d and the Rayleigh length noted in the previous footnote.

We could of course have written L_d instead of L_{nl} because the two are the same when $N = 1$. The final term on the right-hand side implies that the phase velocity of the soliton solution is slightly less than that of a free-propagating wave, which is expected given that the index at the centre of the beam is raised by the nonlinear refractive index term.

Spatial solitons in a slit geometry have been observed experimentally [35,36], and have potential applications in all-optical devices. However, the problem we have been discussing has an exact parallel in the context of optical pulse propagation, where ξ becomes a time variable, and group velocity dispersion takes the role that diffraction plays here. This important topic is treated in Section 7.4, where higher-order solitons are discussed, and pictures of the soliton solutions presented.

To what extent can the results we have obtained be extended to the case of two transverse dimensions? Even without doing any more mathematics, an interesting feature of this new case can be deduced from Eq. (5.27). In circular geometry where the beam widths in the x and y dimensions are the same, W^2 will be a measure of the beam area, and Eq. (5.27) now indicates that the balance between diffraction and self-focusing depends on power![11] Moreover, according to the equation, the critical power for self-focusing depends only on the wavelength of the light, and the nonlinear refractive index of the medium. The standard formula for critical power, which is the same as the right-hand side of Eq. (5.27) apart from a numerical factor, is [37]

$$P_{\text{crit}} = \alpha(\lambda^2/4\pi n_0 n_2). \tag{5.29}$$

The parameter α depends on the precise beam profile, and is usually close to 2.

Typical values of P_{crit} are remarkably modest. For fused silica, which has a notably low value of n_2, P_{crit} is still only a few megawatts, a very small laser power by any standard. This raises an immediate question about what can be done to prevent self-focusing occurring. Fortunately, the distance scale over which self-focusing occurs is linked to L_{nl}, which depends on intensity not on power. The power may exceed P_{crit} but, provided the beam is wide and the intensity low, nothing dramatic will happen for a long distance.

There are several other issues too. As we have seen, stable soliton solutions exist in the case of one transverse dimension. However, the corresponding solutions in two transverse dimensions are unstable if the Kerr effect is the only nonlinear process involved. Stable solutions can, however, exist, at least in principle, if saturable absorption occurs at the same time. As noted at the beginning of this section, rather than whole-beam self-focusing, which is what we have been discussing, what tends to happen in practice is that beam profiles become subject to local filamentation, a process that is driven by a complicated four-wave mixing interaction. The detailed theory of self-trapping is complex, and beyond the scope of the present book.

Whereas intensity-dependent refractive index causes self-focusing when the intensity varies in space, it leads to temporal self-phase modulation (SPM) when intensity varies in

---

[11] Previously, W^2 was the square of the beam width in the x-direction, and the extent of the beam in the y-direction was arbitrary.

time.[12] SPM plays a crucial role in many techniques involving short optical pulses, and the topic is treated in detail in Sections 7.3–7.6 of Chapter 7.

5.3 Stimulated Raman scattering

5.3.1 Raman coupling of laser and Stokes waves

We now move on to consider the process of stimulated Raman scattering (SRS). SRS was discovered accidentally by Woodbury and Ng in 1962 [38–39], while they were working with a ruby laser that was Q-switched using a nitrobenzene Kerr cell. As well as 694.3 nm (431.8 THz) radiation, they found that laser action was also occurring at 765.8 nm (391.5 THz), and noticed that the 40.3 THz frequency difference corresponded exactly to a vibrational resonance frequency of the nitrobenzene molecule. They called the device a Raman laser, and the principle is now in widespread use, especially in fibre Raman systems.

SRS is the stimulated version of spontaneous Raman scattering, a process that was first observed in the 1920s using conventional light sources. In those experiments, the scattered frequency component was called the Stokes wave, and this terminology has been retained for the stimulated process.[13] In practice, SRS is frequently based on vibrational resonances, although rotational or electronic states may also be involved. The terms 'stimulated rotational Raman scattering' and 'stimulated electronic Raman scattering' are sometimes used. A further possibility is that the laser and Stokes waves interact via an acoustic wave; this case is called stimulated Brillouin scattering, and is discussed in Section 5.5 below.

For all the third-order processes discussed earlier in the chapter, the optical field frequencies and their simple combinations (i.e. sums and differences) have been assumed to be far from any material resonances, whatever their origin. As a consequence, the nonlinear coefficients were real, and the energy in the optical fields conserved. To discover how SRS originates, we start by writing Eq. (5.13) for the optical Kerr effect in the modified form

$$\tilde{P}(\omega_S) = \tfrac{3}{2}\varepsilon_0 \chi^{(3)}(\omega_S;\ \omega_L, -\omega_L, \omega_S)\left|\tilde{E}(\omega_L)\right|^2 \tilde{E}(\omega_S). \qquad (5.30)$$

To make things simple, the spatial aspects of the original equation have been dropped, and the polarisation and the fields have been replaced by their slowly varying envelopes.[14] In anticipation of their role in stimulated Raman scattering, the frequencies of the strong and weak waves have been written ω_L and ω_S ($< \omega_L$) where L and S stand respectively for 'laser' and 'Stokes'.

---

[12] The word 'temporal' in this sentence is normally omitted, so 'self-phase modulation' (SPM) invariably refers to the time-dependent case.

[13] The reference is to Sir George Stokes, who first discussed the change of wavelength of light in the context of fluorescence in a famous paper of 1852. The terminology is increasingly used to describe any down-shifted frequency.

[14] The relationship between $\hat{P}$ and $\tilde{P}$, and $\hat{E}$ and $\tilde{E}$ is specified in Eqs (2.4)–(2.5). As explained later, the process is automatically phase matched, so direct replacements can be made.

As long as the third-order coefficient in Eq. (5.30) remains real, the Stokes wave polarisation will be in phase with the Stokes field, and the equation will therefore mediate the intensity-dependent refractive index change characteristic of the optical Kerr effect. But suppose the coefficient becomes complex. What will be the effect of the imaginary part? If we substitute Eq. (5.30) with $\chi^{(3)} = \chi' + i\chi''$ into the general coupled-wave equation (2.8), we obtain

$$\frac{\partial \tilde{E}_S}{\partial z} = \frac{3\omega_S}{4cn_S}(\chi'' - i\chi')\left|\tilde{E}_L\right|^2 \tilde{E}_S \qquad (5.31)$$

where the field frequencies are now indicated by suffices. Clearly the sign of χ'' is critical, since it determines whether the Stokes field grows or decays. To resolve the issue, we anticipate a result from Chapter 9 where the quantum mechanical basis of SRS is discussed. Equation (9.10) indicates that χ'' is positive, in the absence of a population inversion in the nonlinear material. If the population is concentrated in the ground state, we duly conclude that the Stokes wave in Eq. (5.31) will grow, and this is the basis of the SRS process. The corresponding equation governing the growth of the Stokes wave intensity $I_S = \frac{1}{2}n_S c\varepsilon_0 \left|\tilde{E}_S\right|^2$ is readily shown to be

$$\frac{\partial I_S}{\partial z} = \left(\frac{3\omega_S \chi_R'' I_L}{c^2 \varepsilon_0 n_P n_S}\right) I_S = g_S I_S \qquad (5.32)$$

where the second step defines the factor in brackets as the Stokes gain coefficient g_S. The subscript R (for Raman) has been added to the nonlinear coefficient at the same time. If the energy conversion from ω_L to ω_S is small, we can assume that the laser intensity is undepleted, in which case Eq. (5.32) has the straightforward exponential solution

$$I_S(z) = I_S(0) \exp\{g_S z\}. \qquad (5.33)$$

According to this equation, the Raman gain coefficient is proportional to the pump laser intensity, so enormous gain factors are potentially available by increasing I_L.[15] Indeed, the Stokes wave will grow from noise in a cell of length L, once $\exp\{g_S L\}$ gets sufficiently large. When $g_S L \sim 25$, for example, the gain factor is approaching $\sim 10^{11}$, which is likely to be more than sufficient. Once the stimulated Raman process takes off, the assumption that the pump laser field is undepleted is of course likely to break down, and we will consider this aspect of the problem shortly.

A simple energy level diagram of SRS is shown in Fig. 5.4(a). Peak gain occurs when the difference frequency ($\omega_L - \omega_S$) between the laser and Stokes waves coincides with a resonance (at angular frequency Ω) in the medium. As mentioned already, the level associated with the resonance (labelled 2 in Fig. 5.4) is often a vibrational state, but it could also be rotational or electronic in origin.[16] As already noted (and suggested by the figure), the growth of the Stokes field will be accompanied by depletion of the laser field, which is

---

[15] We refer to g_S as the gain *coefficient* and $\exp\{g_S z\}$ as the gain *factor*. It is important to distinguish the two, and good practice to avoid using the word 'gain' on its own.

[16] Strictly speaking, all frequency labels in Fig. 5.4 should be multiplied by $\hbar$ as befits an energy level diagram.

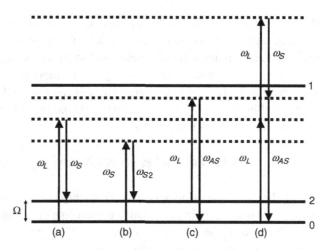

Fig. 5.4 Schematic energy level diagrams for stimulated Raman scattering (a) first Stokes, (b) second Stokes, (c) first anti-Stokes, (d) parametric anti-Stokes.

governed by equations analogous to Eqs (5.31) and (5.32), namely

$$\frac{\partial \tilde{E}_L}{\partial z} = \frac{3\omega_L}{4cn_L} \left(-\chi_R'' - i\chi_R' \right) \left| \tilde{E}_S \right|^2 \tilde{E}_L \tag{5.34}$$

$$\frac{\partial I_L}{\partial z} = -\frac{3\omega_L \chi_R''}{c^2 \varepsilon_0 n_L n_S} I_S I_L. \tag{5.35}$$

For every photon gained by the Stokes wave, one is lost by the laser, but optical energy is not conserved because the population of the Raman level grows at the same time. The sign changes in Eqs (5.34)–(5.35) compared to Eqs (5.31)–(5.32) arise because the nonlinear coefficient governing the polarisation at ω_L is the conjugate of the coefficient in Eq. (5.30) since

$$\chi^{(3)}(\omega_L;\ \omega_S, -\omega_S, \omega_L) = \chi^{(3)*}(\omega_S;\ \omega_L, -\omega_L, \omega_S). \tag{5.36}$$

There has been no mention of phase matching in the discussion of SRS simply because it is automatic, as indeed it is for the optical Kerr effect. Had the spatial aspects of the travelling waves been shown explicitly in Eq. (5.30), that equation would have read

$$\tilde{P}(\omega_S)\exp\{i(\omega_S t - \mathbf{k}_S.\mathbf{r})\} = \tfrac{3}{2}\varepsilon_0 \chi^{(3)} \left| \tilde{E}(\omega_L) \right|^2 \tilde{E}(\omega_S)\exp\{i(\omega_S t - \mathbf{k}_S.\mathbf{r})\} \tag{5.37}$$

where

$$\left| \tilde{E}(\omega_L) \right|^2 = \left| \tilde{E}(\omega_L)\exp\{i(\omega_L t - \mathbf{k}_L.\mathbf{r})\} \right|^2. \tag{5.38}$$

All the space-time terms therefore cancel out, which is another way of saying that the wave vector mismatch is zero, namely

$$\Delta\mathbf{k} = \mathbf{k}_S - (\mathbf{k}_L - \mathbf{k}_L + \mathbf{k}_S) = 0. \tag{5.39}$$

This equation also highlights the fact that, since the SRS process is automatically phase matched, the Stokes wave is free to grow in any direction it chooses. Of course, the laser field is normally in the form of a beam, and so Stokes waves propagating in roughly the same direction will receive preferential growth simply because that gives them the best opportunity to feed off the laser energy. But this criterion could equally be met by a Stokes wave travelling in the opposite direction to the pump and, indeed, backward stimulated Raman scattering is a viable process.

5.3.2 Anti-Stokes generation and higher-order effects

It would be more accurate to refer to the frequency component at $\omega_S = \omega_L - \Omega$ as the first Stokes wave because, once generated, it can become the pump for a further Raman process in which a second Stokes wave at $\omega_{S2} = \omega_L - 2\Omega$ is created; see Fig. 5.4(b). A cascade of further down-shifted frequencies may occur.

Up-shifted (anti-Stokes) frequency components can also be generated through the Raman resonance. There are at least two possible ways in which the first anti-Stokes (AS) component at $\omega_{AS} = \omega_L + \Omega$ can be produced. First of all, by analogy with Eq. (5.30), it could arise from the anti-Stokes polarisation term

$$\tilde{P}(\omega_{AS}) = \tfrac{3}{2}\varepsilon_0\chi^{(3)}(\omega_{AS}; \ \omega_L, -\omega_L, \omega_{AS}) \left| \tilde{E}(\omega_L) \right|^2 \tilde{E}(\omega_{AS}). \tag{5.40}$$

As with Eqs (5.30) and (5.31), the key issue is the sign of the imaginary part of the coefficient. Detailed examination of the quantum mechanics (see Chapters 8 and 9) reveals that if most of the population is in the ground state, the wave at ω_{AS} will decay, but if a population inversion exists between states 0 and 2 in Fig. 5.4, the anti-Stokes wave will grow and the Stokes wave will decay. The anti-Stokes process is represented by Fig. 5.4(c); a photon is lost at ω_L, one is gained at ω_{AS}, and the difference in energy is provided by de-excitation of the medium. These conclusions are entirely reasonable. If the population is mostly in the ground state, the process that moves population from state 0 to state 2 dominates, while if the population of state 2 is larger than that of state 0, the reverse process takes precedence.

Up-shifts are also possible through the sequence shown in Fig. 5.4(d), where the anti-Stokes wave is produced by the four-wave mixing combination $\omega_L + \omega_L = \omega_{AS} + \omega_S$, where two laser waves combine to create a Stokes wave and an anti-Stokes wave. This is sometimes referred to as parametric anti-Stokes Raman generation, even though a Stokes wave is driven as well. Whereas the Raman processes discussed so far have all been automatically phase matched, this parametric process[17] involves the phase-matching condition

$$\Delta \mathbf{k}_{AS} = \mathbf{k}_{AS} + \mathbf{k}_S - 2\mathbf{k}_L. \tag{5.41}$$

The corresponding wave vector diagram for phase-matched anti-Stokes generation is shown in Fig. 5.5. If the optical dispersion curve in the region between ω_S and ω_{AS} were a straight line, it is easy to show that the anti-Stokes generation process would be phase matched when all three waves propagated collinearly. Normally, the curvature of the line ensures

_____
[17] See Section 9.4 for a discussion of the word 'parametric'.

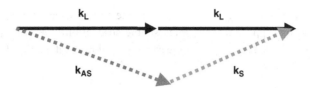

Fig. 5.5   Wave vector diagram for parametric anti-Stokes scattering.

that phase matching occurs when the Stokes and anti-Stokes waves are at a small angle to the laser beam. This gives rise to the ring structure which is characteristic of Raman generation via this (and other) four-wave mixing process.

There are many other possibilities too. For example, the four-wave process $\omega_L + \omega_S = \omega_{AS} + \omega_{S2}$, where collinear laser and Stokes waves create angled anti-Stokes and second Stokes waves, can be phase matched in a similar scheme to Fig. 5.5. And a new and exciting scenario is created if *two* strong waves with frequency separation Ω are inserted into the Raman medium. In this case, a wide frequency comb of waves extending in both the Stokes and anti-Stokes directions is created in a process called ultra-broadband multi-frequency Raman generation [40–42].

5.3.3 Raman waves

In the treatment of Section 5.3.1, the physical origin of the Raman resonance was hidden inside the nonlinear coefficient. We noted that the resonance could be electronic, vibrational, or rotational in origin, but the characteristics of the associated oscillations did not feature anywhere in the analysis. We will now discuss an alternative approach to SRS in which the Raman oscillations appear explicitly in the calculation.

For the purposes of discussion, we will assume that the Raman resonance has its origin in the molecular vibrations in a liquid. An important concept to grasp at the outset is that the vibrational oscillators at different positions in the nonlinear medium are driven by the laser and Stokes waves in such a way that specific phase relationships are established between them. Together, these oscillators create a travelling vibrational wave, with its own wavelength, wave vector, and dispersion characteristics.[18] The idea that disconnected oscillators can be phased to create a travelling wave may be unfamiliar to some readers, but sports fans will immediately recognise the analogy with the 'audience wave' phenomenon, where spectators stand and sit in a controlled time sequence, creating a wave that travels around a large stadium.[19]

We start by going back to first principles and writing the equation for the linear polarisation in the simple form

$$P = \varepsilon_0 N \alpha E \tag{5.42}$$

[18] These have been called 'Raman waves' in the section heading to cover the more general possibility that they might alternatively be rotational or electronic in origin.

[19] This is also known as a 'stadium wave' or (in the UK) as a 'Mexican wave' because it first attracted widespread attention during the 1986 football World Cup in Monterrey, Mexico.

where $N\alpha\ (= \chi^{(1)} = n^2 - 1)$ is the linear susceptibility, and n is the refractive index. Let us suppose that the microscopic susceptibility α of a molecule depends weakly on a coordinate (call it q) that measures the vibrational amplitude such that

$$\alpha(q) = \alpha_0 + \alpha_1 q + \cdots \tag{5.43}$$

where $\alpha_1 = (\partial\alpha/\partial q)|_{q=0}$. Suppose also that, within the liquid, the molecular displacements are so phased that as a function of time and position, q forms a travelling wave of angular frequency Ω and wave vector $\mathbf{K}$ given by

$$q = \tfrac{1}{2}\tilde{q}\exp\{i(\Omega t - \mathbf{K}.\mathbf{r})\} + \text{c.c.} \tag{5.44}$$

with an associated wave of refractive index

$$n(q) \cong n_0 + (\alpha_1/2n_0)\left[\tilde{q}\exp\{i(\Omega t - \mathbf{K}.\mathbf{r})\} + \text{c.c.}\right]. \tag{5.45}$$

Thus, if $E = \tfrac{1}{2}\left(\tilde{E}_L\exp\{i(\omega t - \mathbf{k}_L.\mathbf{r})\} + \text{c.c.}\right)$, it follows from Eqs (5.43)–(5.44) that the polarisation in Eq. (5.42) will include terms such as

$$P = \cdots + \tfrac{1}{2}\varepsilon_0 N\alpha_1\left(\tilde{E}_L\tilde{q}^*\exp\{i(\omega_S t - (\mathbf{k}_L - \mathbf{K}).\mathbf{r}\}\right.$$
$$\left. + \tilde{E}_L\tilde{q}\exp\{i(\omega_{AS}t - (\mathbf{k}_L + \mathbf{K}).\mathbf{r})\}\right) + \cdots \tag{5.46}$$

where $\omega_S = \omega_L - \Omega$ and $\omega_{AS} = \omega_L + \Omega$. These terms serve as source antennae for the first Stokes and first anti-Stokes waves. Inserting the Stokes term into the general coupled-wave equation (2.8) yields

$$\frac{\partial\tilde{E}_S}{\partial z} = -i\frac{\omega_S}{2cn_S}N\alpha_1\tilde{E}_L\tilde{q}^* \tag{5.47}$$

while the analogous equation for the laser field is

$$\frac{\partial\tilde{E}_L}{\partial z} = -i\frac{\omega_L}{2cn_L}N\alpha_1\tilde{E}_S\tilde{q}. \tag{5.48}$$

The corresponding differential equation for q is

$$\frac{\partial^2 q}{\partial t^2} + \frac{2}{\tau}\frac{\partial q}{\partial t} + \Omega_0^2 q = \frac{F}{m} \tag{5.49}$$

where F is the effective force driving the vibrations, m is the mass, Ω_0 the natural resonance frequency, and τ the damping time. An expression for the force can be found from $F = -du/dq$, where $u = -\tfrac{1}{2}\varepsilon_0\alpha(q)E^2$ is the oscillator energy. With the help of Eq. (5.43), we obtain $F = \tfrac{1}{2}\varepsilon_0\alpha_1\overline{E^2}$, where the time-averaging overbar removes the optical frequency terms. We now substitute F into Eq. (5.49) along with Eq. (5.44) for q, and we assume that both the laser and Stokes fields are already present so that

$$E = \tfrac{1}{2}\left(\tilde{E}_L\exp\{i(\omega_L t - \mathbf{k}_L.\mathbf{r})\} + \tilde{E}_S\exp\{i(\omega_S t - \mathbf{k}_L.\mathbf{r})\} + \text{c.c.}\right). \tag{5.50}$$

Neglecting terms in $\ddot{\tilde{q}}$ and $\Gamma\dot{\tilde{q}}$, we obtain

$$\frac{\partial \tilde{q}}{\partial t} + \frac{\tilde{q}}{\tau} = -i\left(\frac{\varepsilon_0 \alpha_1}{4m\Omega}\right) \tilde{E}_L \tilde{E}_S^* \tag{5.51}$$

where we have set $\Omega = \Omega_0$ and $\mathbf{k}_L - \mathbf{k}_S = \mathbf{K}$.

Equations (5.47), (5.48) and (5.51) are the differential equations governing the three waves involved in the stimulated Raman interaction. In the strongly damped case, it is easy to get from here to the results of the previous subsection. When τ in Eq. (5.51) is small, the time derivative can be dropped, and the vibrational amplitude, which is now slaved to the laser and Stokes fields, is

$$\tilde{q} = -i\left(\frac{\varepsilon_0 \alpha_1 \tau}{4m\Omega}\right) \tilde{E}_L \tilde{E}_S^*. \tag{5.52}$$

When this result is substituted into Eqs (5.47) and (5.48), these equations become

$$\frac{\partial \tilde{E}_S}{\partial z} = +\left\{\frac{\omega_S}{2cn_S}\left(\frac{\varepsilon_0 N \alpha_1^2 \tau}{4m\Omega}\right)\left|\tilde{E}_L\right|^2\right\} \tilde{E}_S \tag{5.53}$$

$$\frac{\partial \tilde{E}_L}{\partial z} = -\left\{\frac{\omega_L}{2cn_L}\left(\frac{\varepsilon_0 N \alpha_1^2 \tau}{4m\Omega}\right)\left|\tilde{E}_S\right|^2\right\} \tilde{E}_L \tag{5.54}$$

which are consistent with Eqs (5.32) and (5.35) if

$$\chi_R'' = \frac{N\varepsilon_0 \alpha_1^2 \tau}{6m\Omega}. \tag{5.55}$$

The terms in curly brackets in Eqs (5.53) and (5.54) are the coefficients governing gain at the Stokes frequency, and loss at ω_L. At the beginning of the process, the Stokes field is small, and the laser field $\tilde{E}_L$ is approximately constant. At all stages, the local value of $\tilde{q}$ follows from Eq. (5.52).

Notice how the final few steps have served to hide the vibrational wave from view once again. As soon as Eq. (5.52) is substituted into Eqs (5.47) and (5.48), and the brackets in Eqs (5.53) and (5.54) have been replaced by χ_R'' from Eq. (5.55), all explicit reference to the molecular vibrations disappears.

To gain further physical insight, we now consider the properties of the vibrational wave of Eq. (5.44). We have seen that the wave consists of a spatial distribution of vibrating molecules whose phases are determined by the interference of the laser and Stokes waves according to Eq. (5.52). In fact, the wave vectors of all three waves are linked by $\mathbf{k}_L = \mathbf{k}_S + \mathbf{K}$, which can be represented by the vector triangle of Fig. 5.6. The figure includes a hint of the wavefront structure of the Raman wave, and this raises the question of its dispersion characteristics. In most circumstances, the oscillators participating in SRS interact with each other extremely weakly, and this means that there is virtually no constraint on the magnitude of $\mathbf{K}$, or on the wavelength of the wave. The dispersion curve is therefore essentially a flat horizontal line positioned at frequency Ω for all $|\mathbf{K}|$, as shown schematically in Fig. 5.7. The velocity of the wave is of course Ω/K and, since Ω is essentially fixed while K is free to vary, the velocity varies widely too. Similar attributes apply

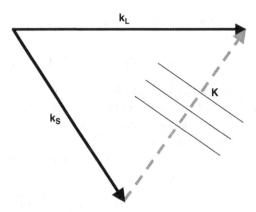

Fig. 5.6   Wavevector triangle for stimulated Raman scattering showing the wave vector **K** of the Raman wave.

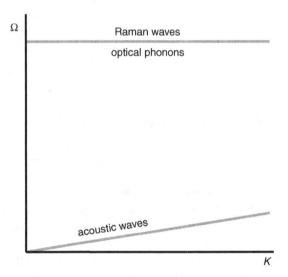

Fig. 5.7   Schematic dispersion diagram (not to scale) for Raman waves (and optical phonons) (top), and acoustic waves (bottom).

incidentally to optical phonons, which mediate the stimulated Raman process in crystalline materials.[20]

Since the magnitude of **K** is unconstrained, the direction of $\mathbf{k}_S$ in Fig. 5.6 is not fixed, and this confirms the fact that the Stokes wave is free to grow in any direction it chooses. This is another way of saying that the stimulated Raman process is automatically phase matched in all directions.

---

[20] Acoustic phonons in a solid are basically sound waves, in which neighbouring atoms are generally displaced in the same direction. For optical phonons, on the other hand, neighbouring atoms are displaced in opposite directions; the oscillations are therefore of a very different kind, and have very different dispersive properties.

Figure 5.6 highlights another interesting aspect of the Raman wave approach, namely that it makes SRS look like a three-wave process, whereas the viewpoint of Section 5.3.1 made it seem like a four-wave process. We have seen the transition from one viewpoint to the other in Eqs (5.52)–(5.55). The distinction between three- and four-wave processes is evidently less clear-cut than we might have thought. Earlier in the book, three-wave processes have normally involved three *electromagnetic* waves coupled by a $\chi^{(2)}$ nonlinearity, and we have become accustomed to the idea that these can only be supported in media lacking a centre of symmetry. As we have seen, however, the Raman waves we are dealing with here are waves of vibration or rotation or (in the case of electronic states) waves of excitation, created in the interference between the laser and Stokes waves. There is no 'one-wayness' in them, and so no violation of symmetry principles.

5.4 Interaction of optical and acoustic waves

5.4.1 Physical basis

In their simplest form, acoustic waves are waves of compression and rarefaction, waves therefore in which the density varies in space and time about its mean value. Since refractive index depends on density, acoustic waves are associated with waves of refractive index, and we have already seen in Section 5.3.3 how a refractive index wave (see Eq. 5.45) couples the laser and the Stokes waves in stimulated Raman scattering. This suggests that light and acoustic waves can interact too, and acousto-optics is the name given to this field of study.

The dictionary definition of 'acoustic' is 'relating to sound', but it must be stressed at the outset that the acoustic frequencies we will be dealing with in this section are far higher than 20 kHz, which is normally regarded as the upper limit of hearing for a young person. We shall rather be considering the frequency range from a few MHz up to a few GHz, which is at the high end of the ultrasound spectrum.

A number of different physical processes are potentially involved in the interaction of light and ultrasound, and it is worth listing some of them and summarising their key features.

The photoelastic effect: Also known as the piezo-optic effect, this concerns the change in the dielectric properties of a material in the presence of strain. Since photoelasticity is often measured by recording changes in birefringence, it is sometimes thought to occur only in anisotropic media. But in fact the symmetry properties of the photoelastic tensor are identical to those of the third-order coefficients described in Section 5.2.1 above, and so photoelasticity can certainly occur in amorphous media.

Electrostriction: This is a universal property of dielectrics in which stress and strain are created in the presence of an electric field. The effect varies as the square of the applied field (so it is not reversed when the sign of the field is reversed), and it is sometimes known as quadratic electrostriction for this reason. Thermodynamic arguments indicate that electrostriction and photoelasticity are closely interrelated.

Piezoelectricity: The piezoelectric effect describes the appearance of an electrical polarisation in a material when a strain is applied. Although this sounds a little like photoelasticity, it differs insofar as piezoelectricity occurs only in non-centrosymmetric crystals. Indeed the symmetry properties of the piezoelectric tensor are identical to those governing second-order nonlinear effects such as second harmonic generation.

The inverse piezoelectric effect:[21] As the name implies, this is piezoelectricity in reverse. Instead of stress/strain causing the medium to become polarised, the inverse piezoelectric effect relates to the *creation* of stress/strain in a non-centrosymmetric medium by the application of an electric field. Unlike electrostriction, the effect is directly proportional to the applied field, and the sign changes when the field is reversed. In broad terms, inverse piezoelectricity is to electrostriction as the Pockels effect is to the DC Kerr effect.

Optical absorption: If two optical beams interfere in an absorbing medium, the material will tend to heat up in the region of the anti-nodes, and the associated thermal expansion can drive an acoustic wave.

   The two most important processes involved in the coupling of optical and acoustic waves are usually photoelasticity and electrostriction. The first describes how ultrasound affects light through the effect of stress and strain on refractive index, while the second describes how light *creates* stress and strain, enabling two optical waves to drive an acoustic wave. It is worth noting that processes of these two kinds were implicated in stimulated Raman scattering; see Eqs (5.42)–(5.45) for the first kind and the discussion after Eq. (5.49) for the second. As we shall discover, a Raman-type process known as stimulated Brillouin scattering (SBS) exists in which an acoustic wave plays the role of the vibrational wave of Section 5.3.3. However, there are important differences between SBS and SRS, which arise because acoustic frequencies are much lower than vibrational frequencies, and because the dispersion characteristics of ultrasound are very different from those of Raman waves. For the vibrational waves of Section 5.3.3, we have seen that frequency is virtually independent of wavelength, so the schematic dispersion relation in Fig. 5.7 is essentially horizontal. In contrast, the dispersion relation for an acoustic wave is much closer to a straight line through the origin, and this is represented by the lower plot in Fig. 5.7; the gradient of the line is basically the velocity of an acoustic wave. It must be stressed that the figure is purely schematic, and completely out of scale. The ultrasound frequencies we will be considering are four to five orders of magnitude smaller than typical vibrational frequencies so, if Fig. 5.7 were drawn to scale and the Raman wave line were

---

[21] Also known as the 'converse', 'reciprocal', or 'reverse' piezoelectric effect.

visible at the top, the acoustic dispersion line would be indistinguishable from the K-axis at the bottom.

In Sections 5.4.2 and 5.4.3, we will consider two situations where a laser signal is diffracted from an acoustic wave generated externally, that is itself essentially unaffected by the diffraction process. Then, in Section 5.5, we will move on to stimulated Brillouin scattering, in which the acoustic wave plays the role of the Raman wave in stimulated Raman scattering.

5.4.2 Diffraction of light by ultrasound

Consider an acoustic wave of angular frequency Ω_A travelling in the y-direction. The longitudinal acoustic displacement Q can be written

$$Q = \tfrac{1}{2} \left(\tilde{Q} \exp\{i(\Omega_A t - K_A y)\} + \text{c.c.} \right), \qquad (5.56)$$

where $\tilde{Q}$ is the slowly varying complex amplitude, $K_A = \Omega_A V$ is the angular wave number, and V is the wave velocity. Since in one dimension, strain $S = \partial Q / \partial y$, the associated acoustic strain wave is given by

$$S = \tfrac{1}{2} \left(-i K_A \tilde{Q} \exp\{i(\Omega_A t - K_A y)\} + \text{c.c.} \right). \qquad (5.57)$$

The presence of strain influences the propagation of optical fields through changes in the dielectric constant given by

$$\delta\varepsilon = -\gamma_e S \qquad (5.58)$$

where γ_e is the electrostrictive constant.[22] It follows that

$$\delta\varepsilon = \tfrac{1}{2}\gamma_e \left(i K_A \tilde{Q} \exp\{i(\Omega_A t - \mathbf{K}_A.\mathbf{r})\} - i K_A \tilde{Q}^* \exp\{-i(\Omega_A t + \mathbf{K}_A.\mathbf{r})\} \right) \qquad (5.59)$$

where the equation has been generalised to allow for propagation in an arbitrary direction. The change in the dielectric constant leads to a nonlinear polarisation $P^{\mathrm{NL}} = \varepsilon_0 \delta\varepsilon E$. If the field in this equation is an incident laser beam given by $E = \tfrac{1}{2} \left(\tilde{E}_L \exp\{i(\omega_L t - \mathbf{k}_L.\mathbf{r})\} + \text{c.c.} \right)$ and we use Eq. (5.58) for $\delta\varepsilon$, we obtain

$$P^{\mathrm{NL}} = \tfrac{1}{4}\varepsilon_0 \gamma_e K_A \left(i \tilde{E}_L \tilde{Q} \exp\{i(\omega_L + \Omega_A)t - (\mathbf{k}_L + \mathbf{K}_A).\mathbf{r}\} \right.$$

$$\left. - i \tilde{E}_L \tilde{Q}^* \exp\{i(\omega_L - \Omega_A)t - (\mathbf{k}_L - \mathbf{K}_A).\mathbf{r}\} + \text{c.c.} \right). \qquad (5.60)$$

The terms in this equation clearly represent optical antennae driving up-shifted and down-shifted fields in new directions, in other words the diffraction of light by ultrasound. A schematic diagram of the interaction is shown in Fig. 5.8, where the acoustic wave is

---

[22] The electrostrictive constant is defined by $\gamma_e = (\rho \partial\varepsilon / \partial\rho)_{\rho=\bar{\rho}}$ where $\bar{\rho}$ is the mean value of the density ρ.

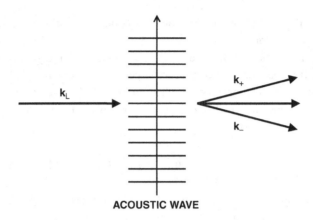

ACOUSTIC WAVE

Fig. 5.8 Diffraction of an optical wave by an acoustic wave.

shown propagating at right angles to the incident laser wave, creating diffracted waves in the directions of $\mathbf{k}_L \pm \mathbf{K}_A$. The process is used in acousto-optic modulators and deflectors.

The diffraction angles drawn in the figure are roughly $\pm 15°$, but a simple calculation shows that deflections of this magnitude are hard to achieve. The angular wave number of an optical wave at 1 μm is $2\pi \times 10^6 \, \mathrm{rad\,s^{-1}}$ and, since $\tan 15° \cong 0.25$, the magnitude of $\mathbf{K}_A$ needs to be about $\frac{1}{2}\pi \times 10^6 \, \mathrm{rad\,s^{-1}}$. Since the velocity of ultrasound in solids is around $4000\mathrm{m\,s^{-1}}$, the necessary acoustic frequency is

$$\frac{\Omega_A}{2\pi} = \frac{10^6 \times 4000}{4} = 1 \, \mathrm{GHz}. \tag{5.61}$$

While acousto-optical devices at a few GHz are now available, frequencies in the 20–200 MHz range are more common.

Although 1 GHz is a high frequency for ultrasound, it is very small compared to typical optical frequencies of around 300 THz. It follows that the relative optical frequency shifts in acousto-optical diffraction are minute, and so the angular wave numbers of the incident and scattered beams are virtually the same (i.e. $k_+ \cong k_- \cong k_L$). This is relevant to the phase matching conditions. Figure 5.9(a) shows that, if the acoustic wave is perpendicular to the incident light wave, the wave vectors cannot form a closed triangle. A simple calculation (see Problem 5.7) shows that the wave vector mismatch in this case is $\Delta k \cong \frac{1}{2}k_L\theta^2$, and so the coherence length for the process is

$$L_{\mathrm{coh}} \cong \frac{\lambda_L}{\theta^2} \cong \frac{\Lambda_A^2}{\lambda_L} \tag{5.62}$$

where λ_L and Λ_A are the respective optical and acoustic wavelengths. This works out as around 16 μm for $\lambda_L = 1$ μm and $\theta = 15°$, although the figure rises to around 150 μm at 5°.

Acousto-optic devices in which the interaction length is substantially smaller than L_{coh} are said to operate in the Raman–Nath regime. Longer interaction lengths can be realised under phase-matched conditions. Phase matching is easily achieved by angling the light and acoustic waves so that one of the triangles is closed, as shown in Fig. 5.9(b). The figure

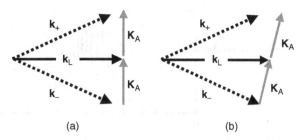

(a)   (b)

Fig. 5.9 Wavevector triangles for (a) the acoustic wave perpendicular to the optical wave (as in Fig. 5.8), (b) a slanted acoustic wave to ensure the phase matching of the $\mathbf{k}_-$ wave.

shows the case where

$$\mathbf{k}_L = \mathbf{k}_- + \mathbf{K}_A. \tag{5.63}$$

Notice that, since $k_L \cong k_-$, the closed triangle in Fig. 5.9(b) is effectively isosceles. Equation (5.63) is similar to the Bragg condition in X-ray diffraction, and indeed the case where phase matching is achieved in acousto-optics is known as the Bragg regime. Whereas in X-ray diffraction, the 'grating' arises from a stationary spatial pattern of atoms in a crystal, here it characterises a travelling acoustic wave. In fact, it is precisely because the acoustic wave presents a refractive index grating moving slightly away from the incident laser beam in Fig. 5.9(b), that the $\mathbf{k}_-$ beam is slightly down-shifted (by Ω_A) in angular frequency. The frequency change is consistent with the Doppler shift from a moving mirror. An even closer analogy with X-ray diffraction occurs in a process called stimulated Rayleigh scattering where the laser beam scatters from an essentially stationary refractive index pattern in the nonlinear medium.

5.4.3 The acousto-optic dispersive filter

The acousto-optic programmable dispersive filter (AOPDF) devised by Tournois [43], commonly known by its trade name (the Dazzler), is designed to control and customise the structure of femtosecond optical pulses. Consider, in the simplest case, a pulse of acoustic waves given by

$$Q(z,t) = \tfrac{1}{2} \left(\tilde{Q}(z,t) \exp\{i(\Omega_A t - K_A z)\} + \text{c.c.} \right). \tag{5.64}$$

The pulse travels through a birefringent crystal in the z-direction, collinearly with two optical pulses, the input (1) and the output (2). One of the optical pulses (say the input) travels as an ordinary wave, and the other (the output) as an extraordinary wave. A schematic diagram is shown in Fig. 5.10.

Phase matching is crucial to the operation of the device. The key idea is to tailor the acoustic pulse so that the phase-matching condition determines the point in the crystal at which the input channel couples to the output channel. Because all three waves are collinear, the condition can be written in scalar form as $k_2 = k_1 + K_A$. The frequency condition $\omega_2 = \omega_1 + \Omega_A$ naturally applies too and, since acoustic frequencies (typically

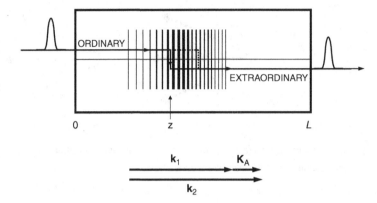

Fig. 5.10   Collinear configuration for an acousto-optic dispersive filter.

$\sim$50 MHz in an AOPDF) are tiny compared to optical frequencies, ω_1 and ω_2 will be the same to about 1 part in 10^7. However, k_1 and k_2 will be significantly different because they are associated with ordinary and extraordinary waves, which have different refractive indices. The relationship between the three angular wave numbers under phase-matched conditions is shown in the lower part of Fig. 5.10.

The velocity of acoustic waves is typically more than five orders of magnitude smaller than the velocity of light, so the acoustic pulse is effectively frozen in the crystal as far as the optical pulses are concerned, as indicated in the figure. Hence we can remove the time dependence from Eq. (5.64) and write

$$Q(z) = \tfrac{1}{2} \left(\left| \tilde{Q}(z) \right| \exp\{-i K'_A(z)z\} + \text{c.c.} \right) \tag{5.65}$$

where the phase characteristics of the acoustic pulse have been absorbed into K'_A, which is now a function of z. Each spectral component of the input pulse will therefore couple into the output channel at the point in the acoustic wave where the phase-matching condition is satisfied, in other words when[23]

$$K'_A(z) = \frac{\omega \left| \tilde{n}_e(\omega) - n_o(\omega) \right|}{c}. \tag{5.66}$$

Once the dispersion characteristics of the crystal are known, this equation can be solved numerically to give z as a function of ω. For a crystal of length L, the frequency-dependent phase accumulated by the optical pulses in the crystal is easily shown to be

$$\phi(\omega) = -\frac{\omega}{c}\left(n_o z(\omega) + \tilde{n}_e(\omega)(L - z(\omega))\right) \tag{5.67}$$

which controls the phase of the transfer function $T(\omega)$ from input to output. However, the process also allows the *amplitude* of the transfer function to be controlled, because this is determined by $\left| \tilde{Q}(\omega) \right|$. Overall we can write

$$\tilde{E}_2(\omega) = T(\omega)\tilde{E}_1(\omega) \tag{5.68}$$

---

[23] $\tilde{n}_e(\omega)$ is the refractive index of the extraordinary wave; see Section 3.3.3.

which implies that amplitude and phase modulation of the optical pulse can be effected simultaneously provided an acousto-optic pulse with the appropriate amplitude and phase structure can be constructed. The problem in practice is of course to find the acoustic profile needed to deliver a desired $T(\omega)$, while compensating for the group velocity dispersion introduced by the device itself at the same time.[24] This non-trivial task has to be done using a specially designed software package.

In summary, the AOPDF enables the structure of a femtosecond pulse to be managed in a programmable way through the mediation of a precisely tailored acoustic signal. However, the details are usually rather more complicated than the simple treatment given here suggests. For example, the acoustic pulse might be a shear wave rather than a compression wave and, for various reasons, a non-collinear beam configuration is necessary in the popular acousto-optic crystal TeO_2. Moreover, since acousto-optic crystals tend to be mechanically as well as optically anisotropic, the acoustic Poynting vector will not necessarily be parallel to $\mathbf{K}_A$, and so acoustic walk-off will need to be taken into account as well.

5.5 Stimulated Brillouin scattering

5.5.1 Introduction

The previous two sections dealt with the diffraction of light by ultrasound generated externally by an acoustic transducer. We now consider a process, similar to stimulated Raman scattering, in which an acoustic wave plays the role of the Raman wave, and the Stokes and acoustic waves are driven by the laser. First observed in 1964 [44], the process might logically be called stimulated acoustic Raman scattering, but it is in fact known as stimulated Brillouin scattering (SBS).

There are close parallels between SBS and SRS, but significant differences too. The most important is that, for reasons connected with the dispersion characteristics of acoustic waves, the Stokes wave in SBS is normally observed in the backward direction. In the typical configuration shown in Fig. 5.11, the laser beam is focused into the nonlinear medium where

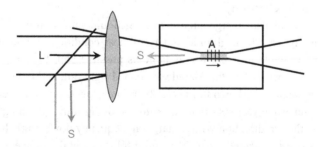

Fig. 5.11   Schematic diagram of backward stimulated Brillouin scattering.

[24] The group velocity dispersion of optical pulses is treated in Chapters 6 and 7.

the SBS interaction takes place in the region of the focus. The laser drives a forward-going acoustic wave and a backward-directed Stokes wave, which is picked off by the partially reflecting mirror beyond the lens.

The most common process by which two optical waves drive an acoustic wave is based on the longitudinal electrostrictive effect. The existence of an electrostrictive force can be deduced from Eq. (5.58) by inferring that the stored electrostatic energy density is changed in the presence of strain by $\delta u_e = -\frac{1}{2}\varepsilon_0\gamma_e S E^2$. This in turn implies the existence of a pressure $p = \delta u_e/S = -\frac{1}{2}\varepsilon_0\gamma_e E^2$, from which it follows that the force on unit volume of material resulting from optical fields travelling along the z-axis is

$$F_v = -\frac{\partial p}{\partial z} = \frac{1}{2}\varepsilon_0\gamma_e\frac{\partial(E)^2}{\partial z}. \tag{5.69}$$

The equation of motion for the acoustic displacement Q in response to the force is

$$\frac{1}{2}\varepsilon_0\gamma_e\frac{\partial(E)^2}{\partial z} + \frac{1}{C}\frac{\partial^2 Q}{\partial z^2} - \Gamma\frac{\partial Q}{\partial t} = \rho\frac{\partial^2 Q}{\partial t^2} \tag{5.70}$$

where the force terms on the left-hand side respectively represent electrostriction, elasticity, and loss. The parameter C is the compressibility of the medium, ρ is the mass density, and Γ is the acoustic damping coefficient.

It has been mentioned already that the Stokes wave in SBS travels in the backward direction, and we now need to consider why that is. It is in fact quite easy to see that collinear propagation in which all waves travel in the same direction is not a viable option. As for the programmable dispersive filter (PDF) of Section 5.4.3, the conditions to be satisfied are $\omega_L = \omega_S + \Omega_A$, and $\mathbf{k}_L = \mathbf{k}_S + \mathbf{K}_A$ and, as in that case, the laser and Stokes frequencies (ω_L and ω_S) will be virtually identical. For the PDF, the magnitudes of $\mathbf{k}_L$ and $\mathbf{k}_S$ (the angular wave numbers k_L and k_S) were significantly different because one wave was ordinary and the other extraordinary. But SBS is normally observed in isotropic materials, so k_L and k_S will now be virtually equal, while the magnitude of $\mathbf{K}_A(=K_A)$ will be determined by the angle in the vector triangle of Fig. 5.12(a). Clearly, if the laser and Stokes waves co-propagate, $\mathbf{k}_L$ and $\mathbf{k}_S$ will be parallel, K_A will be extremely small, the acoustic wavelength will be extremely large, and $\Omega_A(= K_A V)$, which was already small, will be effectively zero.

It will also emerge later that the Brillouin gain is proportional to K_A, so the gain would in any case be negligible under collinear conditions. Anyway, with this point in mind, it is desirable to make K_A as large as possible and, from Fig. 5.12(a), it is clear that this is achieved when the triangle is opened right out, and the laser and Stokes wave counter-propagate, as shown in Fig. 5.12(b). The acoustic wave travels in the $+z$-direction and the laser wave scatters off it, as from a moving mirror, leading to a Stokes wave travelling in the $-z$-direction with a marginal frequency down-shift. It is clear from the figure that $K_A = k_L + k_S \cong 2k_L \cong 2k_S$, and it follows that the frequency shift is given by

$$\frac{\Omega_A}{2\pi} \simeq \frac{2n_L V}{\lambda_L^{vac}} \tag{5.71}$$

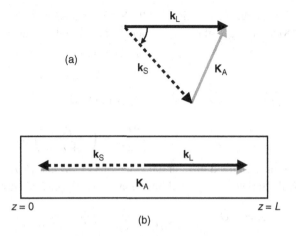

Fig. 5.12 Wave vector orientations in stimulated Brillouin scattering: (a) general angle; (b) the backward case.

where $V (= \Omega_A / K_A)$ is the speed of ultrasound in the nonlinear medium, and λ_L^{vac} is the vacuum wavelength of the laser. Values of $\Omega_A / 2\pi$ around 5 GHz are typical for liquids, with rather lower frequencies in gases, and rather higher frequencies (15–25 GHz) in solids.

5.5.2 The coupled-wave equations of stimulated Brillouin scattering

To analysis the SBS process, we write the optical and acoustic waves in the form

$$E = \tfrac{1}{2} \left(\tilde{E}_L \exp\{i(\omega_L t - k_L z)\} + \tilde{E}_S \exp\{i(\omega_S t + k_S z)\} + \text{c.c.} \right), \tag{5.72}$$

$$Q = \tfrac{1}{2} \left(\tilde{Q} \exp\{i(\Omega_A t - K_A z)\} + \tilde{Q}^* \exp\{-i(\Omega_A t - K_A z)\} \right) \tag{5.73}$$

where the first equation includes the incident laser beam and the counter-propagating Stokes wave; notice the plus sign in the Stokes wave argument. We substitute Eqs (5.72) and (5.73) into Eq. (5.70) and, after some algebra, we obtain

$$\frac{\partial \tilde{Q}}{\partial z} + \frac{\Gamma \tilde{Q}}{2\rho V} = -\frac{\varepsilon_0 \gamma_e}{8\rho V^2} \tilde{E}_L \tilde{E}_S^* \tag{5.74}$$

where we have used $\omega_L - \omega_S = \Omega_A$ and $k_L + k_S = K_A$. The proof uses the relation $V = (C\rho)^{-1/2}$, and several first and second space and time derivatives of $\tilde{Q}$ have been dropped.

Equation (5.74) governs the process in which the optical fields drive the acoustic wave; notice particularly the presence of the electrostrictive constant with the laser and Stokes fields on the right-hand side. The equation for the Stokes field can be found from the second term on the right-hand side of Eq. (5.60). The nonlinear polarisation that drives the wave at ω_S is

$$P_S^{NL} = \tfrac{1}{2} \left(-\tfrac{1}{2} i \varepsilon_0 \gamma_e K_A \tilde{Q}^* \tilde{E}_L \right) \exp\{i(\omega_S t + k_S z)\} + \text{c.c.}$$

$$= \tfrac{1}{2} \tilde{P}_S \exp\{i(\omega_S t + k_S z)\} + \text{c.c.} \tag{5.75}$$

where the second step defines $\tilde{P}_S$. We again customise the standard coupled-wave equation (2.8) to the wave at ω_S and, when $\tilde{P}_S$ is inserted into the right-hand side, the result is

$$\frac{\partial \tilde{E}_S}{\partial z} = \left(\frac{\omega_S \gamma_e K_A}{4 c n_S} \right) \tilde{E}_L \tilde{Q}^*. \tag{5.76}$$

The analogous equation for the laser field (based on $\tilde{P}_L = \frac{1}{2} i \varepsilon_0 \gamma_e K \tilde{E}_S \tilde{Q}$) is

$$\frac{\partial \tilde{E}_L}{\partial z} = \left(\frac{\omega_L \gamma_e K_A}{4 c n_L} \right) \tilde{E}_S \tilde{Q}. \tag{5.77}$$

The three coupled differential equations (5.74) and (5.76)–(5.77) govern the evolution of the three fields in the Brillouin interaction. If we now introduce the new parameters

$$\tilde{A} = -\tilde{Q}, \quad a = \frac{\Gamma}{2 \rho V}, \quad b = \frac{\varepsilon_0 \gamma_e}{4 V \Gamma}, \quad c_S = \frac{\omega_S \gamma_e K_A}{4 c n_S} \text{ and } c_L = \frac{\omega_L \gamma_e K_A}{4 c n_L},$$

the equations take on the more manageable appearance

$$\frac{\partial \tilde{A}}{\partial z} + a \tilde{A} = a b \tilde{E}_L \tilde{E}_S^* \tag{5.78}$$

$$\frac{\partial \tilde{E}_S}{\partial (-z)} = c_S \tilde{E}_L \tilde{A}^* \tag{5.79}$$

$$\frac{\partial \tilde{E}_L}{\partial z} = -c_L \tilde{E}_S \tilde{A}. \tag{5.80}$$

We have defined $\tilde{A}$ as the negative of the acoustic displacement since this makes the signs more convenient, and we have also included a negative sign in the derivative of Eq. (5.79) to highlight the fact that the Stokes wave propagates in the backward direction.

5.5.3 Approximations and solutions

Solutions to Eqs (5.78)–(5.80) can be obtained within several different approximations. First of all, the equation for $\tilde{A}$ can be simplified if one or other of the terms on the left-hand side can be ignored. Discarding the second term amounts to neglecting acoustic damping which, though rarely appropriate, leads to an interesting solution, which we discuss later. In the case of strong acoustic damping, which is normally more realistic, the spatial derivative is dropped, and this leads immediately to

$$\tilde{A} = b \tilde{E}_L \tilde{E}_S^*. \tag{5.81}$$

This means that the acoustic wave is slaved to the laser and Stokes waves, just like the vibrational wave in the SRS case in Eq. (5.52). When Eq. (5.81) is substituted into Eqs (5.79) and (5.80), we obtain

$$\frac{\partial \tilde{E}_S}{\partial (-z)} = + \left\{ b c_S \left| \tilde{E}_L \right|^2 \right\} \tilde{E}_S \tag{5.82}$$

$$\frac{\partial \tilde{E}_L}{\partial z} = - \left\{ b c_L \left| \tilde{E}_S \right|^2 \right\} \tilde{E}_L \tag{5.83}$$

from which it is easy to show that

$$\frac{1}{\hbar\omega_S} \frac{\partial I_S}{d(-z)} = -\frac{1}{\hbar\omega_L} \frac{\partial I_L}{dz}. \tag{5.84}$$

This means that for each Stokes photon gained in the $-z$-direction, a laser photon is lost in the $+z$-direction.

The equations can be simplified further if depletion of the laser field is negligible. Equation (5.83) can then be discarded, and $\left|\tilde{E}_L\right|^2$ treated as a constant in Eq. (5.82), which now has a solution in which the Stokes wave grows exponentially in the $-z$-direction, with a gain coefficient given by

$$g_S = bc_S \left|\tilde{E}_L\right|^2 = \frac{\omega_S \gamma_e^2 K_A \varepsilon_0 \left|\tilde{E}_L\right|^2}{16 c n_S V \Gamma}. \tag{5.85}$$

In a cell of length L, as suggested in Fig. 5.12(b), the solution is

$$\tilde{E}_S(z) = \tilde{E}_S(L) \exp\{g_S(L - z)\} \tag{5.86}$$

where $\tilde{E}_S(L)$ is the Stokes field at the rear surface, which would be the input field in the case of a Brillouin amplifier. The form of the solution is plotted in Fig. 5.13, where $\tilde{E}_S(L) = 1$ (arbitrary units), and $g_S L = 5$. The Stokes field grows from right to left, and because the acoustic wave is tied to the Stokes wave through Eq. (5.81), $\tilde{A}$ is strongest at $z = 0$ and falls away exponentially thereafter. This is despite the fact that the acoustic wave propagates in the $+z$-direction. As for SRS, the Stokes wave will grow from noise if the gain coefficient, which depends on the laser intensity through Eq. (5.83), is sufficiently high.

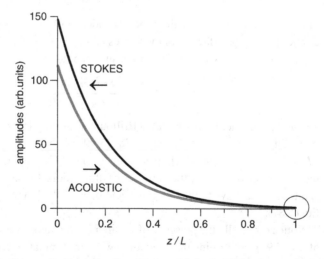

Fig. 5.13   Stokes and acoustic wave amplitudes as a function of distance for stimulated Brillouin scattering in the case of strong acoustic damping. Laser depletion is ignored.

An analytical solution is also possible when laser depletion and acoustic damping are both ignored. As noted earlier, the latter assumption is usually unrealistic, but it is instructive to pursue this case because it has some interesting features. As an added bonus, the system comes with the impressive title of 'contraflow Hermitian parametric coupling' [45], so it is clearly worthy of attention! The equations to be solved are now

$$\frac{\partial \tilde{A}}{\partial z} = r \tilde{E}_S^*; \quad \frac{\partial \tilde{E}_S}{\partial (-z)} = s \tilde{A}^* \tag{5.87}$$

where $r = ab\tilde{E}_L$ and $s = c_S \tilde{E}_L$. It is easy to show that a solution of the type $\tilde{A}(z) = \tilde{A}(0) \exp\{gz\}$ and $\tilde{E}_S(z) = \tilde{E}_S(0) \exp\{gz\}$ exists. However, g is now imaginary, and given by

$$g = \pm i \sqrt{rs^*} = \pm i\kappa \tag{5.88}$$

where

$$\kappa = \sqrt{abc_S |E_L|^2} = \sqrt{\frac{\omega_S \gamma_e^2 K_A \varepsilon_0 |E_L|^2}{32 c n_S \rho V^2}}. \tag{5.89}$$

This implies that the solutions are oscillatory, but there turns out to be a singularity after a quarter of a cycle, so complete oscillations are never seen.

Once again, we consider the evolution of $\tilde{A}$ and $\tilde{E}_S$ in the configuration of Fig. 5.12(b). As before, the pump wave enters at $z = 0$, while the Stokes wave develops in the backward direction. If we set the boundary condition that $A(0) = 0$, the solution is

$$\tilde{A}(z) = \frac{r \tilde{E}_S^*(L)}{\kappa} \frac{\sin \kappa z}{\cos \kappa L}; \quad \tilde{E}_S(z) = \tilde{E}_S(L) \frac{\cos \kappa z}{\cos \kappa L} \tag{5.90}$$

where

$$\frac{r}{\kappa} = e^{i\phi_L} \sqrt{\frac{c \varepsilon_0 n_S}{2\omega_S \rho K_A V^2}}.$$

The key feature of this result is the dependence of both amplitudes on $(\cos \kappa L)^{-1}$. This implies that the gain factors of the Stokes and acoustic waves become infinite when $\kappa L \rightarrow \pi/2$, which occurs when the laser intensity reaches

$$I_L = \frac{4\pi^2 c^2 n_S n_L \rho V^2}{\omega_S \gamma_e^2 K_A L^2}. \tag{5.91}$$

This is the threshold for stimulated Brillouin scattering in this albeit unrealistic case.

A plot of the solutions given in Eq. (5.90) is shown in Fig. 5.14 for the case where $\kappa L = 0.99 \times (\pi/2)$ for which $(\cos \kappa L)^{-1} = 63.7$. The field units are arbitrary and we have set $\tilde{E}_S(L) = r/\kappa = 1$ for the purposes of illustration. The acoustic wave grows to the right and the Stokes wave to the left, each field being linked to the gradient of the other through Eq. (5.87).

Notice especially the presence of K_A in the numerator of Eq. (5.89) and the denominator of Eq. (5.91), confirming that the Stokes wave in stimulated Brillouin scattering grows most strongly when K_A is maximised, which is in the backward direction according to Fig. 5.12.

Recent review material on SBS can be found in [46, 47].

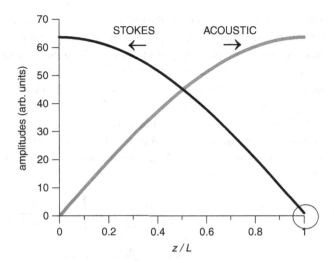

Fig. 5.14   As for Fig. 5.13, but with acoustic damping neglected.

5.6 Optical phase conjugation

5.6.1 Phase conjugate beams

When an optical beam passes through an aberrating medium, it emerges with a distorted wavefront. If it is then reflected back through the medium by a normal mirror, the damage is compounded. By contrast, phase conjugate reflection creates a beam that retraces its steps through the medium as in a time-reversed movie so that the beam emerges with its aberrations removed. Phase conjugate beams have important applications in laser engineering and, in this section, we will consider how they can be created.

Before discussing how a phase conjugate wave can be generated, we need to understand what it is, and how it differs from a normal wave. As usual, we start by considering a beam (call it wave 1) travelling broadly in the $+z$-direction, namely

$$E_1(\mathbf{r}, t) = \tfrac{1}{2} \left(\hat{E}_1(\mathbf{r}) \exp\{i\omega t\} + \hat{E}_1^*(\mathbf{r}) \exp\{-i\omega t\} \right). \tag{5.92}$$

Here $\hat{E}_1(\mathbf{r}) = \tilde{E}_1(\mathbf{r}) \exp\{-ik_1 z\}$, and $\tilde{E}_1(\mathbf{r})$ is slowly varying with respect to $\mathbf{r} = (x, y, z)$.[25] As the name suggests, the corresponding 'phase conjugate' wave $E_C(\mathbf{r}, t)$ is obtained by replacing $\hat{E}_1(\mathbf{r})$ in Eq. (5.92) by its complex conjugate $\hat{E}_1^*(\mathbf{r})$ so that

$$E_C(\mathbf{r}, t) = \tfrac{1}{2} \left(\hat{E}_1^*(\mathbf{r}) \exp\{i\omega t\} + \hat{E}_1(\mathbf{r}) \exp\{-i\omega t\} \right). \tag{5.93}$$

But the terms could equally be written in reverse order, in which case we would have

$$E_C(\mathbf{r}, t) = \tfrac{1}{2} \left(\hat{E}_1(\mathbf{r}) \exp\{-i\omega t\} + \hat{E}_1^*(\mathbf{r}) \exp\{i\omega t\} \right) = E_1(\mathbf{r}, -t) \tag{5.94}$$

---

[25] We are assuming a continuous (CW) beam. More generally, the fields would of course be time dependent.

where the final step shows that the phase conjugate wave is identical to the 'time reversed' version of the original wave, which is obtained by changing t to $-t$ in Eq. (5.92). The conclusion is that replacing $\hat{E}_1(\mathbf{r})$ in Eq. (5.92) by its conjugate is equivalent to reversing the sign of t, so it is immaterial whether one regards the phase conjugate field as being $\hat{E}_1^*(\mathbf{r})$ running in forward time, or $\hat{E}_1(\mathbf{r})$ running in reverse time. A movie tracing the evolution of $E_C(\mathbf{r}, t)$ would be identical in every respect to a movie of $E_1(\mathbf{r}, t)$ run backwards.

A trivial first example is the plane wave. While $\hat{E}_1(z) = A \exp\{-ik_1z\}$ travels in the $+z$-direction, its conjugate $\hat{E}_1^*(z) = A \exp\{+ik_1z\}$ travels in the $-z$-direction, and is equivalent to $\hat{E}_1(z)$ under time reversal. As a second example, consider the Gaussian beam, which is given (see Section 2.5) by

$$\hat{E}_1(r, z) = \left(\frac{A}{1 - iz/z_R} \right) \exp \left\{ -\frac{r^2}{w_0^2(1 - iz/z_R)} - ik_1z \right\} \tag{5.95}$$

where $r^2 = x^2 + y^2$, z_R is the Rayleigh length, and $w^2(z) = w_0^2(1 + (z/z_R)^2)$. Once again, taking the complex conjugate is equivalent to replacing z by $-z$, so the direction of travel (or the direction of time if you prefer) is reversed.

It is now easy to see that phase conjugate reflection is fundamentally different from what happens when a beam encounters a conventional mirror. The two cases are contrasted in Fig. 5.15. The focus of the Gaussian beam is at $z = 0$, and the solid black line shows the spherical wavefront (labelled 1) as it approaches the normal mirror M at $z = z_m$. After reflection, the beam continues to expand in the $-z$-direction, as shown by the dotted line RR, which is now curved in the opposite sense. The reflected wave is in fact equivalent to a leftward travelling Gaussian beam centred at $z = 2z_m$; see Problem 5.9. By contrast, the phase conjugate wave simply retraces the path of the original wave, a possibility that is represented in the figure by the dotted line CC, which is curved in the same sense as the original wave, and is contracting back towards $z = 0$. The difference in the width and curvature of RR and CC highlights the difference between normal reflection and phase conjugated reflection.

As noted at the start of this section, interest in phase conjugation is centred on the potential cancellation of optical aberrations, so we now need to consider how aberrated beams behave after reflection at a phase conjugate mirror. The key point always to keep in mind is that

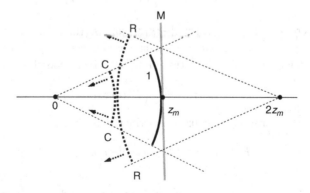

Fig. 5.15 Conventional reflection (RR) and phase conjugate reflection (CC) for a simple spherical wavefront.

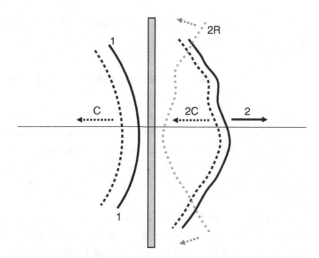

Fig. 5.16 Spherical wavefront (1) is distorted (2) after propagating through an aberrating plate. Conventional reflection creates wavefront 2R, which will be further aberrated by its backward passage through the plate. The phase conjugated wavefront 2C will have its aberrations removed as it returns through the plate creating wavefront C.

the phase conjugate wave is the same as the time-reversed wave so, if a beam acquires aberrations in (say) an optically imperfect window, its phase conjugate reflection will pass back through the window and emerge with the aberrations removed. A double pass through a laser amplifier using a phase conjugate mirror to return the beam is clearly an attractive prospect in laser engineering. However, it is important to appreciate that the procedure only cancels linear optical aberrations, and that distortions stemming from nonlinear optical phenomena are added rather than subtracted in the second pass.

To illustrate these points, consider the spherical wavefront labelled 1 in Fig. 5.16, where it is shown approaching an aberrating plate from the left. As it passes through the plate, it picks up phase distortions $\phi(x, y)$, and emerges as the distorted wavefront 2 given by

$$\hat{E}_2(\mathbf{r}) = \hat{E}_1(\mathbf{r}) \exp\{i\phi(x, y)\}. \tag{5.96}$$

The beam is shown expanding away to the right with ripples in its wavefront. If it were reflected back by an ordinary mirror, it would return as wavefront 2R, which is curved in the opposite sense. After passing through the plate a second time, the beam would emerge doubly distorted, and would continue to expand away to the left. On the other hand, a phase conjugate reflector would return wavefront 2 to the plate as wavefront 2C given by

$$\hat{E}_2^*(\mathbf{r}) = \hat{E}_1^*(\mathbf{r}) \exp\{-i\phi(x, y)\}. \tag{5.97}$$

This has exactly the same shape as wavefront 2, but is now travelling in the opposite direction. Moreover, since it is equivalent to a time-reversed version of wavefront 2, it retraces the steps of that wavefront, and emerges on the left as wavefront C with its aberrations removed. The operation can be represented by the equation

$$\hat{E}_C(\mathbf{r}) = \hat{E}_1^*(\mathbf{r}) \exp\{-i\phi(x, y)\} \exp\{i\phi(x, y)\} = \hat{E}_1^*(\mathbf{r}). \tag{5.98}$$

The phase distortions cancel out, and a phase conjugate replica of the original clean beam is recovered.

5.6.2 Phase conjugation by four-wave mixing

Various ways of generating phase conjugate beams have been studied. As we shall discover, the process is closely analogous to holography, but holography performed in real time through a nonlinear interaction, rather than via the creation and development of a photographic plate. Any medium in which a refractive index grating can be imprinted and instantaneously read out is a potential candidate for mediating the phase conjugation process.

Let us assume (see Fig. 5.17) that the wave to be phase conjugated (again denoted wave 1) travels in the $+z$-direction, while a pair of waves (2 and 3) travel along a different line but with oppositely directed wave vectors $\mathbf{k}_2$ and $-\mathbf{k}_2$. The three waves can be written

$$E_1 = \tfrac{1}{2}\left(\tilde{E}_1(\mathbf{r})\exp\{i(\omega t - k_1 z)\} + \tilde{E}_1^*(\mathbf{r})\exp\{-i(\omega t - k_1 z)\}\right) \tag{5.99}$$

$$E_2 = \tfrac{1}{2}\left(\tilde{E}_2(\mathbf{r})\exp\{i(\omega t - \mathbf{k}_2.\mathbf{r})\} + \text{c.c.}\right) \tag{5.100}$$

$$E_3 = \tfrac{1}{2}\left(\tilde{E}_3(\mathbf{r})\exp\{i(\omega t - (-\mathbf{k}_2.\mathbf{r}))\} + \text{c.c.}\right). \tag{5.101}$$

Even without going into the fundamental mechanism of the process, the possibility of phase conjugate wave generation can be spotted within the basic formalism of third-order nonlinear processes. For if one works out the cube of the total field $(E_1 + E_2 + E_3)$, a term appears in the nonlinear polarisation of the form

$$P_4^{\text{NL}}(\mathbf{r}) \sim \tilde{E}_2(\mathbf{r})\tilde{E}_3(\mathbf{r})\exp\{i2\omega t\} \times \tilde{E}_1^*(\mathbf{r})\exp\{-i(\omega t - k_1 z)\} \tag{5.102}$$

which comes from the product of the second term in Eq. (5.99) and the first terms in Eqs (5.100) and (5.101). If waves 2 and 3 are plane waves, then $\tilde{E}_2(\mathbf{r})\tilde{E}_3(\mathbf{r}) = A_2 A_3$, and

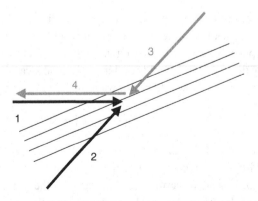

Fig. 5.17 Phase conjugation by four-wave mixing. Wave 1 and plane wavefront 2 write a grating in the nonlinear medium. The reverse plane wavefront 3 interacts with the grating to create the phase conjugated beam 4.

we have

$$P_4^{\mathrm{NL}}(\mathbf{r}) \sim A_2 A_3 \tilde{E}_1^*(\mathbf{r}) \exp\{i(\omega t + k_1 z)\}. \tag{5.103}$$

This polarisation antenna will therefore radiate a field (wave 4) in the backward direction, with a wavefront that is the phase conjugate of the original wave. The interaction could be mediated by the non-resonant four-wave mixing process defined by[26]

$$\hat{P}_4^{\mathrm{NL}}(\mathbf{r}) = \tfrac{3}{2}\varepsilon_0 \chi^{(3)}(\omega;\ \omega, -\omega, \omega) \hat{E}_2(\mathbf{r})\hat{E}_1^*(\mathbf{r})\hat{E}_3(\mathbf{r}) \tag{5.104}$$

where $\hat{E}_2(\mathbf{r}) = \tilde{E}_2(\mathbf{r})\exp\{-i\mathbf{k}_2.\mathbf{r}\}$, etc. in accordance with the definitions of Eqs (2.4)–(2.5).

It is instructive to divide the process into two distinct stages. In the first stage, the interference of wave 1 and wave 2 (assumed plane) writes a stationary pattern in the medium with the structure

$$\rho_{12}(\mathbf{r}) \sim A_2 \tilde{E}_1^*(\mathbf{r}) \exp\{i(k_1\hat{\mathbf{z}} - \mathbf{k}_2).\mathbf{r}\}. \tag{5.105}$$

This comes from the product of the second term in Eq. (5.99) with the first term in Eq. (5.100). In the second stage, wave 3 interacts with this pattern, creating a term in the polarisation

$$P_4^{\mathrm{NL}}(\mathbf{r}) \sim A_3 \rho_{12}(\mathbf{r}) \exp\{i(\omega t - (-\mathbf{k}_2.\mathbf{r}))\} = A_3 A_2 \tilde{E}_1^*(\mathbf{r}) \exp\{i(\omega t + k_1 z)\} \tag{5.106}$$

which is the form given by Eq. (5.103). In Fig. 5.17, the first stage is displayed in black and the second stage in grey. The fine lines indicate the orientation of the stationary interference pattern. Although these lines are drawn straight in the figure, in practice they will be distorted by aberrations in wave 1, and it is of course precisely these distortions that wave 3 reads to create the phase conjugate wave 4.

The roles of waves 2 and 3 could equally be reversed. In that case, instead of Eqs (5.105) and (5.106), we would have

$$\rho_{13}(\mathbf{r}) \sim A_3 \tilde{E}_1^*(\mathbf{r}) \exp\{i(k\hat{\mathbf{z}} + \mathbf{k}_2).\mathbf{r}\} \tag{5.107}$$

and

$$P_4^{\mathrm{NL}}(\mathbf{r}) \sim A_2 \rho_{13}(\mathbf{r}) \exp\{i(\omega t - (\mathbf{k}_2.\mathbf{r}))\} = A_2 A_3 \tilde{E}_1^*(\mathbf{r}) \exp\{i(\omega t + k_1 z)\} \tag{5.108}$$

which is the same as Eq. (5.106). The corresponding diagram is shown in Fig. 5.18. Notice that the interference pattern is at right angles to the one in Fig. 5.17, and the lines forming the pattern are also drawn closer together because the wavelength will be smaller at these angles.

The patterns of Eqs (5.105) and (5.107) take different physical forms depending on the type of interaction involved. In the case of four-wave mixing envisaged in Eq. (5.104),

---

[26] It is in fact necessary only for waves 2 and 3 to be conjugates of each other for the phase structure of the waves to cancel out, and for Eq. (5.103) to apply. The pre-factor comes from $6 \times \left(\tfrac{1}{2}\right)^2$ because six cross-terms appear when the cube is taken.

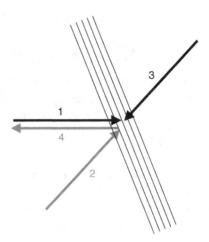

As Fig. 5.17, but with the roles of wavefronts 2 and 3 reversed.

$\rho_{12}(\mathbf{r})$ and $\rho_{13}(\mathbf{r})$ would be second-order terms in the density matrix (see Chapter 8), and the symbol has been chosen with that in mind. The interference patterns would certainly register on a photographic plate, which points to the parallel with conventional holography. In that case, the intention is normally the *re-creation* of the original wavefront, rather than the production of a phase conjugated wave. The initial stage of the process would then involve the combination of the first term in Eq. (5.99) with the second term in the 'reference wave' of Eq. (5.100) to produce the pattern

$$\rho_{12}(\mathbf{r}) \sim \tilde{E}_1(\mathbf{r}) A_2 \exp\{i(\mathbf{k}_2 - k_1\hat{\mathbf{z}}).\mathbf{r}\}. \tag{5.109}$$

The second stage would involve the illumination of this pattern with the reference wave to re-create the original wavefront through the combination

$$P^{NL}(\mathbf{r}) \sim \rho_{12}(\mathbf{r}) A_2 \exp\{i(\omega t - \mathbf{k}_2.\mathbf{r})\} = A_2^2 \tilde{E}_1(\mathbf{r}) \exp\{i(\omega t - k_1 z)\}. \tag{5.110}$$

There would of course be an interval between the two stages, during which the photographic development of the three-dimensional hologram would take place. By contrast, the two stages of phase conjugation by four-wave mixing occur simultaneously, and the term *real-time holography* has been coined for that reason.

5.6.3 Phase conjugation by stimulated Brillouin scattering

Phase conjugation can also be effected by stimulated Brillouin scattering. In early experiments on SBS, it was noticed that the backward Stokes wave was so similar in structure to a time-reversed version of the pumping wave that it worked its way back into the laser source and interfered with its performance. It was soon recognised that the phase conjugating property was an inherent feature of SBS.

At first sight, this phenomenon is unexpected, because there is nothing in Eq. (5.82) to suggest that the Stokes wavefront $\tilde{E}_S$ should have any specific relationship to the conjugate of the laser wavefront $\tilde{E}_L^*$. On the other hand, Fig. 5.11 (shown enlarged in Fig. 5.19),

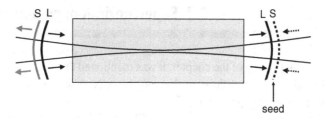

S L   L S

seed

Fig. 5.19   Schematic diagram of phase conjugation by stimulated Brillouin scattering.

indicates that the Stokes beam, propagating in the backward direction from the focal region where the interaction is strongest, is likely to be far closer to $\tilde{E}_L^*$ than to $\tilde{E}_L$ for purely geometrical reasons. After all, we have seen in Fig. 5.15 the difference between the reflection from an ordinary mirror and from a phase conjugate mirror. A mirror placed at the focal point in Fig. 5.19 would certainly cause the incoming beam broadly to retrace its steps, but this would not be the case for any other mirror position. In particular, if we wanted to seed the Stokes beam by launching a weak signal into the cell from the right, the wavefront (shown as the dotted line S on the right of Fig. 5.19) should clearly be curved to match the rightward propagating wavefront $\tilde{E}_L$ (labelled L in the figure). It should therefore correspond to the time reversal of $\tilde{E}_L$, in other words to $\tilde{E}_L^*$. After passing through the SBS interaction region, the Stokes beam wavefront will emerge on the left, curved in the same sense as the laser wave as it approaches the focal region from the left, as shown in the figure.

The issue is therefore not so much whether the structure of the Stokes beam is closer to the conjugate of $\tilde{E}_L$ than to $\tilde{E}_L$ itself – it clearly will be. Rather it is whether the wavefront of $\tilde{E}_S$ will match $\tilde{E}_L^*$ *sufficiently accurately* to display all the attributes of a phase conjugate beam. The evidence of experiment is clearly that it does, so the burden is on the theorist to explain why that is the case.

A qualitative argument runs broadly as follows.[27] If the laser wavefront is aberrated, the spatial structure of the beam throughout the focal region will exhibit corresponding spatial complexity. If one then imagines a range of possible backward-propagating Stokes beams, all competing for the energy of the incident laser through the SBS process, the winner of the competition will be the beam that can march most accurately backwards through the pattern laid down by the laser. This will be the beam that is closest to the phase conjugate of $\tilde{E}_L$.

A question remains. Does the fact that the wavelength of the Stokes beam is slightly higher than that of the laser affect the argument? Does it compromise the accuracy needed if the Stokes beam is to negotiate its way back along the path laid down by the laser? The wavelengths (and frequencies) of the two differ in SBS typically by 1 part in 2×10^4 so, for a laser at 1 µm, serious mismatch between the two will occur over a distance of around 20 mm. But this is far longer than the effective length of the interaction region in a typical SBS experiment, and suggests that the difference is not an issue.

Reviews of phase conjugate optics can be found in [49–51].

---

[27] A semi-quantitative argument along these lines is offered by Boyd [1]. A more complicated picture of phase conjugation in SBS, based on Brillouin-assisted four-wave mixing, is presented by He and Liu [48].

5.7 Supercontinuum generation

At the beginning of the chapter, it was mentioned that when one third-order process occurs, several others often do so at the same time. Nowhere is the tendency more evident than in supercontinuum generation.[28]

In supercontinuum generation (SCG), an intense narrow-band incident pulse undergoes dramatic spectral broadening through the joint action of several third-order processes, resulting in an output pulse of essentially white light. The effect was first observed by Alfano and Shapiro [33] in various glasses, and output spectra spanning the visible spectrum were recorded. The spectral broadening was attributed to self-phase modulation (SPM) but, since it was ten times stronger than anything previously observed, it became known as a supercontinuum [52–53]. In bulk media, the threshold for SCG coincides with the threshold for self-focusing, and the increase in intensity that accompanies beam collapse promotes a family of third-order processes including self-phase modulation, cross-phase modulation, self-steepening, and plasma formation. The mechanism of self-focusing is complicated enough in itself but, despite this, a detailed model of SCG in bulk materials was presented by Gaeta in 2000 [54].

Understanding SCG in optical fibres is significantly simpler because the complexities of self-focusing are absent. Even there, however, a whole family of third-order processes conspires together to deliver the strong spectral broadening, including stimulated Raman scattering and associated four-wave mixing processes, as well as self- and cross-phase modulation [55]. For a recent book on supercontinuum generation in optical fibres, see [56].

When driven by a mode-locked pulse train, SCG is the basis for the generation of wide-band frequency combs that have applications in optical frequency metrology. If combs at the fundamental and second harmonic frequencies are present at the same time, and both span a substantial fraction of an octave, the two will overlap, and this is the basis of an important technique in laser technology for the phase stabilisation of femtosecond pulses; this is explained in more detail in Section 7.10. More generally, SCG can be used as a broadband light source in a wide variety of contexts including spectroscopy, optical coherence tomography, and telecommunications.

Problems

5.1 Show that, with respect to the axes defined in Fig. 5.1, the effective third harmonic generation coefficient for a plane-polarised fundamental field is $\chi_{\text{eff}} = \frac{1}{2}(\chi_1 + \chi_2 + \chi_3 + \chi_4)$ (Eq. 5.8).

5.2 Show that third harmonic radiation cannot be generated using circularly polarised light.

5.3 Find the half-wave voltage for a Kerr cell based on nitrobenzene ($K = 4.4 \times 10^{-12}\,\text{m V}^{-2}$). The cell is 10 cm long, and the plate separation is 1 cm.

---

[28] I am indebted to Dr Louise Hirst for providing me with some useful background material on supercontinuum generation

5.4 Verify Eq. (5.18) starting from the relevant equations in Appendix C.

5.5 Show that the phase velocity of an $N = 1$ soliton is approximately $(cn_2 I_{pk}/2n_0^2)$ lower than its low-intensity value in the same medium.

5.6 Show that Eq. (5.35) follows from Eq. (5.34), and that Eqs (5.32) and (5.35) together ensure that the photon flux (photons/area/time) is conserved.

5.7 Prove Eq. (5.62) for the coherence length of the acousto-optic process diagrammed in Fig. 5.8.

5.8 Equations (5.90) give the solutions to Eqs (5.87) for the boundary condition $\tilde{A}(0) = 0$. Show that for the alternative boundary condition $\tilde{E}_S(L) = 0$, the analogous solutions are

$$\tilde{A}(z) = \tilde{A}(0)\frac{\cos \kappa(L-z)}{\cos \kappa L}; \quad \tilde{E}_S(z) = \frac{\kappa \tilde{A}^*(0)}{r}\frac{\sin \kappa(L-z)}{\cos \kappa L}.$$

5.9 Show that when the Gaussian beam of Eq. (5.95) encounters the ordinary mirror at $z = z_m$ in Fig. 5.15, the reflected wave corresponds to a backward-travelling Gaussian beam centred at $z = 2z_m$, as indicated in the figure.

Dispersion and optical pulses

6.1 Introduction and preview

Just as Chapter 3 on crystal optics served as a prelude to Chapter 4, so the present chapter lays the foundations for Chapter 7 on nonlinear optics in ultrashort pulse technology. The central topic is the phenomenon of dispersion. Although itself a linear effect, dispersion plays a vital role in many nonlinear interactions, and in many techniques used for short pulse generation too. We have already seen how dispersion in the phase velocity leads to phase mismatching in frequency-mixing processes and how, in combination with birefringence, it can also become part of the solution. However, whenever optical pulses are involved, dispersion in the group velocities invariably becomes important, a feature that will crop up repeatedly in the following chapter.

In Section 6.2, we consider the relationship between phase velocity dispersion (PVD) and group velocity dispersion (GVD). The ideas are then applied and elaborated in Section 6.3 where we demonstrate how, in the presence of GVD, a Gaussian pulse broadens and develops a frequency sweep. Hyperbolic secant pulses are contrasted with Gaussians in Section 6.4, and a discussion of chirped pulse compression using gratings and prisms follows in Section 6.5. A partial differential equation governing pulse propagation in the presence of GVD is developed in Section 6.6 for applications in Chapter 7.

6.2 Dispersion in the phase and group velocities

In Section 1.2, we showed that the linear susceptibility is related to refractive index through the equation $n(\omega) = \sqrt{1 + \chi^{(1)}(\omega)}$. But the susceptibility, and hence also the refractive index, are frequency dependent (as indicated), and this underlies the phenomenon of *dispersion*, a word that means 'distribution in different directions' or 'spreading out'. Dispersion describes the separation of white light into colours by a prism, and equally the temporal spreading of an optical pulse as it propagates through a dispersive material. These and similar phenomena arise from the effect of the frequency dependence of n on two key characteristics of light propagation: the phase velocity and the group velocity

For a simple travelling wave of the form $\cos\{\omega t - kz\}$, the phase velocity is

$$v_{\text{phase}} = \frac{\omega}{k} = \frac{c}{n(\omega)} \qquad (6.1)$$

where $k = n(\omega)\omega/c$. On the other hand, the group velocity is given by

$$v_{\text{group}} = \frac{d\omega}{dk}. \tag{6.2}$$

For an optical pulse, one can say in general terms that v_{phase} is the speed of the high-frequency optical oscillations (the optical carrier wave), while v_{group} is the speed of the pulse envelope. So one might expect a propagating pulse to have a profile of the form $E(z,t) \simeq A(t - z/v_{\text{group}}) \cos\{\omega(t - z/v_{\text{phase}})\}$. The analytic result derived in Eq. (6.14) below broadly confirms this conclusion, but also shows that the envelope broadens, and the pulse develops a frequency sweep, known as a 'chirp'.

The phase and group velocities can be conveniently represented on the plot of ω against k, shown schematically in Fig. 6.1. This is called a *dispersion diagram* because it embodies the frequency dependence of the refractive index, which is the origin of dispersion in both v_{phase} and v_{group}. It is clear from Eqs (6.1) and (6.2) that the gradient of OA represents the phase velocity, while the gradient of the tangent TT to the dispersion curve at A represents the group velocity. In the case shown, the group velocity is lower than the phase velocity. If one could watch the propagation of an optical pulse under these conditions, the optical carrier wave oscillations would be seen to be travelling faster than the envelope, growing as they approached the peak from the rear, and dying away as they departed towards the front. There are points on the dispersion curve where the situation is reversed, and the group velocity is higher than the phase velocity.

The excursions of the dispersion curve in Fig. 6.1 are in fact grossly exaggerated. Graphs of the phase and group velocities in fused silica, shown dotted in Fig. 6.2, indicate that the difference between the two (and their individual variation over the wavelength range shown) is barely more than 1%. The eye would therefore find it hard to distinguish a scale drawing of a typical dispersion curve from a straight line through the origin.

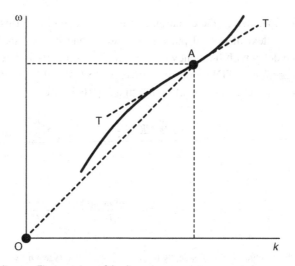

Fig. 6.1   Schematic dispersion diagram. The excursions of the dispersion curve are exaggerated.

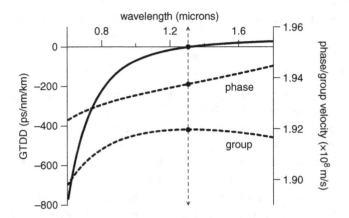

Fig. 6.2 Group time delay dispersion characteristic of fused silica (solid line). The phase and group velocities (plotted against the right-hand axis) are shown dotted.

It is evident from Eq. (6.1) that the frequency dependence of the refractive index leads directly to variation in the phase velocity. This is known as *phase velocity dispersion* (PVD) or sometimes *chromatic dispersion*. In regions of the spectrum when $n(\omega)$ rises with frequency ($dn(\omega)/d\omega > 0$), the dispersion is said to be 'normal', because this is the case for most transparent optical materials in the visible part of the spectrum. The opposite case ($dn(\omega)/d\omega < 0$) is referred to as 'anomalous' dispersion.[1]

The characteristics of group velocity dispersion (GVD) are conveniently expressed in terms of the *group time delay* (GTD), defined as the time it takes a wave group to propagate a distance z, namely

$$\tau_{\text{group}} = \frac{z}{v_{\text{group}}} = z\frac{dk}{d\omega} = \frac{z}{c}\left(n + \omega\frac{dn}{d\omega}\right) = \frac{z}{c}\left(n - \lambda\frac{dn}{d\lambda}\right). \qquad (6.3)$$

The final step in the equation expresses the result in terms of the vacuum wavelength. In the context of optical pulse propagation, one is interested in the *variation* of the group time delay with frequency or wavelength, a property that is known as the *group time delay dispersion* (GTDD), and is equivalent to GVD in the information it provides. The key formulae, obtained by differentiating Eq. (6.3), are

$$\frac{d\tau_{\text{group}}}{d\omega} = z\frac{d^2k}{d\omega^2} = \frac{z}{c}\left(2\frac{dn}{d\omega} + \omega\frac{d^2n}{d\omega^2}\right) \qquad (6.4a)$$

and

$$\frac{d\tau_{\text{group}}}{d\lambda} = -\frac{z\lambda}{c}\frac{d^2n}{d\lambda^2}. \qquad (6.4b)$$

---

[1] These definitions are frequently quoted in terms of the (vacuum) wavelength rather than the frequency. The index falls with increasing wavelength in the case of normal dispersion.

Table 6.1 Signs of parameters controlling PVD and GVD.

	$\dfrac{dn}{d\lambda}$	$\dfrac{dv_{phase}}{d\lambda}$	$\dfrac{d^2n}{d\lambda^2}$	$\dfrac{dv_{group}}{d\lambda}, \beta_2$	$D_\lambda, \dfrac{d\tau_{group}}{d\lambda}$	Chirp
Phase velocity (chromatic) dispersion						
Normal	−	+				
Anomalous	+	−				
Group velocity dispersion						
Positive			+	+	−	Up/+
Negative			−	−	+	Down/−

Notice the simplicity of the group time delay dispersion (GTDD) when expressed in terms of the vacuum wavelength. The GTDD is commonly characterised (especially in optical fibres) by the dispersion parameter

$$D_\lambda = -\frac{\lambda}{c}\frac{d^2n}{d\lambda^2}. \tag{6.5}$$

Convenient units for D_λ are picoseconds (of pulse spread) per nanometre (of bandwidth) per kilometre (of fibre).

The wavelength dependence of D_λ in fused silica is plotted as a solid line in Fig. 6.2. Below 1.3 μm, $d^2n/d\lambda^2$ is positive, so D_λ and $d\tau_{group}/d\lambda$ are both negative. The group time delay falls with increasing wavelength, and the group velocity correspondingly rises, as shown in the figure. Close to 1.3 μm, $d^2n/d\lambda^2$ passes through zero and, at higher wavelengths, $d^2n/d\lambda^2$ goes negative, while D_λ and $d\tau_{group}/d\lambda$ become positive.

The GVD is said to be 'positive' or 'negative' according to the sign of $d^2n/d\lambda^2$. Occasionally people speak of 'normal' and 'anomalous' GVD, but this usage is best avoided because it encourages the erroneous belief that dispersion in the group velocity neatly parallels dispersion in the phase velocity. But Fig. 6.2 makes it quite clear that there is no necessary correspondence between the two kinds of dispersion. The phase velocity of fused silica rises with wavelength (and the refractive index therefore falls) throughout the range displayed, and so the PVD is 'normal' everywhere. However, the group velocity rises on the left-hand side of the figure ('positive' GVD), reaches a maximum near 1.3 μm, and falls away thereafter ('negative' GVD).

The signs of the various dispersion parameters are summarised in Table 6.1, which highlights the point that the signs of PVD and GVD are determined by different characteristics.

The fact that the GVD goes to zero near 1.3 μm makes this an attractive wavelength for high bit-rate optical communications in optical fibres. The zero dispersion wavelength can also be shifted to some extent by adjusting the fibre geometry and, in practice, it is normally moved to around 1.55 μm where InGaAsP lasers operate and the transmission of fused silica is maximised.

6.3 Dispersion and chirping of a Gaussian pulse

6.3.1 Propagation analysis

Group velocity dispersion (the fact that different frequency groups travel at different speeds) has two important consequences: the pulse envelope broadens, and the carrier wave within the envelope acquires a frequency sweep. For positive GVD, low frequencies travel faster than high frequencies, and so arrive earlier. The carrier frequency duly rises through the pulse, which is said to have acquired a positive 'chirp' (or an 'up-chirp'). In the opposite case, the frequency sweep is negative, and the pulse exhibits a 'down-chirp'.

We will now perform a simple exercise on a Gaussian pulse to see how it changes during propagation in a dispersive medium.[2] We use the general definition of the real field

$$E(z,t) = \tfrac{1}{2}(\hat{E}(z,t) \exp\{i\omega_0 t\} + \text{c.c.}) \tag{6.6}$$

where ω_0 is the angular frequency of the optical carrier wave before dispersion is introduced. The field envelope $\hat{E}(z,t)$ is generally complex, but we will assume that initially (at $z = 0$) it has the real Gaussian form

$$\hat{E}(0,t) = A \exp\{-t^2/2t_0^2\}. \tag{6.7}$$

The full width at half maximum intensity (FWHM) of the intensity profile is easily shown to be $\Delta t_0 = t_0\sqrt{4\ln 2} \simeq 1.665 t_0$.

To discover how the pulse changes as it propagates, we take the Fourier transform of the analytic signal $V(0,t) = \hat{E}(0,t) \exp(i\omega_0 t)$, namely[3]

$$\bar{V}(0,\omega) = \frac{1}{2\pi} \int\limits_{-\infty}^{\infty} \hat{E}(0,t) \exp\{-i(\omega - \omega_0)t\}dt = \frac{At_0}{\sqrt{2\pi}} \exp\left\{-\tfrac{1}{2}(\omega - \omega_0)^2 t_0^2\right\} \tag{6.8}$$

which is a Gaussian centred on the carrier frequency ω_0.

The evolution of the pulse as it propagates through the dispersive medium is governed by a factor of the form $\exp\{-ikz\}$ within $\hat{E}(z,t)$; see Eq. (2.4). The effect of dispersion can therefore be found by applying a phase change of $\phi(\omega) = -k(\omega)z$ to $\bar{V}(0,\omega)$ namely

$$\bar{V}(z,\omega) = \bar{V}(0,\omega) \exp\{-ik(\omega)z\}. \tag{6.9}$$

We expand the phase $\phi(\omega)$ as a Taylor series about ω_0 namely

$$\phi(\omega) \equiv -k(\omega)z = \phi_0 + \left.\frac{d\phi}{d\omega}\right|_0 (\omega - \omega_0) + \tfrac{1}{2}\left.\frac{d^2\phi}{d\omega^2}\right|_0 (\omega - \omega_0)^2 + \cdots \tag{6.10}$$

where $\phi_0 = -k(\omega_0)z$, and it follows from Eqs (6.3) and (6.4) that the derivatives are

$$\left.\frac{d\phi}{d\omega}\right|_0 = -\tau_{\text{group}}; \qquad \left.\frac{d^2\phi}{d\omega^2}\right|_0 = -\frac{d\tau_{\text{group}}}{d\omega}. \tag{6.11}$$

---

[2] Readers who are not interested in the details of the following proof can skip to Eq. (6.14).
[3] See Appendix E for a discussion of the analytic signal.

It will also be convenient for later use to write the analogous expansion for $k(\omega)$ in the form

$$k(\omega) = k(\omega_0) + \beta_1(\omega - \omega_0) + \tfrac{1}{2}\beta_2(\omega - \omega_0)^2 + \cdots \qquad (6.12)$$

where the expansion coefficients, which will feature in several later formulae, are

$$\beta_1 \equiv \left.\frac{dk}{d\omega}\right|_0 = v_{\text{group}}^{-1} = \frac{\tau_{\text{group}}}{z}; \quad \beta_2 \equiv \left.\frac{d^2k}{d\omega^2}\right|_0 = \frac{1}{z}\frac{d\tau_{\text{group}}}{d\omega} = -\frac{D_\lambda \lambda^2}{2\pi c}. \qquad (6.13)$$

Substituting Eq. (6.10) into Eq.(6.9), and transforming back to the time domain yields

$$\hat{E}(z,t)\exp(i\omega_0 t) = \int_0^\infty \bar{V}(z,\omega)\exp\{i\omega t\}d\omega$$

$$= \underbrace{\frac{A}{\sqrt{1+i\xi}}}_{\text{term 1}} \underbrace{\exp\{i\omega_0 t_{\text{phase}}\}}_{\text{term 2}} \underbrace{\exp\left\{-\frac{t_{\text{group}}^2}{2t_0^2(1+\xi^2)}\right\}}_{\text{term3}} \underbrace{\exp\left\{\frac{i\xi t_{\text{group}}^2}{2t_0^2(1+\xi^2)}\right\}}_{\text{term 4}}$$

$$(6.14)$$

where $t_{\text{phase}} = t - z/v_{\text{phase}}$, $t_{\text{group}} = t - \tau_{\text{group}}$, and ξ is a dimensionless distance given by

$$\xi = \frac{\beta_2 z}{t_0^2} = -\left(\frac{D_\lambda \lambda^2}{2\pi c t_0^2}\right)z. \qquad (6.15)$$

Of the four terms on the right-hand side of Eq. (6.14), two (terms 1 and 3) affect the pulse envelope, and two (terms 2 and 4) affect the phase of the carrier. The result has five key features:

- the optical carrier wave (term 2) moves at the phase velocity since $t_{\text{phase}} = t - z/v_{\text{phase}}$;
- the envelope (term 3) moves at the group velocity since $t_{\text{group}} = t - z/v_{\text{group}}$;
- the pulse width (term 3) is increased by the factor $\sqrt{1+\xi^2}$;
- the pulse peak intensity (term 1) is reduced by the factor $\sqrt{1+\xi^2}$, which follows when one works out the modulus squared;
- the pulse acquires a frequency *chirp* (term 4), which means that the optical carrier frequency changes through the pulse.

These issues are reviewed in the following subsections.

6.3.2 Characteristic dispersion distance

The first question to ask is: how far must the pulse travel through the dispersive medium before $|\xi| = 1$? This *characteristic dispersion distance* L_{disp} marks the point at which the peak intensity has fallen by $\sqrt{2}$ and the pulse duration has increased by $\sqrt{2}$. From $\xi = (\beta_2/t_0^2)z$, it follows that

$$L_{\text{disp}} = \frac{t_0^2}{|\beta_2|} = \frac{\Delta t_0^2}{4\log 2\,|\beta_2|}. \qquad (6.16)$$

We conclude that the larger $|\beta_2|$ the smaller L_{disp} (as expected), and that L_{disp} increases as the *square* of the initial pulse duration. The square arises because a pulse of twice the duration needs to travel twice as far to be broadened by $\sqrt{2}$, but also has half the bandwidth to be dispersed.

6.3.3 The bandwidth theorem and the time–bandwidth product

Next, we consider the relationship between the pulse duration and the spectral bandwidth. From Eq. (6.8), the FWHM of the spectral intensity is easily shown to be

$$\Delta \nu = \frac{1}{\Delta t_0} \frac{2 \log 2}{\pi} = \frac{0.441}{\Delta t_0}. \tag{6.17}$$

Moreover, we have seen from Eq. (6.14) that dispersion causes the pulse duration to increase according to

$$\Delta t = \Delta t_0 \sqrt{1 + \xi^2}. \tag{6.18}$$

Hence the *time–bandwidth product* is

$$\Delta t \Delta \nu = 0.441 \sqrt{1 + \xi^2}. \tag{6.19}$$

This is a special case of the *bandwidth theorem*, which states that

$$\Delta t \Delta \nu \geq K \tag{6.20}$$

where K, a number of order unity, is 0.441 for Gaussian profiles. When the minimum time–bandwidth product is realised, which occurs at $\xi = 0$ in the present example, a pulse is said to be *bandwidth-limited* or *transform-limited*.

6.3.4 Chirp

The magnitude of the frequency chirp can be determined by writing term 4 of Eq. (6.14) as $\exp\{i\psi\}$, and remembering that $d\psi/dt$ represents the shift in the carrier frequency. The chirp (in rad s^{-2}) is therefore

$$\frac{d\omega}{dt} = \frac{d^2\psi}{dt^2} = \frac{\xi}{t_0^2(1 + \xi^2)}. \tag{6.21}$$

For long propagation distances ($|\xi| \gg 1$), this simplifies to

$$\frac{d\omega}{dt} \cong \frac{1}{t_0^2 \xi} = \frac{1}{\beta_2 z} = \frac{2\pi c}{\lambda^2 D_\lambda z}; \quad \frac{d\nu}{dt} \cong \frac{1}{2\pi \beta_2 z} = \frac{c}{\lambda^2 D_\lambda z} \quad (|\xi| \gg 1) \tag{6.22}$$

where the result has been quoted in both rad s^{-2} and in Hz s^{-1}.

It is worth trying to understand the physics behind these formulae. First of all, Eqs (6.21) and (6.22) indicate that, if $\beta_2 > 0$, the chirp is positive. Positive β_2 means that the GTD increases with frequency, so higher frequencies suffer a greater delay than lower frequencies, which is consistent with an increase in carrier frequency through the pulse.

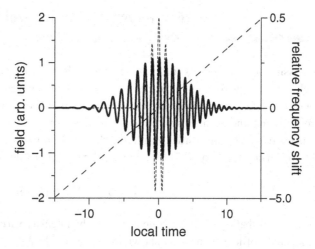

Fig. 6.3 Profile of a chirped Gaussian pulse with $\xi = +3$. The parent pulse is shown dotted. The slanting dotted line is the relative frequency shift (plotted against the right-hand axis), and confirms that the chirp is linear.

Secondly, Eq. (6.21) indicates that, as the pulse propagates through the dispersive medium, the chirp increases until $|\xi| = 1$, and decreases thereafter. The initial increase occurs as the pulse spectrum begins to be spread out, but the chirp ultimately falls as the fixed bandwidth is distributed over an increasingly long time.

Let us follow this line of thought a little further. For $|\xi| \gg 1$, the pulse duration is $\Delta t = |\xi| \, \Delta t_0$ from Eq. (6.18), so dividing the bandwidth from Eq. (6.17) by the pulse duration yields

$$\frac{\Delta \nu}{\Delta t} = \frac{1}{|\xi| \, \Delta t_0^2} \frac{2 \log 2}{\pi} = \frac{c}{\lambda^2 \, |D_\lambda| \, z} \tag{6.23}$$

with the help of Eq. (6.15). This is very reassuring, because it is in perfect agreement with Eq. (6.22), apart from the sign which has not been considered.

An example of a pulse chirped by dispersion is shown in Fig. 6.3. In order to make the chirping clearly visible, a pulse only two optical cycles in duration has been chosen as the parent (shown dotted). The value of ξ is +3, and careful inspection of the figure shows that the pulse is indeed up-chirped; the carrier frequency is clearly higher for $t_g > 0$ than for $t_g < 0$.

The slanting dotted line (plotted against the right-hand axis in the figure) records the frequency shift relative to that of the optical carrier. This ranges from around -0.3 on the leading edge to around $+0.3$ on the trailing edge which, in real world terms, are high numbers. However, had more realistic values been used, the chirp would have been hard to see by eye.

Finally, notice that dispersion has reduced the amplitude of the pulse because of term 1 in Eq. (6.14). If a quick glance at the figure suggests that energy is not conserved, remember that the field rather than the intensity is plotted in the figure.

6.4 Hyperbolic secant profiles

Gaussian profiles possess some convenient mathematical properties; for example, no other bell-shaped function would have allowed the calculation leading to Eq. (6.14) to have been performed so simply. However, hyperbolic secant profiles arise in the analysis of optical solitons, and indeed most real optical signals are probably closer to sech than to Gaussian pulses. It is therefore worth considering the basic properties of sech profiles.

The analogue of Eq. (6.7) for a hyperbolic secant profile is

$$E_0(0, t) = A \operatorname{sech}\{t/t_0\}. \tag{6.24}$$

Remember that if the field profile is a sech, the intensity varies as sech^2, the FWHM of the intensity profile is then easily shown to be $\Delta t = t_0 \ln\{3 + \sqrt{8}\} \simeq 1.763\, t_0$.

Just as the Fourier transform of a Gaussian is another Gaussian, so the Fourier transform of a sech is another sech, namely

$$\bar{V}(0, \omega - \omega_0) = A t_0 \sqrt{\pi/2} \, \operatorname{sech} \left\{ \tfrac{1}{2}\pi(\omega - \omega_0) t_0 \right\}. \tag{6.25}$$

The time–bandwidth product comes out to be $\Delta t \Delta \nu = \pi^{-2} \left(\ln\{3 + \sqrt{8}\} \right)^2 \simeq 0.315$, slightly less than the figure of 0.441 for Gaussians (see Eq. 6.20).

The key difference between Gaussian and sech (or sech^2) profiles lies in the behaviour in the wings. This is best appreciated by viewing the profiles on a log scale as in Fig. 6.4. Whereas the sech^2 function (curve S) has exponential wings, which appear as straight lines on the log scale, the wings of the Gaussian (curve G) plunge increasingly steeply

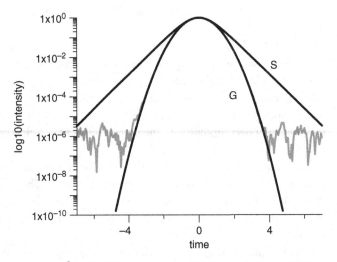

Fig. 6.4   Comparison of Gaussian and sech^2 profiles of the same intensity FWHM plotted on a logarithmic intensity scale. In reality, the wings are always embedded in a noise background (shown in grey).

downwards. Naturally, in any experimental situation, the wings ultimately end in the noise background, as indicated in grey.[4]

6.5 Compression using gratings and prisms

The pulse broadening and chirping that result from GVD could easily be reversed if a material with the opposite GVD characteristics could be found. However, materials exhibiting negative GVD are needed to reverse the effects of positive GVD, and these are virtually unobtainable. Fortunately, the same effect can be realised by using arrangements of diffraction gratings or prisms; indeed the grating arrangement shown in Fig. 6.5, first proposed by Treacy in 1969 [57], has been widely used ever since for compressing up-chirped pulses.

A simple geometrical argument shows that the optical path of a plane wave negotiating the Z-shaped path between the gratings is shorter for short wavelengths and longer for long. As the wavelength increases, the diffraction angle δ in Fig. 6.5 increases too; points B and C will lie higher in the diagram (see arrows), and the path length ABC will increase. Hence, the lower wavelengths (higher frequencies) at the rear of an up-chirped pulse will take a shorter path, and hence will catch up the lower frequencies at the front. In terms of the frequency-dependent phase, we can write

$$\phi(\omega) = \underbrace{-\tfrac{1}{2}(\beta_2 z)(\omega - \omega_0)^2}_{\text{GVD}} + \underbrace{\tfrac{1}{2}(\gamma_2 p)(\omega - \omega_0)^2}_{\text{grating pair}} \tag{6.26}$$

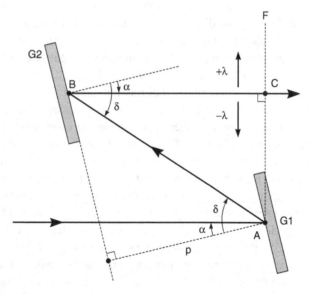

Fig. 6.5   Geometry of a grating pair.

---

[4] The fluctuating noise background shown in Fig. 6.4 is numerically generated random noise.

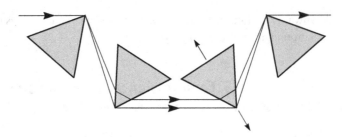

Fig. 6.6   Four-prism compressor.

where the first term comes from Eqs (6.10) and (6.12), and p is the perpendicular distance between the gratings. If $\gamma_2 p$ is positive, the second term counters the quadratic term in $\beta_2 z$ in the case of normal GVD (see Section 6.3).

To find γ_2 in Eq. (6.26), it is necessary to calculate the group time delay τ_{group} of different frequency groups travelling from the point of incidence A on grating G1 in Fig. 6.5 to B on the second grating G2, and thence to C. The result (see Appendix F for details) is

$$\frac{d\tau_{\text{group}}}{d\omega} = -\gamma_2 p = -\frac{4\pi^2 cp}{\Lambda^2 \omega^3 \cos^3 \bar{\delta}} \tag{6.27}$$

where Λ is the ruling spacing of the gratings, and $\bar{\delta}$ is the mean value of angle δ in Fig. 6.5. Notice that $\gamma_2 p$ is necessarily positive, so it is correctly signed to offset positive $\beta_2 z$ in Eq. (6.26). On the other hand, $d\tau_{\text{group}}/d\omega$ is negative, which means that higher frequencies overtake lower frequencies. These are the characteristic features of negative GVD.

Grating pairs are often used in a double-pass configuration. The output is either reflected back through the system by a plane mirror, or two grating pairs are used in series. Apart from doubling the dispersion, both schemes have the advantage that different frequency components are not spatially separated in the output beam.

Gratings can deliver substantial amounts of negative GVD, but they introduce significant loss at the same time. An alternative technique based on chromatic dispersion in sets of prisms offers much higher transmission, and is therefore suitable for dispersion control inside laser cavities.[5] Prisms are also less expensive that gratings, and prism systems can deliver dispersion of either sign too. On the other hand, the amount of GVD they can introduce is smaller than for gratings, basically because chromatic dispersion in glass is much weaker than diffractive dispersion from gratings. The four-prism configuration shown in Fig. 6.6 is a popular arrangement; as for gratings, placing two prism pairs in series ensures that different frequency components are not spatially separated in the output beam.

The properties of a single prism pair can be understood from the geometry of Fig. 6.7 in which a typical ray follows the path shown from A to B. As pointed out by Martinez *et al.* [58], the frequency-dependent phase introduced by refraction through the prisms can be measured between *any* pair of points on the input and output wavefronts, so the line between the prism vertices at V and W can be used as a convenient reference in this case.[6]

---

[5] Another option for intra-cavity dispersion control is to use chirped dielectric mirrors.
[6] This technique could equally have been used to analyse the grating pair (see Problem 6.6).

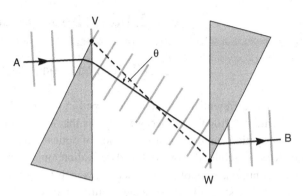

Fig. 6.7 Geometry of a prism pair.

Fig. 6.8 Schematic grating stretcher.

If the distance VW is given by s, and the angle between VW and the ray path is θ, we have

$$\phi(\omega) = -\frac{\omega s \cos \theta(\omega)}{c} \tag{6.28}$$

and hence

$$\frac{d\tau_{\text{group}}}{d\omega} = -\frac{d^2\phi}{d\omega^2} = -\frac{\omega s \cos\theta}{c}\left(\frac{d\theta}{d\omega}\right)^2 - \frac{s \sin\theta}{c}\left(2\frac{d\theta}{d\omega} + \omega\frac{d^2\theta}{d\omega^2}\right). \tag{6.29}$$

The key conclusion is that $d\tau_{\text{group}}/d\omega$ will be negative if the first term is dominant, which will be the case for small θ. Under these conditions, the prism pair, like the grating pair, exhibits the characteristic of negative dispersion. However, an arbitrary amount of positive dispersion can easily be included by increasing the distance a pulse travels in the body of one or both of the prisms. Moreover, if one of the prisms is translated along a line perpendicular to its base, as suggested by the arrows on the third prism in Fig. 6.6, the amount of positive GVD can be controlled.

Finally, we note that placing a telescope between a pair of gratings, as shown in Fig. 6.8, reverses the sign of the dispersion, and this kind of arrangement is routinely used to introduce positive GVD for stretching pulses in chirped-pulse amplification systems. Mirrors are often used instead of the lenses shown in the figure; indeed placing a single spherical mirror of the correct strength between a pair of gratings in a reflective double-pass arrangement offers an elegant solution.

6.6 Optical pulse propagation

6.6.1 Pulse propagation equation

The simple differential equations developed in Chapter 2 (see e.g. Eq. 2.8) were suitable for tracing the growth and decay of signals that are essentially monochromatic. To study the propagation and nonlinear interaction of optical pulses, more sophisticated differential equations are required. For a detailed derivation, one needs to go back to Maxwell's wave equation, but it is possible to advance a reasonable argument in favour of the correct result by developing some of the formulae obtained in this chapter in a simple hand-waving way.

For an incremental propagation distance δz, Eq. (6.9) can be written in the differential form $\delta \bar{V}(\omega) = -ik(\omega)\bar{V}(\omega)\delta z$. With the help of Eq. (6.12), this result can be used to construct the differential equation

$$\frac{d\bar{V}(z,\omega)}{dz} = -i(\beta_0 + \beta_1(\omega - \omega_0) + \tfrac{1}{2}\beta_2(\omega - \omega_0)^2 + \cdots)\bar{V}(z,\omega). \tag{6.30}$$

This equation can now be Fourier transformed, term by term, into the time-domain equation

$$\frac{\partial \hat{E}(z,t)}{\partial z} = \left(-i\beta_0 - \beta_1\frac{\partial}{\partial t} + \tfrac{1}{2}i\beta_2\frac{\partial^2}{\partial t^2} + \cdots\right)\hat{E}(z,t) \tag{6.31}$$

where standard formulae for transforming differential operators have been used. Since $\beta_1 = v_{\text{group}}^{-1}$, Eq. (6.31) can be rewritten in the form

$$\left(\frac{\partial}{\partial z} + \frac{1}{v_{\text{group}}}\frac{\partial}{\partial t}\right)\hat{E}(z,t) = i\left(-\beta_0 + \tfrac{1}{2}\beta_2\frac{\partial^2}{\partial t^2} + \cdots\right)\hat{E}(z,t) \tag{6.32}$$

which contains all the ingredients needed to produce the features seen in Eq. (6.14). The directional derivative on the left-hand side represents the rate of change of the field with distance as seen by an observer travelling with the pulse, and ensures that the pulse envelope travels at the group velocity. The first term on the right-hand side ensures that the carrier wave travels at the phase velocity, while the term in β_2 underlies the development of chirp and the change in the pulse shape.

6.6.2 Transformation to local time

A number of refinements to Eq. (6.32) can now be introduced. First of all, the phase velocity term can be removed by writing $\hat{E}(z,t) = \tilde{E}(z,t)\exp\{-ik_0z\}$ where $k_0 = \beta_0$. The reduces the equation to

$$\left(\frac{\partial}{\partial z} + \frac{1}{v_{\text{group}}}\frac{\partial}{\partial t}\right)\tilde{E}(z,t) = \tfrac{1}{2}i\beta_2\frac{\partial^2\tilde{E}(z,t)}{\partial t^2}. \tag{6.33}$$

A further very common procedure is to transform to the so-called *local time frame*, where local time is measured relative to the time of arrival of a particular point on the pulse. This removes the motion of the pulse from the equations, which has the advantage that changes to the pulse profile are highlighted, and that a steady-state pulse is represented by a function with zero derivatives in the direction of propagation.

The local time coordinates (Z, T) are duly defined by $Z = z$, $T \equiv t_{group} = t - z/v_{group}$. By using the chain rule, it is easy to show that

$$\frac{\partial}{\partial z} + \frac{1}{v_{group}} \frac{\partial}{\partial t} \Rightarrow \frac{\partial}{\partial Z} \quad \text{and} \quad \frac{\partial}{\partial t} \Rightarrow \frac{\partial}{\partial T},$$

and Eq. (6.33) therefore becomes

$$\frac{\partial \tilde{E}'(Z, T)}{\partial Z} = \tfrac{1}{2} i \beta_2 \frac{\partial^2 \tilde{E}'(Z, T)}{\partial T^2} \tag{6.34}$$

where the field $\tilde{E}'(Z, T)$ in transformed coordinates is related to $\tilde{E}(z, t)$ by

$$\tilde{E}(z, t) = \tilde{E}'(Z, T) \equiv \tilde{E}'(z, t - z/v_{group}). \tag{6.35}$$

Strictly speaking, $\tilde{E}$ and $\tilde{E}'$ are different mathematical functions, a point that is often glossed over. The distinction can be understood by considering the trivial case of a pulse travelling at the group velocity without change of shape, i.e. with $\beta_2 = 0$. A schematic representation of $\tilde{E}(z, t)$ in this case is shown in the lower part of Fig. 6.9, where the slanting lines are contours of $\tilde{E}(z, t)$, and thicker lines represent higher field values. Notice how the peak of

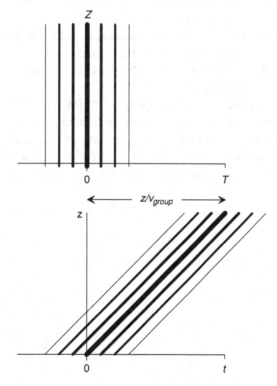

Fig. 6.9   Pulse propagation in (top) the local time frame (Z-T), and (bottom) in the normal z-t frame.

the pulse moves from $t = 0$ to $t = z/v_{\text{group}}$ in propagating a distance z. This should be compared with the upper part of the figure, which shows the corresponding representation of $\tilde{E}'(Z, T)$. The contours now run parallel to the Z-axis, at fixed values of local time T. The difference between the two parts of Fig. 6.9 is precisely the difference between $\tilde{E}$ and $\tilde{E}'$.

For future purposes, we will set these niceties aside, and write Eq. (6.34) in the form

$$\frac{\partial \tilde{E}}{\partial z} = \tfrac{1}{2} i \beta_2 \frac{\partial^2 \tilde{E}}{\partial T^2} \tag{6.36}$$

where the distinction between $\tilde{E}$ and $\tilde{E}'$ has been ignored, and Z has been replaced by z on the left-hand side on the understanding that the derivative is to be performed at constant T, not at constant t. The parameter T has been retained on the right-hand side as a reminder that this is a local time-frame equation. Equation (6.36) will be used to construct the differential equation governing short pulse propagation at the beginning of the next chapter.

Problems

6.1 Prove the final step of Eq. (6.3).

6.2 It appears from Fig. 6.2 that the zero-dispersion wavelength (where the group time delay dispersion changes sign) might perhaps correspond to a point of inflexion in the phase velocity as a function of wavelength. Is this suggestion correct or incorrect?

6.3 (a) Prove the formula $\Delta t_0 = t_0 \sqrt{4 \ln 2} \simeq 1.665 t_0$ for the intensity FWHM of the Gaussian pulse defined in Eq. (6.7).

　　(b) A transform-limited Gaussian pulse at 1 μm has an intensity FWHM of 1 ps. Find its bandwidth in both THz and nm.

　　(c) What is the maximum chirp (in GHz/ps) that can be realised by dispersing the pulse of part (b)?

6.4 Prove the result for the intensity FWHM of a sech^2 pulse quoted after Eq. (6.24).

6.5 Find the characteristic dispersion distance L_{disp} for a 100 fs Gaussian pulse at 1 μm in a medium with a dispersion parameter of 20 ps nm^{-1} km^{-1}.

6.6 Apply the technique used to find the dispersion characteristics of the prism pair in Eqs (6.28) and (6.29) to derive Eq. (6.27) for the grating pair.

7   Nonlinear optics with pulses

7.1 Introduction and preview

While everything in Chapter 6 came under the umbrella of linear optics, in this chapter we will be looking at a range of nonlinear phenomena associated with optical pulses. We will consider the process of self-phase modulation in considerable detail, and see how it can be used, in combination with dispersion, to stretch and compress optical pulses, and to generate optical solitons. We will examine the adverse effects of group velocity dispersion on second harmonic generation, and examine various ways of optimising the bandwidth in optical parametric chirped pulse amplification. Finally, we will discover how nonlinear optical techniques can be used in the diagnosis of ultrashort pulses and for the stabilisation of the carrier-envelope phase.

7.2 Wave equation for short pulses

A detailed derivation of the differential equation governing the propagation of short optical pulses under nonlinear conditions is long and intricate; see for example [59]. To avoid this, we will take a series of reasonable steps that lead to the correct conclusion.

We start, as in Section 2.2, by substituting Eqs (2.4) and (2.5) into Eq. (2.3), but we now retain three time-dependent terms that were previously discarded. The result is

$$2ik\frac{\partial \tilde{E}}{dz} + k^2\tilde{E} + \frac{2i\omega}{c^2}\frac{\partial \tilde{E}}{dt} - \frac{\omega^2}{c^2}\tilde{E} = \mu_0\left(\omega^2(\tilde{P}^{\mathrm{L}} + \tilde{P}^{\mathrm{NL}}) - 2i\omega\frac{\partial(\tilde{P}^{\mathrm{L}} + \tilde{P}^{\mathrm{NL}})}{dt}\right). \quad (7.1)$$

We will mainly be dealing with signals close to a single frequency ω, so only one frequency component has been included, and the subscripts on $\tilde{E}$ and $\tilde{P}^{\mathrm{NL}}$ have therefore been dropped. We now set $\tilde{P}^{\mathrm{L}} = \varepsilon_0\chi^{(1)}\tilde{E}$ and $k = n_0\omega/c$, where $n_0 = \sqrt{1 + \chi^{(1)}}$ is the refractive index.[1] Equation (7.1) then becomes

$$\frac{\partial \tilde{E}}{dz} + \frac{1}{v_{\mathrm{phase}}}\frac{\partial \tilde{E}}{dt} = -i\frac{\omega}{2c\varepsilon_0 n_0}\left(1 - \frac{2i}{\omega}\frac{\partial}{dt}\right)\tilde{P}^{\mathrm{NL}} \quad (7.2)$$

---

[1] The symbol n_0 is used for the low-intensity value of the refractive index in this chapter. The symbol n will be reserved for the value when intensity dependence effects are taken into account; see Section 7.3.

where $v_{\text{phase}} = c/n_0$. When the complex envelopes have no time variation, this equation reduces to Eq. (2.8).

The most obvious deficiency of Eq. (7.2) is that it takes no account of dispersion. This shortcoming can be remedied by switching to the frequency domain and expanding the angular wave number as a Taylor series in ω, as was done in Sections 6.3 and 6.6. Equation (7.2) then becomes

$$\frac{\partial \tilde{E}}{\partial z} + \frac{1}{v_{\text{group}}} \frac{\partial \tilde{E}}{\partial t} = \tfrac{1}{2} i \beta_2 \frac{\partial^2 \tilde{E}}{\partial t^2} - i \frac{\omega}{2 c \varepsilon_0 n_0} \left(1 - \frac{2i}{\omega} \frac{\partial}{\partial t} \right) \tilde{P}^{\text{NL}} \qquad (7.3)$$

where the dispersion parameter is given by

$$\beta_2 = \frac{1}{z} \frac{d\tau_{\text{group}}}{d\omega}$$

as explained in Chapter 6.

When Eq. (7.3) is transformed to the local time frame $Z = z$, $T = t - z/v_{\text{group}}$ (see Section 6.6.2), it reduces to[2]

$$\frac{\partial \tilde{E}}{\partial z} = \tfrac{1}{2} i \beta_2 \frac{\partial^2 \tilde{E}(z,t)}{\partial T^2} - i \frac{\omega}{2 c \varepsilon_0 n_0} \left(1 - \frac{i}{\omega} \frac{\partial}{\partial T} \right) \tilde{P}^{\text{NL}}. \qquad (7.4)$$

In writing this equation, some of the finer details discussed at the end of the last chapter have been glossed over. Although Z has been replaced by z on the left-hand side, it must be remembered that the derivative is to be performed at constant T, not at constant t. Notice that the factor 2 in the final operator of Eq. (7.3) has disappeared in the local time conversion. In fact, it is frequently a good approximation to ignore this operator altogether and, for much of this chapter, we will abbreviate Eq. (7.4) to

$$\frac{\partial \tilde{E}}{\partial z} = \tfrac{1}{2} i \beta_2 \frac{\partial^2 \tilde{E}}{\partial T^2} - i \frac{\omega}{2 c \varepsilon_0 n_0} \tilde{P}^{\text{NL}}. \qquad (7.5)$$

However, the term that has been omitted plays an essential role in a number of important nonlinear processes, especially when pulses containing very few optical cycles are involved. We shall return to this issue at several points later in this chapter.

7.3 Self-phase modulation

We first encountered the process of intensity-dependent refractive index (IDRI) in Chapter 1 and then (in more detail) in Chapter 5. We now consider the effect of IDRI on the propagation of an optical pulse. Written in terms of slowly varying envelopes, and with the coordinate subscripts dropped, Eq. (5.16) reads

$$\tilde{P}^{\text{NL}} = \tfrac{3}{4} \varepsilon_0 \chi_1^{\text{IDRI}} \left| \tilde{E} \right|^2 \tilde{E}. \qquad (7.6)$$

---

[2] The step from Eq. (7.3) to Eq. (7.4) is not straightforward; see [59] for a detailed treatment.

As shown in Chapter 5, this implies that the refractive index acquires an intensity-dependent term given by

$$n = n_0 + \left(\frac{3\chi_1^{\text{IDRI}}}{4n_0^2 c\varepsilon_0} \right) I = n_0 + n_2 I \tag{7.7}$$

where n_0 is the refractive index at low intensities, and the second step defines the nonlinear index n_2.[3] In fused silica $n_2 \simeq 3 \times 10^{-16}$ cm^2 W^{-1}; in most other materials, the value is one or two orders of magnitude larger. The sign of n_2 is almost always positive, which means that refractive index normally *increases* with intensity.

The most obvious consequence of Eq. (7.7) for an optical pulse is that the phase velocity ($= c/n$) varies with intensity, and this causes the phase of the optical carrier wave to become modulated in the process known as self-phase modulation (SPM). Specifically, since the angular wave number is $k = n\omega/c$, and an optical signal acquires a phase of $-ikz$ as it propagates, a pulse will develop a phase modulation given by $\phi = -\omega n_2 I(t)z/c$. The same conclusion follows from Eq. (7.5). If we ignore the GVD term for the moment, and insert Eq. (7.6) for the nonlinear polarisation, the equation reads

$$\frac{\partial \tilde{E}}{\partial z} = -i \left(\frac{\omega n_2 I}{c} \right) \tilde{E}. \tag{7.8}$$

The solution is clearly $\tilde{E}(z) = \tilde{E}(0) \exp\{i\phi(z)\}$ where

$$\phi(z,t) = -\omega n_2 I(t) z/c. \tag{7.9}$$

Notice that these equations imply that the pulse envelope will not be affected by the nonlinearity, and this is certainly borne out in practice when the IDRI is weak and/or the distance of propagation is small. Under other conditions, the more sophisticated Eq. (7.4) has to be used as the governing equation, and one such case is dealt with in Section 7.5 below.

Equation (7.9) shows that the phase change at the peak of the pulse rises to 1 radian after a propagation distance of $c/\omega n_2 I_{\text{pk}}$, and this provides a convenient definition of the characteristic distance for SPM, namely

$$L_{\text{spm}} = \frac{c}{\omega |n_2| I_{\text{pk}}} \tag{7.10}$$

which is identical to Eq. (5.22). After travelling a distance of (say) $4\pi \times L_{\text{spm}}$, the phase change at the peak will therefore be 4π, and Fig. 7.1 shows what the field profile looks like at this point. Although superficially similar to Fig. 6.3 (the case of GVD), the structure of the frequency modulation is more complicated in the present figure. For SPM, the frequency shift (plotted as a dotted line in the figure) is

$$\frac{\partial \phi}{\partial t} = -\frac{\omega n_2 z}{c} \frac{\partial I}{\partial t}$$

from Eq. (7.9), while the *rate of change* of frequency (the chirp) is

$$\frac{d\omega}{dt} = \frac{d^2\phi}{dt^2} = -\frac{\omega n_2 z}{c} \frac{\partial^2 I}{\partial t^2}.$$

---

[3] The definition of the nonlinear refractive index is sometimes based on the equation $n = n_0 + n_2 |E|^2$.

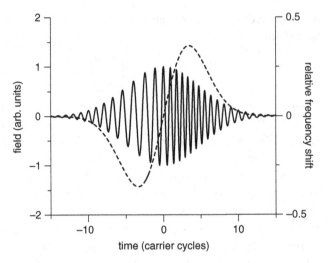

Fig. 7.1 Profile of a pulse chirped by self-phase modulation. The dotted line traces the relative frequency shift. Compare Fig. 6.3.

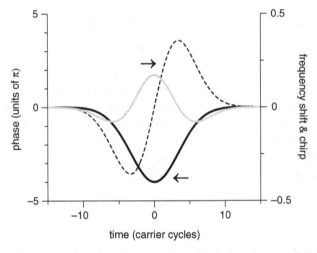

Fig. 7.2 Characteristics of Fig. 7.1: (solid line) time-dependent phase; (dotted line) relative frequency shift (= first phase derivative); (grey line) chirp (= second phase derivative).

Whereas in the dispersive case (shown in Fig. 6.3), the chirp is constant and positive across the entire profile, now it is negative on both wings, but positive in around the peak of the profile.

More detailed information is contained in Fig. 7.2, which shows the phase $\phi(t)$, and curves based on its first and second time derivatives. From Eq. (7.9), $\phi(t) \sim -I(t)$, and this is the solid line in Fig. 7.2. The dotted line is the relative frequency shift from Fig. 7.1, which varies as $d\phi/dt$, while the solid grey line is the chirp. Notice that the

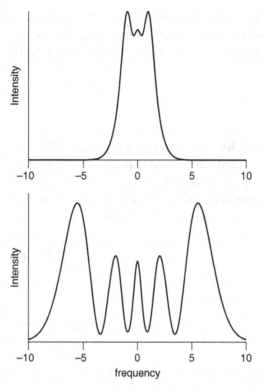

Fig. 7.3   Spectra of pulses chirped by SPM: peak phase shift π (top), 4.5π (bottom).

region of positive chirp lies between the two points of inflexion of $I(t)$, which are close to the half-maximum intensity points.

Figure 7.3 shows spectra of self-phase modulated Gaussian pulses for peak phase shifts of π and 4.5π (top and bottom, respectively). The frequency scale is normalised to the intensity FWHM of the parent pulse spectrum. In general, the spectra consist of a series of lobes, the number corresponding roughly to the peak phase shift in units of π [60, 61]. We will discuss how the increased bandwidth can be exploited to create shorter pulses in Section 7.6 below.

7.4 Self-phase modulation with dispersion: optical solitons

Dispersion has been ignored in Section 7.3, so we must now consider what happens to a pulse when both SPM and GVD occur together. We duly restore the dispersion term to its rightful place in Eq. (7.8), which now reads

$$\frac{\partial \tilde{E}}{\partial z} = \tfrac{1}{2} i \beta_2 \frac{\partial^2 \tilde{E}}{\partial T^2} - i \left(\frac{\omega n_2 I}{c} \right) \tilde{E}. \tag{7.11}$$

We now use the dispersion distance L_{disp} from Eq. (6.16), and the pulse width measure t_0 from Eq. (6.24) to define dimensionless distance and time coordinates $\zeta = z/L_{\text{disp}} = |\beta_2| z/t_0^2$ and $\tau = T/t_0$. In terms of the relative field $\tilde{U} = \tilde{E}/\tilde{E}_{\text{pk}}$, Eq. (7.11) then becomes

$$\frac{\partial \tilde{U}}{\partial \zeta} = i \left(\frac{\text{sgn}\{\beta_2\}}{2} \frac{\partial^2}{\partial \tau^2} - \frac{L_{\text{disp}}}{L_{\text{spm}}} \left| \tilde{U} \right|^2 \right) \tilde{U}. \qquad (7.12)$$

This equation has very different properties depending on the sign of the GVD parameter β_2. As we have seen earlier, both GVD and SPM cause pulse chirping. Since the nonlinear index n_2 is almost invariably positive, SPM generally creates an up-chirp. Positive GVD ($\beta_2 > 0$; see Table 6.1) creates an up-chirp too so, in this case, GVD and SPM reinforce each other; see Section 7.6 below. On the other hand, when $\beta_2 < 0$, the two processes work in opposition, and Eq. (7.12) then reads

$$i \frac{\partial \tilde{U}}{\partial \zeta} = \left(\tfrac{1}{2} \frac{\partial^2}{\partial \tau^2} + N^2 \left| \tilde{U} \right|^2 \right) \tilde{U} \qquad (\beta_2 < 0) \qquad (7.13)$$

where

$$N = \sqrt{\frac{L_{\text{disp}}}{L_{\text{spm}}}} = \sqrt{\frac{\omega n_2 I_{\text{pk}} t_0^2}{c \, |\beta_2|}}. \qquad (7.14)$$

Equation (7.13) is in the form of the nonlinear Schrödinger equation (NLSE), and is essentially identical to Eq. (5.24). In Chapter 5, we were discussing spatial solitons; now we are dealing with temporal solitons – time-dependent pulse solutions of Eq. (7.13). Whereas, before, the dispersive term involved the transverse spatial coordinate ξ, and was associated physically with diffraction, now the corresponding parameter is τ, and the term represents GVD. The scale length L_{diff} (diffraction) is replaced by L_{disp} (GVD), while the definitions of L_{spm} and L_{nl} are identical. Once again, N is the soliton 'order'.

For the lowest-order ($N = 1$) solution of Eq. (7.13), the GVD and SPM terms are in perfect balance, and the soliton propagates as an unchanging solitary pulse. The solution is

$$\tilde{U}(\tau, \zeta) = \text{sech}\{\tau\} \exp\{-\tfrac{1}{2} i \zeta\}. \qquad (7.15)$$

It is easy to verify that this result is a stationary solution of Eq. (7.13); i.e. $\partial \tilde{U}/\partial \zeta = 0$. Since $N = 1$, it follows from Eq. (7.14) that

$$\frac{c}{\omega \, |n_2| \, I_{\text{pk}}} (= L_{\text{spm}}) = \frac{t_0^2}{|\beta_2|} (= L_{\text{disp}}) \qquad (7.16)$$

so the peak intensity of the soliton is related to the duration through the equation

$$I_{\text{pk}} t_0^2 = \frac{c \, |\beta_2|}{\omega \, |n_2|}. \qquad (7.17)$$

The equations imply that if the soliton duration is halved, the strength of the dispersion is quadrupled, and the peak intensity must be quadrupled to compensate if a stationary $N = 1$

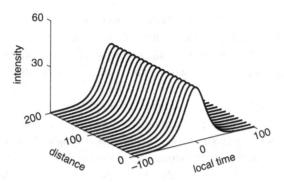

Fig. 7.4   Steady-state propagation of the $N = 1$ soliton.

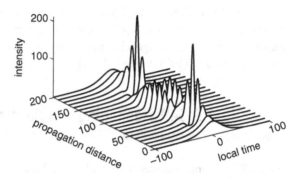

Fig. 7.5   Propagation of the $N = 2$ soliton over the first period.

solution is to be maintained. Rewriting Eq. (7.15) in terms of the original parameters yields

$$\tilde{E}(z, T) = \tilde{E}_{\text{pk}} \, \text{sech}\{T/t_0\} \exp\{-\tfrac{1}{2}iz/L\} \tag{7.18}$$

where $L = L_{\text{disp}} = L_{\text{spm}}$. While the soliton envelope travels at the group velocity, the phase term implies that the phase velocity is reduced from its original value of c/n_0; see Problem 7.3.

A numerical solution to Eq. (7.13) for $N = 1$ is shown in Fig. 7.4. The figure is entirely unexciting visually, but its truly remarkable feature lies precisely in the fact that the profile does not evolve as it propagates. In fact, since we already know that Eq. (7.15) is a stationary solution to the NLSE, the figure is largely a test of the numerics. However, the simulation does provide the additional information that the first-order soliton is stable under propagation.

More complex analytical solutions of Eq. (7.13) exist for higher integer values of N, i.e. when SPM is N^2 times stronger relative to GVD than for $N = 1$. For all higher-order solitons, the solutions are periodic rather than stationary. The profiles evolve cyclically under propagation with a universal period of $z_0 = \tfrac{1}{2}\pi L_{\text{disp}}$ that does not depend on N; in all cases, the profile returns to its initial form at integral multiples of z_0. Figure 7.5 shows the structure of the $N = 2$ soliton over the first complete period. The contortions within the period become increasingly complicated as N increases.

An interesting practical issue arises if the pulse launched into a soliton-supporting system does not exactly match a soliton solution. Broadly speaking, the initial pulse evolves into the soliton whose order N_f is the nearest integer to the initial (non-integer) value N_i given by Eq. (7.14). The peak intensity and pulse width adjust themselves until Eq. (7.14) is satisfied for $N = N_f$, and some fraction of the initial energy is dissipated in the process. If $N_i < \frac{1}{2}$, no soliton is formed, and the pulse simply disperses.[4]

Optical soliton propagation was first reported in 1980 by Mollenauer, Stolen and Gordon, who observed picosecond pulse narrowing and subsequent steady-state propagation in an optical fibre [62]. Because the spreading associated with GVD was perfectly balanced by SPM, solitons immediately became candidates for high bit-rate optical communications. However, the solitons still tended to lengthen under propagation because of energy losses (see Eq. 7.17), and so frequent replenishment of the energy was necessary. In 1988, Mollenauer and Smith demonstrated transmission of 55 ps pulses over 4000 km by supplying Raman gain at 42 km intervals [63]. Despite this, very few soliton-based fibre links are currently in commercial use, largely because of advances in conventional fibre transmission. Further information can be found in [64, 65].

7.5 Self-steepening and shock formation

When the effects of intensity-dependent refractive index are weak, its principal effect is on the phase and spectral structure of an optical pulse. However, when it is strong, the pulse envelope is distorted as well as the optical carrier wave [60,66]. To analyse this case, it is necessary to go back to Eq. (7.4), which includes the differential operator on the right-hand side that has been ignored in the previous two sections. When Eq. (7.6) is inserted for the nonlinear polarisation and the terms rearranged, the equation becomes

$$\left(\frac{\partial}{\partial z} + \frac{(n_{\text{group}} + n_2 I)}{c} \frac{\partial}{\partial t} \right) \tilde{E} = \frac{1}{2} i \beta_2 \frac{\partial^2 \tilde{E}}{\partial t^2} - i \frac{n_2}{c} \left(\omega - i \frac{\partial}{\partial t} \right) I \, \tilde{E}. \qquad (7.19)$$

The most significant feature of this result is that the group index is now intensity dependent, so the group velocity falls with intensity in the same way as the phase velocity. This means that the pulse peak travels more slowly than the wings, so the trailing edge of the pulse envelope will begin to steepen, leading potentially to the formation of a shock.

The simple geometry of Fig. 7.6 enables the propagation distance before a shock forms on the pulse envelope to be calculated. Time runs horizontally in the figure, and distance vertically. The slanting lines trace the paths of two neighbouring points on the pulse profile,[5] which travel at different speeds because of the different local intensities. The figure shows the lines meeting at S after propagating a distance L, at which point the intensity gradient becomes infinite. The trajectory starting at A travels at speed v and reaches S in time t, while the neighbouring trajectory from point B travels at $v - dv$ and takes $t + dt$ to reach

---

[4] See Agrawal [61] for more details.
[5] These lines are known mathematically as the characteristics of the governing differential equation.

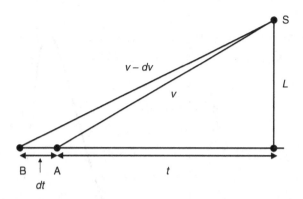

Fig. 7.6   Space-time geometry of self-steepening and shocking.

the same point. Clearly $L = vt = (v - dv)(t + dt)$ and it follows that

$$\frac{dv}{dt} = \frac{v}{t} = \frac{v^2}{L} = \frac{c^2}{L n_{\text{group}}^2} \tag{7.20}$$

where n_{group} is the group index. On the other hand, we also know that

$$v = \frac{c}{n_{\text{group}} + n_2 I},$$

so

$$\frac{dv}{dt} = -\frac{c n_2}{(n_{\text{group}} + n_2 I)^2} \frac{dI}{dt} \simeq -\frac{c n_2}{n_{\text{group}}^2} \frac{dI}{dt}. \tag{7.21}$$

Combining Eqs (7.20) and (7.21) yields

$$L = -\frac{c}{n_2 dI/dt}.$$

The shock first occurs at the minimum value of this function, in other words at the point where the intensity is falling most rapidly. Hence

$$L_{\text{shock}} = \text{Min } L = -\frac{c}{n_2 \text{Min}\{dI/dt\}} = \frac{c\Delta t}{1.83 n_2 I_{\text{max}}}. \tag{7.22}$$

Note that $\text{Min}\{dI/dt\}$ in the denominator refers to the steepest negative gradient; the final step applies to a Gaussian pulse with a full width at half maximum intensity of Δt. Figure 7.7 shows the intensity profile as the point of shocking is approached.

While this analysis is entertaining, one should probably not get too excited by the possibility of optical shock formation. The steepening of the trailing edge of an intense pulse (known as self-steepening) is certainly a real phenomenon, but the analysis of shocking given here is very simplistic. We have for instance ignored the other terms in Eq. (7.19), particularly the one representing group velocity dispersion, which is the principal counter-balance to shock formation. After all, as the profile steepens and a shock is imminent, the bandwidth will be widening rapidly, and dispersive effects will be correspondingly strengthened.

An even more esoteric possibility is that steepening and shock formation might occur not on the envelope of a pulse but within its carrier-wave structure. Surprisingly, the

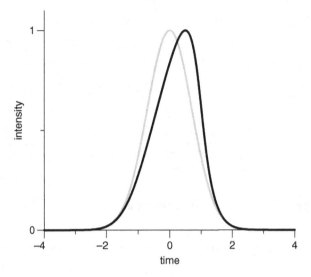

Fig. 7.7   A self-steepened pulse envelope.

theory of this process was studied by Rosen as early as 1965 [67], before work on
envelope shocking had been developed. The issue was revisited in the late 1990s [68],
and carrier wave steepening has been considered as a tool in high harmonic generation
experiments [69].

7.6 Compression of self-phase modulated pulses

We saw in Chapter 6 that when a pulse is subject to *positive* group velocity dispersion
(GVD), the spectral width is not changed, but the temporal profile broadens, and the pulse
acquires a positive frequency sweep. In Section 6.5, we studied ways in which combinations
of gratings and prisms can be used to reverse this chirp, and potentially return the profile to
its original shape. We discovered in Section 7.4 that solitons occur when SPM and *negative*
GVD are judiciously balanced. On the other hand, we saw in Section 7.3 that, when a pulse
is up-chirped by SPM acting *on its own*, the temporal profile is largely unaffected, but the
spectrum is broadened, and the broadening can be dramatic (see Fig. 7.3). But once the
spectral width is increased, the bandwidth theorem (Section 6.3.3) tells us that the potential
pulse duration after compression is decreased, and this suggests that it should be possible
to use gratings and prisms to achieve shorter pulse durations than at the outset. Our aim in
the present section is therefore to discover how to manipulate the spectral phase profile to
compress pulses that have been chirped by SPM.

Although gratings and prisms work well for countering GVD, this is because GVD
creates an essentially linear chirp that is readily reversed. Unfortunately, things are not so
straightforward for SPM. There is no problem with the signs; SPM produces an up-chirp,
and requires negative GVD to reverse it. The difficulty is that, as indicated in Figs 7.1 and

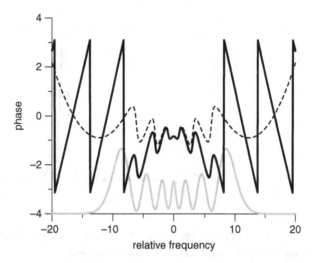

Fig. 7.8   Spectral profiles of a pulse chirped by SPM with a peak phase shift of 6.5π: spectral phase (bold), spectral intensity (grey), spectral phase after optimal quadratic correction (dotted).

7.2, the chirp produced by SPM is not linear. The point is further evidenced in Fig. 7.8, where the phase variation across the spectrum of a self-phase modulated pulse with a peak phase shift of 6.5π is displayed (solid black line); the spectral intensity is shown as a grey line for reference. The complexity of Fig. 7.8 is the manifestation in the frequency domain of the non-uniform frequency sweep within a pulse chirped by SPM; see Fig. 7.1.

So what happens when one tries to compress the pulse of Fig. 7.8? First of all, let us try the effect of an 'ideal' compressor, one that simply eliminates the spectral phase variations in the figure. Such a device is virtually impossible to realise in the laboratory, but the operation is easily performed numerically by a single line of code that sets all the spectral phase to zero. The result is shown as the bold line in Fig. 7.9, where the intensity scale is normalised to the peak intensity of the original pulse (shown dotted grey), and the time-scale is normalised to its FWHM. The peak intensity has increased by a factor of over 11, and the width reduced by roughly the same amount. However, the quality is poor as evidenced by the lobes in the wings.

A more realistic possibility is to try compressing the parent profile using a quadratic phase correction. Of course, the phase variation in Fig. 7.8 is *not* quadratic so, to optimise the process, one has to scan a range of values and choose the one that yields the highest peak power. The corrected phase profile for this optimum condition is shown dotted in Fig. 7.8, and the dotted black line in Fig. 7.9 shows the corresponding temporal profile. The peak intensity is slightly lower than for ideal compression, and the side lobes have been replaced by a broad pedestal supporting the central peak.

The key to improving the quality of the compressed pulse is to combine SPM with an appropriate amount of *positive* GVD. This recipe turns out to produce a broad temporal profile with a much smoother spectrum than when SPM acts alone, and it can therefore be more efficiently compressed. However, the exact way in which SPM and GVD conspire to

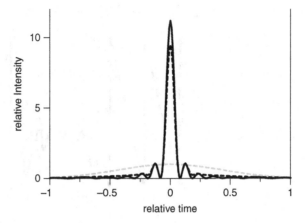

Fig. 7.9 Temporal intensity profiles of a pulse chirped by SPM: under 'ideal' compression (bold), under optimal quadratic compression (dotted), parent (pre-compression) profile (grey).

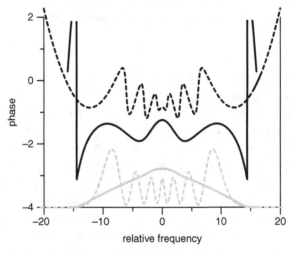

Fig. 7.10 Spectral profiles of a pulse chirped by SPM combined with GVD: spectral phase (bold), spectral intensity (grey). The corresponding profiles without GVD from Fig. 7.8 are shown dotted.

achieve this desirable end is not entirely clear, and whole research papers have been written about how to optimise the process. In practical compressors, one typically introduces SPM and GVD in an optical fibre. Both processes act throughout the length of the fibre, but whereas the effect of GVD increases with distance as the spectral bandwidth gets larger, the SPM process weakens as the pulse spreads out and the peak power drops.

For simulation purposes, it is convenient to use the same scale lengths as in the soliton studies. In the examples that follow, the strength of the initial pulse is $N = 12$, and the distance of propagation is $2.05L_{\text{disp}}(= 1.3z_0$ where z_0 is the soliton distance measure defined at the end of Section 7.4).

The solid lines in Fig. 7.10 show the spectral characteristics of the broadened pulse (before compression) for comparison with those of Fig. 7.8 (shown dotted). In the lower

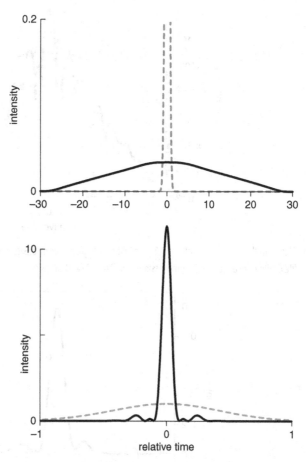

Temporal intensity profiles corresponding to Fig. 7.10: broadened profile before compression (top), compressed profile (bottom). In both frames the original parent profile is shown in grey.

part of the figure (in grey), the multiple peaks in the original spectral intensity profile have been smoothed out, while the ripples in the original phase profile (shown higher up) have largely been eliminated.

The corresponding temporal profiles are shown in Fig. 7.11. The broadened pulse before compression is shown in the upper frame, and the corresponding profile after compression in the lower frame. In both cases, the original parent pulse is shown in dotted grey, and time and intensity are normalised to the characteristics of the parent, which has unit peak intensity and unit FWHM on the scales used. The compression factor (relative to the parent) is about 12, but the pulse still exhibits small side lobes, so it would be wise to display the results on a logarithmic intensity scale to see the wing structure more clearly. After all, experimentalists are interested in pulse 'contrast' (the dynamic intensity range between the pulse peak and its wings), and this cannot be assessed properly in the figure as it stands.

The profiles from Fig. 7.11 are displayed on a log scale in Fig. 7.12. The compressed profile is shown in bold, while the dotted line shows the profile after quadratic compression

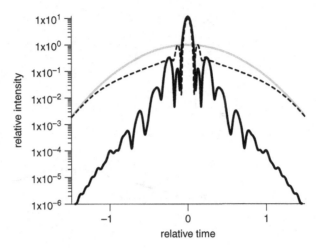

Fig. 7.12 Profiles from the lower frame of Fig. 7.11 plotted on a logarithmic intensity scale. The dotted line corresponds to the profile under optical quadratic compression from Fig. 7.9 (without GVD).

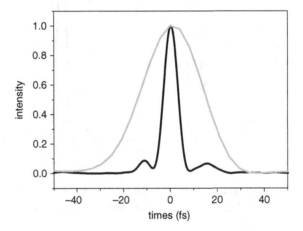

Fig. 7.13 Experimental pulse compression result. A 30 fs parent pulse (grey) is narrowed to under 7 fs (black) using a hollow fibre compressor. (Courtesy of J.W.G. Tisch, Imperial College Attosecond Laboratory.)

from Fig. 7.9. The grey line shows the profile of the original pulse. It can now be seen that the broad pedestal of Fig. 7.9 has been eliminated, and the contrast ratio in the wings has been improved by nearly four orders of magnitude.

The idea of first broadening the bandwidth of an optical pulse, and subsequently compressing it using diffraction gratings or their equivalent, goes back as far as 1969 [70]. However, it was not until the 1980s that the development of optical fibre technology allowed the principles to be put effectively into practice. Some remarkable experimental results with fibre-grating compressors were soon achieved. In 1984, for example, Johnson *et al.* [71] compressed a 33 ps pulse to 410 fs, which corresponds to a compression ratio of 80, an extremely impressive figure for a single-stage compressor. The recent result [72] shown in Fig. 7.13 is on a much more challenging time-scale. The combination of a hollow fibre and a grating was used to compress a 30 fs parent pulse (shown in grey) to under 7 fs (shown in black). Since the pulse energy was around 0.5 mJ, the peak power after compression is approaching 100 GW.

7.7 Group velocity dispersion in second harmonic generation

We now turn from self-action effects, in which a pulse modifies its own propagation characteristics, to inter-pulse effects in which pulses (often at different frequencies) interact. We start by returning to second harmonic generation (SHG), and considering what happens when the process is driven by pulses rather than by quasi-monochromatic beams. Pulses are attractive because the high peak power they offer makes for high conversion efficiency. But there is also a down side if dispersion causes the fundamental and harmonic pulses to travel at different speeds. To understand this problem, which was recognised in the early days of nonlinear optics [73, 74], we must consider once again the relationship between phase velocity and group velocity that we first encountered in Section 6.2.

Phase matching is, of course, as crucial to achieving high-efficiency nonlinear interactions with pulses as it is with monochromatic beams. In the case of SHG, the term implies *phase velocity* matching between the fundamental and second harmonic waves.[6] The question is: what about group velocity matching? In an SHG experiment with optical pulses, what happens if the group velocities of the fundamental and second harmonic pulses are different?

The key point to understand here is that, as explained in Chapter 6, phase and group velocity effects are manifestations of the same phenomenon – dispersion – and hence they are closely interrelated. In particular, when significant group velocity mismatch occurs in SHG, it means that part of the bandwidth covered by the fundamental and harmonic pulses is not phase matched.

To appreciate this important idea, consider the schematic dispersion diagram shown in Fig. 7.14. The centre of the fundamental spectrum lies at A, and the centre of the corresponding second harmonic spectrum at B. If OAB is a straight line (as drawn), it means that the centre frequencies are phase matched, while OB is double the length of OA because SHG doubles the frequency. The dotted line CAD represents a section of the dispersion curve around A, and its gradient corresponds to the group velocity of the fundamental pulse, just as the gradient of OA represents the phase velocity. Now consider second harmonic generation for the frequency component represented by point C. The second harmonic component that is phase matched lies at E, where OCE is a straight line, and OE is twice the length of OC. A similar argument shows that the phase matched second harmonic of component D lies at F. Since it follows from simple geometry that OCD and OEF are similar triangles, the gradient of EBF, which is the section of the dispersion curve near B, is equal to that of

---

[6] As we saw in Chapter 1, a more complicated condition applies in a general three-wave interaction.

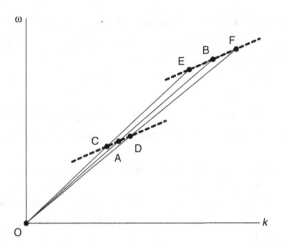

Fig. 7.14 Schematic dispersion diagram showing the phase-matching conditions for second harmonic generation using a short optical pulse.

CAD. So the group velocities of the fundamental and second harmonic pulses must be the same for the pulse to be fully phase matched.

In practice, CAD and EBF are of course dispersion *curves*, and reflect properties of the nonlinear material that cannot be adjusted, except sometimes to a limited extent in optical fibres. Their gradients at A and B will not generally be the same, so the group velocities of the fundamental and harmonic pulses will be different. Hence OCD and OEF will no longer be similar triangles and, if the A and B components are themselves phase matched, the same will not be the case for adjacent components.

It is worth mentioning that, under birefringent phase matching, CAD and EBF would be sections of separate ordinary and extraordinary dispersion curves. For quasi-phase matching, one could perhaps think of EBF as a section of the curve that has been shifted to a lower value of k by the periodic poling.

We now return to the original question: what is the effect of a group velocity mismatch on SHG? As the fundamental pulse propagates through the medium, it starts to generate a pulse of second harmonic in the normal way. But the harmonic pulse travels at a different speed from the fundamental – usually slower – and so it begins to fall behind. Meanwhile, the fundamental pulse continues to generate second harmonic waves as before, and these continue to exude from the back end of the fundamental pulse. The final result is a flat second harmonic signal, stretching out behind the parent pulse over a time interval determined by the group time delay dispersion (GTDD) between the two frequencies accumulated through the nonlinear medium.

A numerical simulation of the behaviour is shown in Fig. 7.15. A 1 ps fundamental pulse (1 nJ at 1560 nm) generates a second harmonic signal at 780 nm in a 75 mm slab of periodically poled lithium niobate (PPLN). The group time delay dispersion between the two signals is 304 fs mm^{-1}, which amounts to 22.8 ps in the entire slab. Centred on zero local time in the figure are the original fundamental pulse (dotted grey), and its

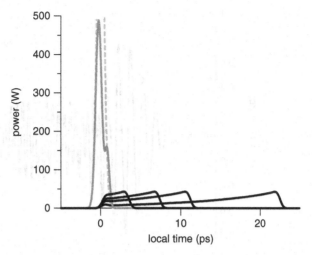

Fig. 7.15 Second harmonic generation by a short optical pulse in the presence of group velocity dispersion.

depleted profile (solid grey) after propagation through the slab, by which time 44% of the original energy has been lost to the harmonic. The harmonic signal at four stages during the interaction (12.5 mm, 25 mm 37.5 mm and 75 mm) is spread out behind the fundamental, as expected, and the extent of the final harmonic profile corresponds to the 22.8 ps group time delay noted earlier. The slanting tops of the harmonic signals are caused by depletion of the fundamental; without depletion the tops would be essentially flat.

Group velocity mismatch not only distorts the shape of the SHG pulse, but it also lowers the overall conversion efficiency too. This is because the second harmonic intensity increases as the square of the distance under phase-matched conditions but, once a section of the second harmonic becomes detached from the parent, its growth is ended and so its effective interaction distance has been reduced. For example, consider a simplistic picture in which the GTDD through the sample was ten times the fundamental pulse duration. The first tranche of second harmonic would detach itself from the parent one-tenth of the way through the nonlinear medium, so its peak intensity would reach only one-hundredth of the level it would have achieved had detachment not occurred. This is partially compensated by the fact that the final second harmonic pulse is ten times as long as the parent, but the overall account shows an energy deficit of a factor of ten. The bottom line is that group velocity mismatch is bad for efficiency, and smears out the second harmonic pulse as well.

Another example of the same phenomenon is shown in Fig. 7.16, this time for a fundamental pulse ($\sim$5 fs duration) containing only a few optical cycles [75]. This result was obtained by the direct solution of Maxwell's equations using a variant of the well-known finite difference time domain (FDTD) method. As in Fig. 7.15, the group velocity of the second harmonic is lower than that of the fundamental, so the high-frequency harmonics lie to the rear (the right) of the fundamental pulse in the figure.

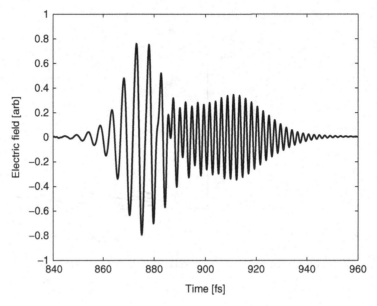

Fig. 7.16   Second harmonic generation by a few-cycle pulse computed by the direct solution of Maxwell's equations; from [75] by permission.

7.8 Optical parametric chirped pulse amplification

7.8.1 The basic principle

Optical parametric chirped pulse amplification (OPCPA) [76, 77] combines the technique of chirped pulse amplification (CPA) with optical parametric amplification (OPA).[7] In chirped pulse amplification [78], a principle originally developed in radar engineering (and known as 'chirp radar') is applied to short-pulse laser physics. The problem to be solved is how to amplify pulses of EM radiation to extreme peak power without damaging the equipment that supplies the energy. The recipe is simple. Start with a short pulse of the desired duration but only modest energy: stretch it in a systematic way, normally by chirping; amplify it to form a broad high-energy pulse; and finally compress the pulse back to its original duration. This creates high-energy pulses that are both extremely short and extremely intense. The peak power needs to be kept within acceptable limits in all system components. In the optical case, both stretcher and compressor are normally based on diffraction gratings; the arrangement of Fig. 6.8 is used for the stretcher, and the standard zigzag configuration of Fig. 6.5 for compression. Damage to the compressor is avoided by the use of large gratings, and by putting the entire compression system under vacuum if necessary.

---

[7] The process is occasionally called chirped pulse optical parametric amplification (CPOPA).

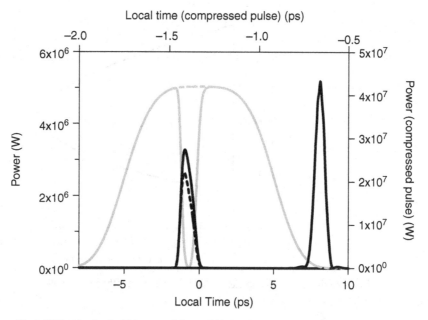

Fig. 7.17 Pulse profiles in OPCPA showing inefficient use of the available bandwidth: signal pulse (bold line), idler pulse (dotted), pump pulse (grey – with parent dotted). The 65 fs compressed signal pulse is shown on the right. [Pump: 50 μJ, 10 ps, 400 nm (centre). Signal pulse: 2 nJ, 7 ps, 714 nm (centre) chirped at 10 THz/ps ($= 17$ nm ps^{-1}). Crystal: 8 mm LBO, all beams in the x-y plane, beam angle 31.14°.]

The attraction of combining chirped pulse amplification with optical parametric amplification is the wide bandwidth offered by parametric amplifiers. The standard procedure is to use a signal pulse with a relatively broad temporal profile and a strong chirp. These characteristics necessarily imply wide spectral bandwidth, and hence the potential for exceptionally short and intense pulses after compression. The potential of pulses with such extreme characteristics in novel scientific applications is immense, and this is why OPCPA is the focus of so much current attention [79]. The pump pulse must naturally be wide enough to amplify the full temporal (and spectral) extent of the signal. The pump itself need not be chirped, although a weak chirp can sometimes be applied to enhance the amplification bandwidth as we shall see shortly.

As a practical example, we demonstrate the OPCPA process for a 10 ps pump pulse driving a 7 ps signal pulse chirped at 10 THz ps^{-1} ($= 17$ nm ps^{-1}) in an 8 mm sample of LBO. Figure 7.17 shows the various profiles after the interaction, the signal in black, the idler in dotted black, the pump in grey, and the initial pump in dotted grey. The separate profile on the right (plotted against the top time axis and the right-hand power axis) shows the 65 fs signal pulse after optimal quadratic compression. A full list of parameter values is given in the figure caption.[8]

---

[8] Further background to the numerical techniques used to create to Fig. 7.17 and similar figures can be found in [80].

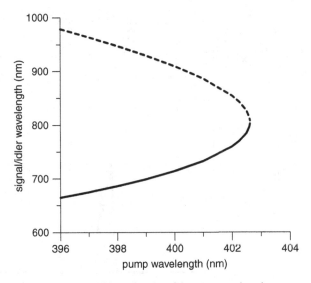

Fig. 7.18   Phase-matched signal/idler wavelengths in LBO as a function of the pump wavelength.

Why are the signal and idler profiles in Fig. 7.17 so narrow? Why has only a thin slice in the centre of the pump pulse been used, and the rest of its energy wasted? The reason is that the signal pulse is strongly chirped, and only the central part of the signal spectrum is phase matched. Time corresponds to frequency in a chirped pulse, and so only a small part of the temporal profile can participate in an efficient nonlinear interaction. Different methods for overcoming this problem are described in the following sections.

7.8.2 Bandwidth enhancement with a chirped pump

Insight into the problem is provided by Fig. 7.18, where the phase matched signal wavelength for LBO is plotted as a function of the pump wavelength. The upper branch of the graph (dotted) traces the idler wavelength. The positive slope of the signal branch near 400 nm indicates that slightly lower (or higher) pump wavelengths are needed to phase-match lower (or higher) signal wavelengths. So the problem could be at least partially solved if the pump pulse carried a chirp that was accurately matched to that of the signal. Careful measurement of Fig. 7.18 reveals that the gradient at 400 nm is around 17, and this suggests that if the pump chirp were $\frac{1}{17}$th that of the signal, or roughly 1 nm ps^{-1} (which converts to 1.875 THz ps^{-1}), phase matching could be maintained over a much wider bandwidth.

Figure 7.19 shows how Fig. 7.17 is transformed when a chirped pump pulse with the specified characteristics is used. A much wider time (and frequency) slice is now involved in the interaction and, with the wider bandwidth, the compressed profile narrows from 65 fs to 18 fs. Further refinements are possible too. One could, for instance, introduce a *nonlinear*

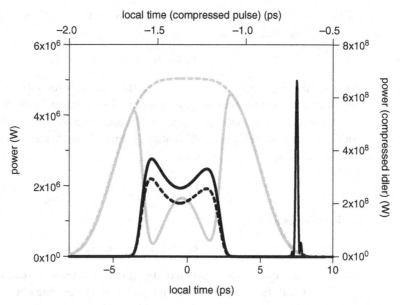

Fig. 7.19   Improved OPCPA bandwidth usage (compared to Fig. 7.17) achieved by chirping the pump pulse at 1.875 THz/ps. See caption to Fig. 7.17 for other details.

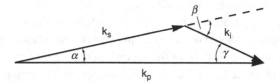

Fig. 7.20   Non-collinear phase matching in OPCPA for bandwidth optimisation.

chirp on the pump or signal, a possibility that has been demonstrated successfully in the laboratory [81].

7.8.3 Bandwidth enhancement with non-collinear beams

Another approach to bandwidth enhancement in OPCPA is to use non-collinear beams. To understand how this might help, consider the case of a narrow-band pump at ω_3 driving a broadband signal at ω_2 [79]. The **k**-vector triangle that ensures phase matching for the signal and idler centre frequencies is shown in Fig. 7.20, where the signal propagates at an angle α to the pump. The question is now whether there is a particular value of α that maximises the phase-matched bandwidth.

We will suppose that the signal and idler are quasi-ordinary waves, while the pump is an extraordinary wave. In this case, the magnitudes of k_2 and k_1 are fixed, but the length of k_3 depends on its angle ϕ to the x-axis.[9] For particular values of ϕ and α that close the triangle, the cosine rule gives

$$2k_3 k_2 \cos \alpha = k_3^2 + k_2^2 - k_1^2. \tag{7.23}$$

Keeping the angle α fixed, we now seek the condition for which the phase matching is unaffected (i.e. the **k**-vector diagram remains closed) when small changes are made to ω_2. As this is done, there will be changes of opposite sign to ω_1, as well as to k_1, k_2, and the angles β and γ in the figure. Differentiating Eq. (7.23) with respect to frequency, and using the fact that $d\omega_2 = -d\omega_1$, yields

$$-k_1 \frac{dk_1}{d\omega_1} = (k_2 - k_3 \cos \alpha) \frac{dk_2}{d\omega_2} \tag{7.24}$$

which reduces to

$$k_3 \cos \alpha = k_2 + k_1 (v_{g2}/v_{g1}) \tag{7.25}$$

where v_{g2} and v_{g1} are the signal and idler group velocities. By combining Eqs (7.23) and (7.25) and with further use of the sine and cosine rules, one obtains for the angle α the expression

$$\alpha = \tan^{-1} \left(\frac{\sin \beta}{k_2/k_1 + \cos \beta} \right) \tag{7.26}$$

where the angle β (defined in the figure) is given by $\beta = \cos^{-1}\{v_{g2}/v_{g1}\}$.

The application of this technique in LBO is illustrated in Fig. 7.21 where the results of simulations based on collinear and non-collinear beams are compared directly. The dotted and solid black lines in the figure show the signal profiles at the end of the interaction in the respective cases, while the corresponding pump profiles are shown in grey. For the parameter values used, which are similar to those in Figs 7.17 and 7.19, the angles are $\alpha = 1.55°$, $\beta \simeq 4.03°$ and $\gamma \simeq 2.48°$. As before, since the signal is strongly chirped, a wider pulse translates immediately to a wider spectrum. The compressed signal pulse widths in the collinear and non-collinear cases are respectively 31.0 and 12.5 fs, which underscores the benefits of non-collinear geometry.

7.8.4 Optical parametric generation in three dimensions

Virtually every example in this book has been based on the plane-wave approximation in which transverse variations in intensity and phase across the wavefronts of the interacting beams are ignored. By contrast, the contour plots displayed in Fig. 7.22 show the transverse intensity profiles of the pump (top), signal (centre), and idler (bottom) after interacting in an optical parametric amplifier. The three beams have been separated purely for display purposes; in reality they lie on top of each other, and the cross-hairs define the common centre line. Once again, the nonlinear medium is LBO, with all beams in the x-y plane.

---

[9] LBO is a biaxial crystal (see Sections 3.5 and 4.6). If, as we are assuming, all beams lie in the x-y plane, those polarised in the z-direction propagate as quasi-ordinary waves.

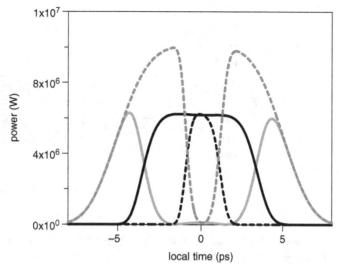

Fig. 7.21 Enhanced OPCPA bandwidth utilisation by non-collinear phase matching: signal pulse (bold line), final pump pulse (grey). Corresponding collinear profiles are shown dotted.
[Pump: 100 μJ, 10 ps, 523.5 nm. Signal: 2 nJ, 7 ps, 850 nm (centre) chirped at 10 THz ps^{-1} ($=$ 24 nm ps^{-1}). Crystal: 8 mm LBO, all beams in the x-y plane, pump propagation direction 15.58°, non-collinear angle $\alpha = 1.545°$.]

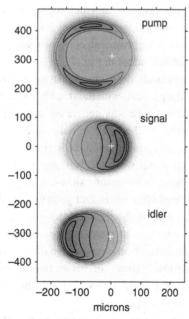

Fig. 7.22 Two-dimensional pump, signal, and idler profiles in OPCPA, with the three profiles separated for display purposes. White crosses mark the common centre line. The seed signal pulse is angled to the right, and the idler runs to the left to ensure phase matching; see Fig. 7.20. Pump walk-off (to the left) is evident.
[Pump 523.5 nm, 1 mJ, 40 ps, 240 μm super-Gaussian. Signal 850 nm, 25 pJ, 20 ps, 240 μm super-Gaussian, 1.3 deg. Crystal: LBO xy, 7 mm, 14.49 deg, 0.85 pm V^{-1}.]

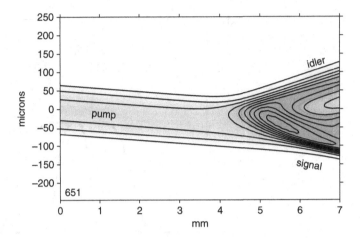

Fig. 7.23 Top view of an OPCPA interaction similar to Fig. 7.22, but with narrower beams to illustrate the separation of signal and idler more clearly.
[Pump: 0.5 mJ, 125 μm super-Gaussian. Signal 0.1 mJ, 125 μm super-Gaussian, −1.3 deg. Other data as Fig. 7.22.]

The signal and idler are quasi-ordinary waves, but the pump is polarised in the x-y plane and so is subject to walk-off, which is clearly visible in its leftward movement of about 60 μm in the 7 mm crystal. The walk-off angle is about 0.5° in this case. A non-collinear beam configuration is used, in which the signal beam is angled rightward at about 1.3°, and this leads to an idler angled leftward at 2.1° (see Fig. 7.20). Over 7 mm, these angles would suggest a ~160 μm rightward shift of the signal, and a ~250 μm leftward shift of the idler, neither of which is evident in Fig. 7.22.

A vivid demonstration of what is happening is provided by Fig. 7.23, which shows an aerial view of the process, with the three beam paths superimposed. To make the point more dramatically, this figure was generated using narrower beams than in Fig. 7.22, and the relative intensities of the three signals were also adjusted for the same reason.[10] For the first 4 mm in the crystal, there is no sign of signal or idler; the pump appears undepleted and, apart from pump walk-off, nothing appears to be happening at all. The parametric generation process is however "slow-cooking". Detailed examination of the data shows that the signal and idler are in fact growing steadily in this region, but this is invisible on the linear scale used in the figure. After all, according to the coupled-wave equations, the signal and idler both need the presence of the other (and of course the pump) in order to grow. Since the interaction would cease if the two parted company, they are forced to cling together for survival. However, about two-thirds of the way through the sample, the pot comes to the boil, and the signal and idler burst out on opposite sides of the pump beam, travelling at the expected angles.[11]

Further results on transverse effects in OPCPA are presented in [82].

---

[10] The angling of the signal and idler beams is reversed in Fig. 7.22, but this does not affect the argument in any way.

[11] Because the axial and transverse scales are different, the beam angles in Fig. 7.22 are magnified by about a factor 10.

7.9 Nonlinear optical diagnosis of short pulses

In the late 1960s, the development of laser mode-locking techniques enabled picosecond optical pulses to be generated for the first time and, since these were beyond the time resolution of oscilloscopes, new measurement techniques had to be devised. Two approaches were explored: streak cameras, and autocorrelation techniques based on nonlinear optics. Versions of the latter approach, highly developed to various levels of sophistication, continue to be used for optical pulse diagnosis today, when pulses in the femtosecond (fs) regime are commonplace, and pulses deep in the attosecond regime are hot topics of current research.[12]

To measure something, one always needs a benchmark. But optical pulses are shorter than all the normal benchmarks, so one resorts to using an optical pulse as its own measure, and this is what is done in autocorrelation methods. The principle is to overlap two replicas of an optical signal with a controllable time displacement between them, and to record some parameter that is sensitive to the degree of overlap as a function of the time displacement. One needs to find a property of the system that behaves differently when the two replicas overlap than when they travel separately (i.e. without overlapping).

A device that immediately springs to mind is the Michelson interferometer, in which an optical signal is divided into two beams that follow paths of different length before being recombined. It is easy to show that the fraction of the pulse energy that reaches the detector under these circumstances is[13]

$$f(\tau) = \tfrac{1}{2}\left(1 + \frac{G_{\mathrm{E}}(\tau)}{G_{\mathrm{E}}(0)}\right) \qquad (7.27)$$

where $G_{\mathrm{E}} = \int\limits_{-\infty}^{\infty} E(t)E(t+\tau)dt$ is the autocorrelation function of the (real) electric field $E(t)$. However, it is easy to show that G_{E} is the Fourier transform of the intensity spectrum, so a Michelson interferometer records the same information as a spectrometer, albeit in a different form.

Additional information can be obtained from a Michelson interferometer if nonlinear optical detection is used, the most common practice being to record the second harmonic of the interferometer output. To analyse this case, we introduce the complex *analytical signal*, which is related to the real field by

$$E(t) = Re V(t) = \tfrac{1}{2}(V(t) + V^*(t)). \qquad (7.28)$$

This equation is actually just a version of Eq. (2.4) in which the electric field is evaluated at a fixed point in space (say $z = 0$), and $V(t)$ includes the complete carrier wave structure.[14]

---

[12] 1 fs $= 10^{-15}$s and 1 as $= 10^{-18}$s.

[13] The rest of the energy is reflected back to the source. Note that, in what follows, $E(t)$ is the field amplitude after two encounters with the 50–50 beam splitting, and therefore one-half that entering the interferometer.

[14] Equation (7.28) does not define $V(t)$ uniquely; a rigorous procedure for doing this is given in Appendix E.

The second harmonic signal depends on the time-averaged fourth power of the total optical field at the output of the interferometer, namely

$$S(\tau) = \overline{(E(t) + E(t + \tau))^4}. \tag{7.29}$$

A laborious calculation now reveals that the relative harmonic signal registered at the output is given by

$$\frac{S(\tau)}{S(\infty)} = \left\{ 1 + \frac{2G_{\mathrm{I}}(\tau) + R(\tau)}{G_{\mathrm{I}}(0)} \right\} \tag{7.30}$$

where $G_{\mathrm{I}}(\tau)$ is the autocorrelation function of the intensity (defined here as $I(t) = |V(t)|^2$) given by

$$G_{\mathrm{I}}(\tau) = \int\limits_{-\infty}^{\infty} I(t)I(t + \tau)dt. \tag{7.31}$$

The background harmonic signal represented by $S(\infty)$ in Eq. (7.30) is the net harmonic signal created by the two replicas travelling separately (i.e. with such a large time separation that there is effectively no overlap at all). Rapidly varying terms contained in $R(\tau)$ in Eq. (7.29) are given by

$$R(\tau) = Re \left\{ e^{i2\omega_0\tau} \int\limits_{-\infty}^{\infty} \tilde{V}^2 \tilde{V}_\tau^2 dt + 2e^{i\omega_0\tau} \int\limits_{-\infty}^{\infty} (\tilde{V}^2 \tilde{V}^* \tilde{V}_\tau^* + \tilde{V} \tilde{V}_\tau \tilde{V}_\tau^{*2})dt \right\} \tag{7.32}$$

where $\tilde{V}(t) = V(t)e^{-i\omega_0 t}$ is the slowly varying part of $V(t)$, and $\tilde{V}$ and $\tilde{V}_\tau$ are abbreviations for $\tilde{V}(t)$ and $\tilde{V}(t + \tau)$, respectively

If the rapidly varying terms in Eq. (7.30) are ignored, it is easy to see that the equation predicts that the signal at the centre of the overlap region ($\tau = 0$) is three times what it is in the wings ($\tau = \infty$). The contrast ratio of the intensity autocorrelation function (IACF) is therefore said to be 3. Unfortunately, however, the mathematics is tending to obscure the physics of what is going on, so we shall adopt a simple-minded approach to try to interpret the result. Consider two square replica pulses of unit width and intensity I'. Outside the overlap region, where the pulses travel separately, $2I'^2$ units of harmonic intensity will be registered, where the 2 arises because there are two pulses in succession, and the square because the second harmonic varies as the square of the fundamental intensity. By contrast, at the centre of the overlap region ($\tau = 0$), the intensity will presumably be doubled, which suggests that the harmonic signal should be $(2I')^2 = 4I'^2$, twice what it is in the wings.

Although this argument indicates correctly that a higher signal is registered within the overlap region, it is wrong in detail because the fine structure in the interference between the two replicas (and contained in $R(\tau)$) has been ignored. Constructive interference doubles the *field* at the anti-nodes, so the anti-node *intensity* rises by $\times 4$ (not $\times 2$), and the corresponding harmonic signal rises to $16I'^2$, or eight times what it is in the wings. Away from the anti-nodes, the interference pattern will be characterised by $\cos^2$ fringes in the intensity, which will lead to $\cos^4$ fringes in the harmonic signal. Since the cycle average of $\cos^4$ is $\frac{3}{8}$, the cycle-averaged harmonic signal is $6I'^2$, which is 50% higher than the crude result

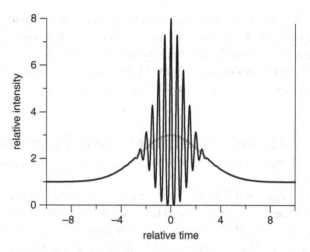

Fig. 7.24   SHG autocorrelation trace showing the fine structure resolved and (grey) unresolved.

obtained in the previous paragraph, and a factor 3 above the $2I'^2$ background value. This confirms the $\times 3$ peak-to-background ratio implied by Eq. (7.30) when the rapidly varying terms are ignored.

Figure 7.24 shows the IACF of the linearly chirped pulse of Fig. 6.3. The full record (including rapidly varying terms) is shown in black, and the cycle-averaged record in grey. Notice that the fine structure does not extend over the entire overlap region, let alone beyond it. This is a consequence of the chirp, which means that strong constructive and destructive interference between the two replicas of the pulse only occurs near the centre of the pattern, where the temporal displacement is small. If one moves sufficiently far enough from the centre, the fine structure is completely smeared out.

The IACF, however accurately measured, does not provide sufficient information to reconstruct even the complete intensity profile of the original pulse, let alone its phase structure [83]. A technique (crude by modern standards) for observing the frequency sweep of a chirped pulse was demonstrated by Treacy as early as 1970 [84]. In Treacy's experiment, pulses were spectrally dispersed, and IACFs of the separate components were recorded simultaneously. The tilted autocorrelation record created by a chirped pulse provided a direct visual demonstration of the frequency sweep.[15]

A number of highly sophisticated techniques of this general type are now available which, in conjunction with suitable numerical tools, enable the complete field structure of the original waveform to be reconstructed. Most of the popular techniques are known by memorable acronyms such as FROG (Frequency-Resolved Optical Gating) [85] and SPIDER (Spectral Phase Interferometry for Direct Electric field Reconstruction) [86]. The field of optical pulse measurement has become increasingly complex, and a full-length book has been written on one single technique [87].

[15] This kind of measurement is known technically as an optical sonogram.

Exhaustive studies of all the available options for optical pulse characterisation can be found in [88, 89]. Most of these methods involve nonlinear optics, although there is no reason in principle why linear techniques cannot be used, provided a non-time stationary filter of some kind is involved.[16] Linear methods would certainly offer increased sensitivity and they continue to be studied for this reason [90].

7.10 Mode-locked pulse trains and carrier envelope phase control

In a typical mode-locked laser, a parent pulse circulating in the laser cavity creates a train of replicas that emerge through the output mirror with a temporal spacing close to the round-trip period of the cavity. Even in a CW laser, there are always sources of jitter, so the parent pulse inevitably changes slightly from one round trip to the next, and one replica pulse is never precisely identical to the one that came before. However, even in the ideal case where the pulse envelope itself is perfectly stable, differences between the phase and group velocities will cause the optical carrier wave to shift under the pulse envelope from transit to transit. This phase slip is highly undesirable in experiments where precise waveform control is needed, and methods are being developed to control it.

The spectrum of a mode-locked pulse train can be found by convolving the spectrum of an individual pulse (e.g. Eq. 6.8) with the longitudinal mode structure of the cavity. In the idealised case, the modes are represented by a simple comb function of the form

$$C(\omega) = \sum_j \delta(\omega - \omega_j) \tag{7.33}$$

where the angular frequencies of the modes ω_j follow from the standing-wave condition $L = j\lambda_j/2$ in a cavity of length L. If we assume the cavity to be filled with a material of refractive index $n_j = n(\omega_j)$, it is easy to show that the mode frequencies are given by $\omega_j = j(\pi c/n_j L)$. The presence of n_j in this formula means that, if the material is dispersive, the modes will not be uniformly spaced, so a pulse circulating freely in the cavity will spread out, just as it would if it propagated through the same length of material under any other circumstances. However, the mode-locking mechanism, be it active or passive, will counter this tendency, enforcing equal mode spacing $\delta\omega$, and enabling us to write

$$\omega_j = j\delta\omega + \delta\omega_0 \tag{7.34}$$

where $\delta\omega_0 (< \delta\omega)$ is a small frequency remainder that takes account of the fact that ω_j will generally not be an integral multiple of $\delta\omega$. Under mode-locked conditions,

$$\delta\omega = \frac{\pi c}{L\bar{n}} \tag{7.35}$$

---

[16] We saw earlier that a basic Michelson interferometer yields nothing more than an intensity spectrum, but no non-time stationary filter is involved in this case.

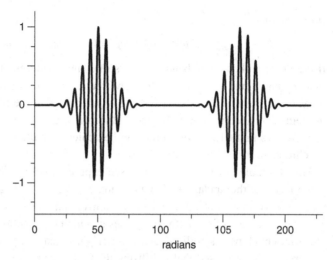

Fig. 7.25   Few-cycle pulses with a carrier-envelope phase slip between them.

where $\bar{n}$ is an effective refractive index that depends on the values of n_j within the bandwidth, and also on the mode-locking mechanism.

The time-dependent real field comes from the sum of all participating modes, namely

$$E(t) = \tfrac{1}{2} \sum_j A_j \exp\{i(\omega_j t + \phi)\} + \text{c.c.}$$

$$= \tfrac{1}{2} \sum_j A_j \exp\{i(j\delta\omega t + \delta\omega_0 t + \phi)\} + \text{c.c.} \tag{7.36}$$

where the amplitudes A_j are real, and ϕ is the common phase of all modes at $t = 0$. Careful thought indicates that the pulse *envelope* repeats with a periodicity of $T = 2\pi/\delta\omega$, while the *phase* of the optical carrier changes from one pulse to the next by $\delta\varphi_{\text{cep}} = T\delta\omega_0 = 2\pi(\delta\omega_0/\delta\omega)$, which is known as the carrier-envelope phase slip (CEP slip).

Figure 7.25 shows an example of the phenomenon described. In the pulse on the left, the peak of the envelope coincides with a peak of the carrier wave, and the pulse structure is 'cosine like', i.e. symmetrical about the centre; in this case the carrier-envelope phase offset is zero. On the other hand, the carrier wave of the pulse on the right has slipped leftward relative to the envelope by 0.4π, and the structure is therefore skewed. If the inter-pulse phase slip were raised to 0.5π, the pulse would be anti-symmetrical (and "sine-like"). In pulses containing many optical cycles, CEP effects are hard to spot by eye, but in few-cycle pulses of the kind shown in Fig. 7.24, they are much more obvious.

Since $\delta\omega$ in Eq. (7.35) depends on L, fine adjustment of the cavity length should enable the frequency comb ω_j to be adjusted, and $\delta\omega_0$ (and hence $\delta\phi_{\text{cep}}$) to be brought to zero. In this case, CEP will not change from pulse to pulse. Various techniques for CEP stabilisation have been devised. Consider, for instance, what happens if a mode-locked signal with a spectrum given by Eq. (7.34) is subject to second harmonic generation. The frequency comb governing the harmonic can be found by summing the frequencies of two fundamental combs so that[17]

$$\omega_l = \omega_j + \omega_k = (j + k)\delta\omega + 2\delta\omega_0. \tag{7.37}$$

If the mode-locked pulse bandwidth is sufficiently wide, the high-frequency wing of the fundamental comb will overlap the low-frequency wing of the second harmonic comb. Now, comparison of Eqs (7.34) and (7.37) indicates that the two combs will be aligned in the region of overlap only if $\delta\omega_0 = 0$. But in that case, the CEP slip will be eliminated, so it follows that CEP stabilisation can be achieved if the two combs can be forced into synchronism.

This desirable outcome can be realised experimentally by heterodyning the frequency components of the fundamental and harmonic waves in the region of spectral overlap, and using a servo loop controlling the cavity mirror spacing L to bring the beat note to zero. This creates a train of mode-locked pulses in which the location of the carrier waves under the pulse envelope is stabilised, and no inter-pulse phase slip occurs [91–93].

It must be emphasised that stabilising the CEP simply stops it varying, and does not of itself determine the absolute carrier phase at the peak of the envelope, which is given by ϕ in Eq. (7.36). However, it is important to recognise that few-cycle pulses are extremely fragile objects and, even if pulses with a specific CEP could be generated on demand, the CEP would be significantly altered if the pulses passed through a few microns of glass; see Problem 7.6. This property can in fact be turned into an opportunity, because it means that the CEP can be tuned by fine adjustment of the path length through a rotatable plate. As explained in Chapter 10, the absolute CEP is of crucial importance in applications such as high harmonic generation and attosecond pulse generation, and CEP tuning is used to optimise the process.

Problems

7.1 Calculate L_{spm} (the characteristic length for self-phase modulation) in fused silica at an intensity of 1 TW cm^{-2}. Express the answer in units of the vacuum wavelength. [n_2 (fused silica) $= 3 \times 10^{-16}$ cm^2 W^{-1}.]

7.2 Verify that Eq. (7.15) is a solution to Eq. (7.13).

7.3 Show that the phase velocity of an $N = 1$ soliton is approximately $(cn_2 I_{pk}/2n_0^2)$ lower than its low-intensity value in the same medium. (This question is essentially the same as Problem 5.5. It is equally applicable to the spatial solitons of Section 5.7 and to the temporal solitons of Section 7.4.)

7.4 Prove the final step of Eq. (7.22) for L_{shock}. Show that the ratio L_{shock}/L_{spm} is roughly four times the number of cycles contained within the intensity FWHM of an optical pulse.

7.5 Try the effect of adding a 3ω component to $\cos \omega t$ and notice how the distortions of the carrier wave depend on the relative phase. This is most conveniently done by plotting the waveforms using a PC.

---

[17] Notice that the 'second harmonic' process includes sum frequency generation among the components of the fundamental.

7.6 CEP phase slip arises in the propagation of an optical pulse through fused silica as a manifestation of the difference between the phase and group velocities. Using the information given in Fig. 6.2 as a guide, estimate the distance the pulse needs to travel before the phase and group time delays differ by a quarter optical cycle (i.e. before a phase slip of $\pi/2$ has occurred). Assume that the vacuum wavelength is 800 nm. Think about the implication of the answer.

8 Some quantum mechanics

8.1 Introduction

A key weakness of the simple approach to nonlinear optics adopted in Chapter 1 was that the physical origin of nonlinearity in the interaction of light and matter was hidden inside the $\chi^{(n)}$ coefficients of the polarisation expansion.[1] This is such a fundamental issue that it is difficult to avoid some mention of how nonlinearity arises within a quantum mechanical framework, even in an introductory text. Unfortunately, the standard technique for calculating the nonlinear coefficients is based on time-dependent perturbation theory, and the expressions that emerge begin to get large and unwieldy even at second order. While every effort has been made in this chapter to provide a gentle lead-in to this aspect of the subject, this is almost impossible to achieve given the inherent complexity of the mathematical machinery.

From a mathematical point of view, Schrödinger's equation is linear in the wave function, but nonlinear in its response to perturbations. At a fundamental level, this is where nonlinear optics comes from. The perturbations of atoms and molecules referred to here arise from external electromagnetic fields. When the fields are relatively weak, the perturbations are relatively small, and the theoretical machinery of time-dependent perturbation theory can be deployed to quantify the effects. This is the regime where the traditional polarisation expansion of Eq. (1.24) applies, indeed the terms in the expansion correspond to successive orders of perturbation theory.

The early part of this chapter is devoted to a brief review of quantum mechanical principles. Perturbation theory techniques are then developed, and expressions for the coefficients governing many of the basic nonlinear processes described elsewhere in this book are derived. The focus is mainly on simple cases, and no attempt is made to obtain expressions for the coefficients in their most general form.

When the perturbations become sufficiently strong and/or the time-scales sufficiently fast, the polarisation expansion breaks down, and other analytical techniques have to be developed. This situation frequently arises close to a resonance, although resonances can sometimes be treated within perturbation theory. Examples of resonant effects, including the stimulated Raman effect, self-induced transparency and electromagnetically-induced transparency are discussed in Chapter 9. High harmonic generation also occurs within the strong-field regime, and this topic is treated in Chapter 10, albeit from an essentially classical standpoint.

---

[1] The other defect was that the tensor nature of the coefficients was ignored.

8.2 Basic ideas

The state of a quantum mechanical system can be specified in several different ways: the *wave function* $\psi(x)$, or the *state vector* $|\psi\rangle$ of Dirac are two obvious examples. The most fundamental approach to quantum mechanics is based on $|\psi\rangle$, which is a vector in an abstract Hilbert space, whereas $\psi(x)$ is one of many possible representations of $|\psi\rangle$ in particular Hilbert spaces [94].[2]

The idea that the state of a system can be represented in different ways should not come as a surprise. After all, the position-dependent wave function $\psi(x)$ is related by Fourier transformation to the momentum space wave function $\Psi(p)$, where $p = \hbar k$, and x and k are transform variables. Since $\psi(x)$ and $\Psi(p)$ both contain complete information about the same state, they are equivalent representations of the system.

Although we will use Dirac formalism in the early part of this chapter, readers who are not familiar with the notation can treat it simply as shorthand for the traditional wave function expressions. As noted by Rae [95], in an equation such as

$$\int \psi_1^* \hat{A} \psi_2 \, dx = \langle 1| \, \hat{A} \, |2\rangle \tag{8.1}$$

the object on the right-hand side can be regarded as a code for the expression on the left-hand side. The *bra vector* $\langle 1|$ translates into the conjugated wave function ψ_1^*, the *ket vector* $|2\rangle$ into ψ_2, and integration is implied.

The words 'represented' and 'representation' have appeared several times in the last few paragraphs. In the early history of quantum mechanics, the words were also used to describe different ways of apportioning the time dependence of the system between the state vectors and the operators. Most modern texts use the word 'picture' in this latter connotation.[3] In the *Schrödinger picture*, the time dependence resides entirely in the state vectors, while in the *Heisenberg picture*, it resides entirely in the operators. On the other hand, in the *interaction picture*, the time dependence is split, with the intrinsic element in the operators and the remainder (which is usually the interesting part) in the state vectors.

8.3 The Schrödinger equation

In state vector notation, the Schrödinger equation reads

$$\frac{\partial |\psi\rangle}{\partial t} = -\frac{i}{\hbar} \hat{H} |\psi\rangle \tag{8.2}$$

where $\hat{H} = \hat{H}_0 + \hat{V}$ includes the unperturbed Hamiltonian $\hat{H}_0$ and a perturbation Hamiltonian $\hat{V}$. In the context of nonlinear optics, $\hat{V}$ is often written in the dipole approximation as $\hat{V} = -\hat{\mu}.\mathbf{E} = e\hat{\mathbf{r}}.\mathbf{E}$, where e is the elementary charge and $\hat{\mu} = -e\hat{\mathbf{r}}$ is the dipole operator.

---

[2] For simplicity, a single space coordinate is applied to wave functions in this section
[3] Throughout Messiah [94], the word 'representation' is put in quotation marks when it is used in this sense.

It is convenient to expand $|\psi\rangle$ in terms of the orthonormal *energy eigenstates* of the system using one of the two alternative forms[4]

$$|\psi\rangle = \sum_n b_n |n_t\rangle = \sum_n a_n |n\rangle. \tag{8.3}$$

Here, the state vector $|n_t\rangle$ contains the intrinsic time dependence of the system, and is related to the time-independent $|n\rangle$ by

$$|n_t\rangle = |n\rangle \exp\{-i\omega_n t\} \tag{8.4}$$

where $\omega_n = E_n/\hbar$ and E_n is the energy eigenvalue. It follows that $a_n = b_n \exp(-i\omega_n t)$ so, in the second form of Eq. (8.3), the intrinsic time dependence is contained in the coefficients a_n, rather than in $|n_t\rangle$. The occupation probabilities of the different energy eigenstates are $|a_n|^2 (= |b_n|^2)$, and these sum to unity so that[5]

$$\sum_n |a_n|^2 = \sum_n |b_n|^2 = 1. \tag{8.5}$$

Written in terms of wave functions, Eq. (8.3) reads

$$\psi = \sum_n b_n \psi_n' = \sum_n a_n \psi_n \tag{8.6}$$

where the *energy eigenfunctions* $\psi_n' = \psi_n \exp\{-i E_n t/\hbar\}$ and the ψ_n are time independent. The orthonormality of the energy eigenstates implies that $\int \psi_n^* \psi_{n'} d\mathbf{r} = \delta_{nn'} = \langle n \mid n' \rangle$ where $d\mathbf{r}$ stands for $dx\,dy\,dz$.

Substituting Eq. (8.3) into Eq. (8.2) allows the time evolution of the quantum system to be expressed is terms of either a_n or b_n via the equations

$$\dot{a}_n = -i\omega_n a_n - \frac{i}{\hbar} \sum_l V_{nl} a_l; \qquad \dot{b}_n = -\frac{i}{\hbar} \sum_l V_{nl}' b_l \tag{8.7}$$

where $V_{nl} = \langle n|\hat{V}|l\rangle$ is the perturbation *matrix element*, $V_{nl}' = V_{nl} \exp\{i\omega_{nl} t\}$, and $\omega_{nl} = \omega_n - \omega_l$. Since $\hat{V} = -\hat{\boldsymbol{\mu}}.\mathbf{E}$, we can write the matrix element out in full as

$$V_{ln} = -(\mu_{ln}^x E_x + \mu_{ln}^y E_y + \mu_{ln}^z E_z) = -(\langle x\rangle_{ln} E_x + \langle y\rangle_{ln} E_y + z_{ln} E_z) \tag{8.8}$$

where the jth Cartesian component of $\boldsymbol{\mu}_{ln}$ is written μ_{ln}^j, or equivalently as $\langle j\rangle_{ln}$ which is slightly easier on the eyes.

The expectation value of an arbitrary operator $\hat{A}$ is given by

$$\langle\psi| \hat{A} |\psi\rangle = \sum_{nl} a_l a_n^* A_{nl} \equiv \sum_{nl} b_l b_n^* A_{nl} \exp\{-i\omega_{nl} t\} \tag{8.9}$$

---

[4] In terms of 'bras' rather than 'kets', the equation reads $\langle\psi| = \sum_n \langle n_t| b_n^* = \sum_n \langle n|a_n^*$.

[5] This is easily proved from Eq. (8.3) using the orthonormality of the energy eigenstates and the fact that the wave vector has unit norm.

where $A_{ln} = \langle l | \hat{A} | n \rangle$. In nonlinear optics, the operator of primary interest is the dipole operator $\hat{\mu} = -e\hat{r}$, because it leads directly to the polarisation via the equation

$$\mathbf{P} = N \langle \mathbf{\mu} \rangle = N \sum_{nl} \mathbf{\mu}_{nl} a_l a_n^*. \qquad (8.10)$$

8.4 The density matrix

8.4.1 Definition of the density matrix

Later in the chapter, we will treat a range of nonlinear optical problems using state vectors and wave functions as in the previous section. However, an alternative formalism based on the density operator (or density matrix) offers significant advantages in some contexts, particularly when resonant effects are involved and/or damping becomes important. We will review the density matrix approach in this section.

The Schrödinger equation (8.2) is entirely equivalent to the equation

$$\frac{\partial \hat{\rho}}{\partial t} = -\frac{i}{\hbar} \left[\hat{H}, \hat{\rho} \right] \qquad (8.11)$$

where $\hat{\rho} = |\psi\rangle\langle\psi|$ is known as the *density operator*, and the commutator $\left[\hat{H}, \hat{\rho} \right] = \hat{H}\hat{\rho} - \hat{\rho}\hat{H}$. Using Eq. (8.3), we can write the density operator in the form

$$\hat{\rho} = |\psi\rangle\langle\psi| = \sum_{ml} a_m a_l^* |m\rangle\langle l| \qquad (8.12)$$

and define the *density matrix* as

$$\rho_{nn'} = \langle n | \hat{\rho} | n' \rangle = a_n a_{n'}^*. \qquad (8.13)$$

Readers who are not familiar with Dirac notation should feel reassured by the second form of Eq. (8.13), which defines the density matrix in terms of the probability amplitudes from Eq. (8.3). Indeed, for the simplest two-level system (shown in Fig. 8.1), the density matrix becomes simply

$$\hat{\rho} = \begin{pmatrix} \rho_{00} & \rho_{01} \\ \rho_{10} & \rho_{11} \end{pmatrix} = \begin{pmatrix} |a_0|^2 & a_0 a_1^* \\ a_1 a_0^* & |a_1|^2 \end{pmatrix} \qquad (8.14)$$

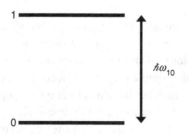

Fig. 8.1   Two-level system.

in which the diagonal elements $\rho_{nn} = |a_n|^2$ are just the population probabilities of the energy eigenstates 0 and 1. The sum of the diagonal elements is therefore unity. The population density N_n of level n is given by $N\rho_{nn}$ where N is the total atomic/molecular number density. As we shall discover shortly, the off-diagonal elements govern the dipole moment of the atoms, and hence the polarisation of the medium. Notice that when (for example) $a_0 = 1$, it follows that $a_1 = 0$, from Eq. (8.5), and that the off-diagonal elements of $\hat{\rho}$ are zero.

The time evolution of ρ can be found by differentiating Eq. (8.13) and using Eqs (8.7). The result is

$$\dot{\rho}_{nn'} = -i\omega_{nn'}\rho_{nn'} - \frac{i}{\hbar}\sum_q \left(V_{nq}\rho_{qn'} - \rho_{nq}V_{qn'}\right) \tag{8.15}$$

where the first term on the right-hand side indicates that the intrinsic time dependence of the off-diagonal element $\rho_{nn'}$ is $\exp\{-i\omega_{nn'}t\}$. However, it is easy to remove this intrinsic motion by defining $\sigma_{nn'} = \rho_{nn'}\exp\{i\omega_{nn'}t\} = b_n b_{n'}^*$. The time dependence of $\sigma_{nn'}$ is then given by

$$\dot{\sigma}_{nn'} = -\frac{i}{\hbar}\sum_q \left(V_{nq}'\sigma_{qn'} - \sigma_{nq}V_{qn'}'\right) \tag{8.16}$$

where $V_{nq}' = V_{nq}\exp\{i\omega_{nq}t\}$, etc. The switch from ρ to σ represents a transformation to the *interaction picture* where the intrinsic time dependence has been transferred to the operators, and the remaining time dependence in $\sigma_{nn'}$ is the result of external perturbations.

At this point, an important issue must be raised: if the density matrix method is entirely equivalent to the state vector method of Section 8.2, what is its advantage? Why not stay with the more familiar state vectors and wave functions? The answer is that the density matrix approach offers clear benefits when one is dealing with *mixed* (as opposed to *pure*) states. If all members of an ensemble of atoms/molecules are in the same quantum state, the state is said to be 'pure'. However, it is far more common for different atoms/molecules to be subject to different stochastic influences, and it is much easier to incorporate these effects within a density matrix treatment than in the state vector approach. Using the two-level case as an example, we can adopt the generalised definition of the density matrix given by

$$\hat{\rho} = \begin{pmatrix} \overline{|a_0|^2} & \overline{a_0 a_1^*} \\ \overline{a_1 a_0^*} & \overline{|a_1|^2} \end{pmatrix} \tag{8.17}$$

where the overbars imply averages over all members of the representative ensemble. Stochastic processes affecting the diagonal elements of the density matrix (e.g. spontaneous emission) redistribute the population between the different states, of which there are just two in this simple case. But effects also exist that randomise the phase relationships between the states but have no effect on the populations, and these cause the off-diagonal elements of $\hat{\rho}$ to decay on a faster time-scale.

We can take these stochastic effects into account by adding phenomenological damping terms to Eq. (8.15). For off-diagonal elements of the generalised density matrix, the equation becomes

$$\dot{\rho}_{nn'} = -i\tilde{\omega}_{nn'}\rho_{nn'} - \frac{i}{\hbar}\sum_q \left(V_{nq}\rho_{qn'} - \rho_{nq}V_{qn'}\right) \qquad (n \neq n') \qquad (8.18)$$

where $\tilde{\omega}_{nn'} = \omega_{nn'} - i\gamma_{nn'}$ and $\gamma_{nn'}$ is the rate of loss of phase coherence between states n and n'. For diagonal elements, the general form is

$$\dot{\rho}_{nn} = \sum_m \alpha_m \rho_{mm} - \beta\rho_{nn} - \frac{i}{\hbar}\sum_q \left(V_{nq}\rho_{qn} - \rho_{nq}V_{qn}\right) \qquad (8.19)$$

where the first term represents a population arriving from other levels, and the second population departing to lower levels. These results could equally have been written in terms of $\sigma_{nn'}$.

8.4.2 Properties of the density matrix

It is clear from its definition that the density matrix is Hermitian, which means that $\rho_{nn'} = \rho_{n'n}^*$, and that its trace is unity (see Eqs 8.5 and 8.14). But it also possesses the key property that the expectation value of any operator $\hat{A}$ is $\left\langle\hat{A}\right\rangle = \text{Tr}\{\hat{\rho}\hat{A}\}$. In the context of nonlinear optics, this enables one to find the dipole moment from $\langle\hat{\mu}\rangle = \text{Tr}\{\hat{\rho}\hat{\mu}\} = \sum_{nn'} \rho_{nn'}\mu_{n'n}$, and once the dipole moment is known, the polarisation follows immediately from $\mathbf{P} = N\langle\boldsymbol{\mu}\rangle$, which is consistent with Eq. (8.10). Typically, only the off-diagonal elements of the dipole moment matrix are non-zero so, for the simplest two-level model, we have

$$\hat{\mu} = \begin{pmatrix} 0 & \mu_{01} \\ \mu_{10} & 0 \end{pmatrix}. \qquad (8.20)$$

It follows that

$$\mathbf{P} = N\langle\boldsymbol{\mu}\rangle = N\left(\rho_{01}\boldsymbol{\mu}_{10} + \rho_{10}\boldsymbol{\mu}_{01}\right) \qquad (8.21)$$

in this special case. The result highlights the important point mentioned earlier that the polarisation is associated with the *off-diagonal elements* of the density matrix. If $\mathbf{P}$ is non-zero, it necessarily implies that the system is in a superposition state; otherwise, all but one of the a_n would be zero, and hence all the off-diagonal elements of the density matrix would be zero too.

Notice that $\langle\boldsymbol{\mu}\rangle$ is real because $\boldsymbol{\mu}_{10} = \boldsymbol{\mu}_{01}^*$. We shall in fact assume that the matrix elements themselves are real, in which case $\boldsymbol{\mu}_{10} = \boldsymbol{\mu}_{01}$; this can usually be arranged without loss of generality. As already noted, we use the notation $\langle j\rangle_{10}$ to represent the jth Cartesian component of $\boldsymbol{\mu}_{10}$.

8.5 Introduction to perturbation theory

In the context of nonlinear optics, perturbation theory involves the successive approximation of the solution for the quantum mechanical state of the system. The procedure can be implemented using either the probability amplitudes a_n (or b_n), or the density matrix $\rho_{nn'}$. It is good to be familiar with both approaches, because each has its particular merits and demerits. The density matrix method includes stochastic effects in a natural way, which is an obvious advantage. But the associated algebra is often more awkward, and one needs to keep a careful watch out for spurious cancellations – situations where terms appear to cancel, but only if causality is ignored. We shall encounter some of the problems that may arise in Section 8.8.2 below.

The application of perturbation theory proceeds in the same way whichever approach is used. Starting from a zeroth-order solution, the relevant evolution equation is used to calculate a first-order correction. The modified solution is then reinserted into the equation, and a second-order correction duly emerges. The operation can in principle be repeated *ad infinitum*, but the algebraic labour increases at an alarming rate. In practice, perturbation theory is rarely extended to fourth order (at least in nonlinear optics), but the complexity at third order is already severe.

In standard perturbation theory notation, one writes $\hat{H} = \hat{H}_0 + \lambda \hat{V}$, and the successive corrections to the probability amplitudes in the form

$$a_n = a_n^{(0)} + \lambda a_n^{(1)} + \lambda^2 a_n^{(2)} + \lambda^3 a_n^{(3)} + \cdots \tag{8.22}$$

The parameter λ (which is nothing to do with wavelength) is used to keep track of terms of different order, and is set to unity at the end of the calculation. In nonlinear optics, however, there are numerous other indicators of term order (the number of matrix elements, for example), and so we simply write

$$a_n = a_n^{(0)} + a_n^{(1)} + a_n^{(2)} + a_n^{(3)} + \cdots \tag{8.23}$$

or the analogous density matrix equation

$$\rho_{nn'} = \rho_{nn'}^{(0)} + \rho_{nn'}^{(1)} + \rho_{nn'}^{(2)} + \rho_{nn'}^{(3)} + \cdots \tag{8.24}$$

From Eq. (8.13), we have $\rho_{nn'} = a_n a_{n'}^*$, and so it is tempting to assume (for example) that

$$\rho_{21}^{(2)} = a_2^{(0)} a_1^{(2)*} + a_2^{(1)} a_1^{(1)*} + a_2^{(2)} a_1^{(0)*} \tag{8.25}$$

in which orders of the amplitudes in the right-hand side terms always sum to 2. One must be a little cautious here, because the density matrix potentially includes damping, while the probability amplitudes do not. However, in the absence of damping, Eq. (8.25) can be verified by taking the time derivatives of both sides and using the relevant evolution equations; see Problem 8.1.

8.6 First-order perturbation theory

In deriving perturbation theory formulae for the linear and nonlinear susceptibilities, we will focus mostly on simple cases where only two or three levels are involved. Readers wishing to see the results in their most generalised form should consult books such as Boyd [1] or Butcher and Cotter [96].

First-order perturbation theory in a two-level model system is very simple, and leads to a formula for the linear susceptibility $\chi^{(1)}$, a parameter that we first encountered in Chapter 1. In the state vector approach, we take the zeroth-order solution as $a_0^{(0)} = 1$, $a_1^{(0)} = 0$, which means that all the atoms are in the ground state. Equations (8.7) then read

$$\dot{a}_1^{(1)} + i\omega_{10}a_1^{(1)} = -\frac{i}{\hbar}V_{10} \tag{8.26}$$

where $a_1^{(1)}$ is the first-order contribution to a_1. If the atoms are subject to a (real) perturbing field whose components are

$$E_j = \tfrac{1}{2}\left(\hat{E}_j \exp\{i\omega t\} + \text{c.c.}\right) \quad (j = x, y, z) \tag{8.27}$$

then

$$V_{10} = -\tfrac{1}{2}\sum_j \langle j\rangle_{10}\left(\hat{E}_j \exp\{i\omega t\} + \text{c.c.}\right) \tag{8.28}$$

where $\langle j\rangle_{ln} \equiv \mu_{ln}^j$ and $j = x, y, z$. Equation (8.26) can now be integrated to give

$$a_1^{(1)} = \frac{1}{2\hbar}\sum_j \langle j\rangle_{10}\left(\frac{\hat{E}_j \exp\{i\omega t\}}{\omega_{10} + \omega} + \frac{\hat{E}_j^* \exp\{-i\omega t\}}{\omega_{10} - \omega}\right). \tag{8.29}$$

The analogous result within the density matrix approach is

$$\rho_{10}^{(1)} = \rho_{01}^{(1)*} = \frac{1}{2\hbar}\sum_j \langle j\rangle_{10}\left(\frac{\hat{E}_j \exp\{i\omega t\}}{\tilde{\omega}_{10} + \omega} + \frac{\hat{E}_j^* \exp\{-i\omega t\}}{\tilde{\omega}_{10} - \omega}\right) \tag{8.30}$$

which is identical to Eq. (8.29) apart from the presence of damping via the complex frequency $\tilde{\omega}_{10} = \omega_{10} - i\gamma_{10}$.

The polarisation of the medium is

$$\begin{aligned}
P_i &= \tfrac{1}{2}\hat{P}_i \exp\{i\omega t\} + \text{c.c.} \\
&= N\left(a_0 a_1^* \langle i\rangle_{10} + a_1 a_0^* \langle i\rangle_{01}\right) \\
&= N\left(\rho_{01}^{(1)} \langle i\rangle_{10} + \rho_{10}^{(1)} \langle i\rangle_{01}\right) \\
&= \frac{N}{2\hbar}\sum_j \left(\frac{\langle j\rangle_{01}\langle i\rangle_{10}}{\tilde{\omega}_{10}^* - \omega} + \frac{\langle i\rangle_{01}\langle j\rangle_{10}}{\tilde{\omega}_{10} + \omega}\right)\hat{E}_j \exp\{i\omega t\} + \text{c.c.} \tag{8.31}
\end{aligned}$$

where the more general form of the result with damping included has been quoted. Notice that terms in $\exp\{i\omega t\}$ from both $\rho_{10}^{(1)}$ and its conjugate $\rho_{01}^{(1)}$ have contributed to the final result. From $\hat{P}_i = \varepsilon_0 \sum_j \chi_{ij}^{(1)} \hat{E}_j$, it follows that the linear susceptibility is

$$\chi_{ij}^{(1)} = \frac{N}{\varepsilon_0\hbar}\left(\frac{\langle j\rangle_{01}\langle i\rangle_{10}}{\tilde{\omega}_{10}^* - \omega} + \frac{\langle i\rangle_{01}\langle j\rangle_{10}}{\tilde{\omega}_{10} + \omega}\right). \tag{8.32}$$

For a many-level system, Eq. (8.30) generalises to

$$\rho_{n0}^{(1)} = \frac{1}{2\hbar} \sum_j \langle j \rangle_{n0} \left(\frac{\hat{E}_j \exp\{i\omega t\}}{\tilde{\omega}_{n0} + \omega} + \frac{\hat{E}_j^* \exp\{-i\omega t\}}{\tilde{\omega}_{n0} - \omega} \right) \tag{8.33}$$

and Eq. (8.32) then becomes

$$\chi_{ij}^{(1)} = \frac{N}{\varepsilon_0 \hbar} \sum_n \left(\frac{\langle j \rangle_{0n} \langle i \rangle_{n0}}{\tilde{\omega}_{n0}^* - \omega} + \frac{\langle i \rangle_{0n} \langle j \rangle_{n0}}{\tilde{\omega}_{n0} + \omega} \right). \tag{8.34}$$

In summary, the applied electric field of Eq. (8.27), acting via the perturbation matrix elements in Eq. (8.28), creates first-order components $a_1^{(1)}$ in Eq. (8.29), or equivalently the off-diagonal density matrix elements $\rho_{01}^{(1)}$ and $\rho_{10}^{(1)}$ of Eq. (8.30). These lead directly to corresponding terms in the polarisation of the medium (Eq. 8.31), and hence in the linear susceptibility (Eqs 8.32 and 8.34). The resonance structure of Eq. (8.34) will be explored in Section 9.2.

As in any forced oscillator system, the correction terms oscillate at the applied frequency rather than the natural resonance frequency, and this fact can be represented by adding 'virtual' levels at $\pm\hbar\omega$ into the energy level scheme of Fig. 8.1, as shown in Fig. 8.2. We will offer an interpretation of Eqs (8.32) and (8.34) in Section 8.7.2, where the significance of the virtual levels is considered in greater detail.

Finally, it is worth remembering that the linear susceptibility is directly related to the dielectric constant matrix through the equation $\varepsilon_{ij} = 1 + \chi_{ij}^{(1)}$, which we encountered in Chapter 3; see the discussion surrounding Eqs (3.5) and (3.6). In general, ε_{ij} is Hermitian ($\varepsilon_{ij} = \varepsilon_{ji}^*$) but, away from resonance, the elements are real and the matrix becomes symmetric. It is this property that allows it to be diagonalised (Eq. 3.6), and the principal dielectric axes defined. However, Eq. (8.34) indicates that, if χ_{ij} is to be real, $\langle j \rangle_{0n} \langle i \rangle_{n0}$ must be real too. We have, of course, assumed the individual matrix elements themselves to be real, so the issue does not arise. But had we not done so, the need for the product to be real when $i \neq j$ raises some interesting issues. When $i = j$, the product is real in any case.

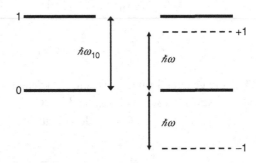

Fig. 8.2 Two-level system showing virtual levels accessed in first-order perturbation theory.

8.7 Second-order perturbation theory

8.7.1 Second harmonic generation

We have just played round 1 of perturbation theory, and we now move on to round 2. First of all, we add a third level (2) to the two-level scheme of Fig. 8.2, to create the three-level system shown in Fig. 8.3; the stack of virtual levels to be discussed later is shown on the right. We will now work out the second harmonic generation coefficient for this system on the assumption that the dipole matrix elements between all three pairs of levels are non-zero, as indicated by the double arrows in the top-left-hand section of the figure. We ignore damping (setting $\tilde{\omega}_{nn'} = \omega_{nn'}$), and we use the probability amplitude approach because it turns out to be algebraically simpler.

At first order, we already have $a_1^{(1)}$ from Eq. (8.29) and, since levels 1 and 2 are functionally equivalent, we have by analogy

$$a_2^{(1)} = \frac{1}{2\hbar} \sum_j \langle j \rangle_{20} \left(\frac{\hat{E}_j \exp\{i\omega t\}}{\omega_{20} + \omega} + \frac{\hat{E}_j^* \exp\{-i\omega t\}}{\omega_{20} - \omega} \right). \qquad (8.35)$$

The second-order corrections can now be found from

$$\dot{a}_1^{(2)} + i\omega_{10}a_1^{(2)} = -\frac{i}{\hbar} V_{12}a_2^{(1)} = \frac{i}{2\hbar} \sum_k \langle k \rangle_{12} \left(\hat{E}_k e^{i\omega t} + \text{c.c.} \right) a_2^{(1)}. \qquad (8.36)$$

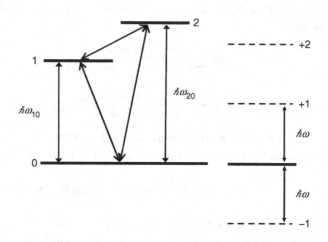

Fig. 8.3   Three-level system showing virtual levels accessed in second-order perturbation theory

Substituting Eq. (8.35) into Eq. (8.36) yields

$$a_1^{(2)} = \frac{1}{4\hbar^2} \sum_{j,k} \langle k \rangle_{12} \langle j \rangle_{20} \left(\frac{\hat{E}_j \hat{E}_k \exp\{i2\omega t\}}{(\omega_{10} + 2\omega)(\omega_{20} + \omega)} + \frac{\hat{E}_j^* \hat{E}_k^* \exp\{-i2\omega t\}}{(\omega_{10} - 2\omega)(\omega_{20} - \omega)} + \cdots \right)$$

$$(8.37)$$

where only terms at $\pm 2\omega$ are shown explicitly. The analogous result for $a_2^{(2)}$ is obtained by the exchange $1 \leftrightarrow 2$. The second harmonic polarisation can now be found from[6]

$$P_i^{\text{SHG}} = \tfrac{1}{2} \hat{P}_i \exp\{i2\omega t\} + \text{c.c.}$$

$$= N \left(a_2^{(2)} a_0^{(0)*} \langle i \rangle_{02} + a_1^{(2)} a_0^{(0)*} \langle i \rangle_{01} + a_1^{(1)} a_2^{(1)*} \langle i \rangle_{21} + \text{c.c.} \right) \quad (8.38)$$

and the final result is[7]

$$\hat{P}_i^{2\omega} = \frac{N}{2\hbar^2} \sum_{j,k} \left(\frac{\langle j \rangle_{01} \langle k \rangle_{12} \langle i \rangle_{20}}{(\omega_{10} - \omega)(\omega_{20} - 2\omega)} + \frac{\langle j \rangle_{02} \langle k \rangle_{21} \langle i \rangle_{10}}{(\omega_{20} - \omega)(\omega_{10} - 2\omega)} \right.$$

$$+ \frac{\langle k \rangle_{01} \langle i \rangle_{12} \langle j \rangle_{20}}{(\omega_{10} - \omega)(\omega_{20} + \omega)} + \frac{\langle k \rangle_{02} \langle i \rangle_{21} \langle j \rangle_{10}}{(\omega_{20} - \omega)(\omega_{10} + \omega)}$$

$$\left. + \frac{\langle i \rangle_{01} \langle j \rangle_{12} \langle k \rangle_{20}}{(\omega_{10} + 2\omega)(\omega_{20} + \omega)} + \frac{\langle i \rangle_{02} \langle j \rangle_{21} k_{10}}{(\omega_{20} + 2\omega)(\omega_{10} + \omega)} \right) \hat{E}_j^\omega \hat{E}_k^\omega. \quad (8.39)$$

The same result can of course be obtained using the density matrix approach, but the algebra is less straightforward. The problem is that when $\hat{P}_i^{2\omega}$ is evaluated in that case, the result has eight terms rather than six, and four contain factors of the type $(\omega_{21} \pm \omega)$, etc. in the denominator. These combine to form terms 3 and 4 in Eq. (8.39), but the step is best avoided.

Equation (8.39) is readily generalised to a many-level system and made more compact at the same time. By introducing summations over multiple intermediate levels, the formula becomes

$$\hat{P}_i^{2\omega} = \frac{N}{2\hbar^2} \sum_{jk} \sum_{nn'} \left(\frac{\langle j \rangle_{0n} \langle k \rangle_{nn'} \langle i \rangle_{n'0}}{(\omega_{n0} - \omega)(\omega_{n'0} - 2\omega)} + \frac{\langle k \rangle_{0n} \langle i \rangle_{nn'} \langle j \rangle_{n'0}}{(\omega_{n0} - \omega)(\omega_{n'0} + \omega)} \right.$$

$$\left. + \frac{\langle i \rangle_{0n} \langle j \rangle_{nn'} \langle k \rangle_{n'0}}{(\omega_{n0} + 2\omega)(\omega_{n'0} + \omega)} \right) \hat{E}_j^\omega \hat{E}_k^\omega. \quad (8.40)$$

Equation (8.39) is just the special case of Eq. (8.40) where there are just two intermediate levels, and the terms with $(n, n') = (1,2)$ and $(2,1)$ are written separately.

8.7.2 Discussion and interpretation

Equations (8.39)–(8.40) merit detailed discussion. In fact, it will be helpful to start by revisiting the first-order case, and relating the terms in the linear susceptibility (Eq. 8.32)

---

[6] Note that all six terms in the second line contain terms at $+2\omega$.

[7] Since j and k are equivalent dummy indices in Eqs (8.39) and (8.40), they can be exchanged at will within the matrix element products, to make the expressions look neat.

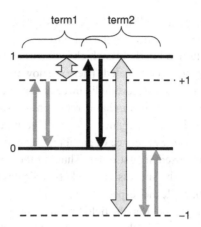

term1   term2

Fig. 8.4   Diagrammatic representation of the structure of Eq. (8.32).

to Fig. 8.2. As noted in Section 8.5, this diagram represents the two real levels (0 and 1) that feature in the equation by solid lines, and the two 'virtual' levels at $\pm\hbar\omega$ by dotted lines. Both terms in the equation use the same real level sequence $0 \rightarrow 1 \rightarrow 0$, and the matrix elements in $\langle j\rangle_{01}\,\langle i\rangle_{10}/(\tilde{\omega}_{10}^* - \omega)$ and $\langle i\rangle_{01}\,\langle j\rangle_{10}/(\tilde{\omega}_{10} + \omega)$ have been ordered so that the subscript sequence 01–10 reflects this. On the other hand, the first term involves the virtual level at $+\hbar\omega$ and the second term the virtual level at $-\hbar\omega$. This can be seen by inspecting the respective denominators. If the complex nature of $\tilde{\omega}_{10}$ is ignored, $(\tilde{\omega}_{10}^* - \omega)$ becomes $(\omega_{10} - \omega)$, which is the angular frequency separation between the real level 1 and the virtual level +1; equally, $(\omega_{10} + \omega)$ is the separation between the same real level and the virtual level -1.

Figure 8.4 uses these ideas to construct a more detailed diagrammatic representation of the mathematics. The two solid arrows in the centre represent the real level sequence $0 \rightarrow 1 \rightarrow 0$, which is common to both terms. The two grey arrows on the left indicate the virtual level sequence $0 \rightarrow +1 \rightarrow 0$ used by the first term, while the grey arrows on the right indicate the sequence $0 \rightarrow -1 \rightarrow 0$ used by the second term. The centre-left portion of Fig. 8.3 therefore represents the first term (term 1), while the centre-right portion represents the second term (term 2) as indicated. The broad double arrows indicate the energy (or frequency) differences between the participating real and virtual levels in each case, a short double arrow implying a strong term. Term 1 (with the negative sign in the denominator and the smaller detuning) will naturally dominate term 2 in this case. In a multi-level model, the situation will naturally be more complicated, but it often remains true that a good approximation to the susceptibility can be obtained from a single dominant term.

It is also worth writing the dominant term in Eq. (8.31) in a slightly different way in order to highlight the structure of the result. In terms of $\tilde{P}_i$, the equation reads

$$\tilde{P}_i = N\,\langle i\rangle_{10} \sum_j \left(\frac{\langle j\rangle_{01}\,\hat{E}_j\,\exp\{i\omega t\}}{\hbar(\tilde{\omega}_{10}^* - \omega)} \right) \qquad (8.41)$$

where the term in brackets is the (dimensionless) ratio of an interaction energy (on the top) and the detuning energy (on the bottom).

We now turn to Eq. (8.39), which is the product of second-order perturbation theory and so naturally more complicated. There are now two intermediate states in the real and virtual level sequences and, accordingly, three matrix elements in the numerator of each term and two brackets in the denominator representing successive angular frequency differences. The equation has been organised so that reading the brackets in the denominator from left to right reveals the sequence of participating real and virtual levels, by direct extension of the procedure used at first order. Thus, in the first term of Eq. (8.39), the appearance of ω_{10} and ω_{20} in the successive brackets indicates that the sequence of real levels is $0 \rightarrow 1 \rightarrow 2 \rightarrow 0$, while the respective appearance of $-\omega$ and -2ω implies that the removal of a fundamental photon from the field first takes the system to virtual state $+\hbar\omega$ and the removal of a second then raises the system further to $+2\hbar\omega$. Finally, a second harmonic photon is emitted, which returns the system to its ground state. Notice that the matrix elements in the numerator have been arranged so that the same real level sequence appears when the indices are read from right to left, namely 01–12–20.

The six terms in Eq. (8.39) represent two different real level sequences each with three different virtual level sequences. Terms 1, 3 and 5 (the three on the left) are for the real level sequence $0 \rightarrow 1 \rightarrow 2 \rightarrow 0$, while the other three are for the sequence $0 \rightarrow 2 \rightarrow 1 \rightarrow 0$. A diagrammatic representation of the first three is presented in Fig. 8.5, which is based on the same ideas as Fig. 8.4. The black arrows (on the far left and right) indicate the real level sequence, and the three possible virtual level sequences are shown in the central columns. The sequences for each of the six terms are summarised in Table 8.1.

As in Fig. 8.4, the double arrows in Fig. 8.5 indicate the values of the frequency denominators, and there are, of course, now two per term whose product affects the term strength. The equation, the table, and the figure have all been arranged so that term 1 is the strongest and term 6 the weakest. The matrix elements have been ordered to reflect the level sequences.

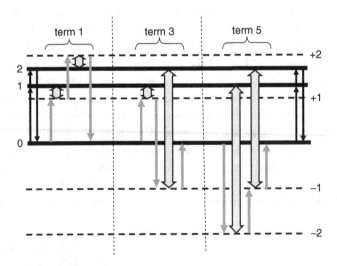

Fig. 8.5   Diagrammatic representation of the structure of Eq. (8.39).

Table 8.1 Real and virtual level sequences in Eq. (8.39).

Term	Real sequence	Virtual sequence
1	$0 \to 1 \to 2 \to 0$	$0 \to +\hbar\omega \to +2\hbar\omega \to 0$
2	$0 \to 2 \to 1 \to 0$	$0 \to +\hbar\omega \to +2\hbar\omega \to 0$
3	$0 \to 1 \to 2 \to 0$	$0 \to +\hbar\omega \to -\hbar\omega \to 0$
4	$0 \to 2 \to 1 \to 0$	$0 \to +\hbar\omega \to -\hbar\omega \to 0$
5	$0 \to 1 \to 2 \to 0$	$0 \to -\hbar\omega \to -2\hbar\omega \to 0$
6	$0 \to 2 \to 1 \to 0$	$0 \to -\hbar\omega \to -2\hbar\omega \to 0$

Consider, for example, term 4, which involves $\langle k \rangle_{02} \langle i \rangle_{21} \langle j \rangle_{10}/((\omega_{20} - \omega)(\omega_{10} + \omega))$. Reading both top and bottom from left to right, the interpretation is that in the first step, a fundamental photon is absorbed and the system goes to real state 2, in the second a harmonic photon is emitted and the system goes to state 1, and in the final step, a further fundamental photon is absorbed and the system returns to the ground state.

The most important point to come out of this discussion is probably the existence of a 'dominant path' determined by real levels that lie close to the virtual levels at $+\hbar\omega$ and $+2\hbar\omega$. Taking the dominant term on its own, Eq. (8.39) reduces to

$$\hat{P}_i^{2\omega} = \frac{N}{2\hbar^2} \sum_{jk} \left(\frac{\langle j \rangle_{01} \langle k \rangle_{12} \langle i \rangle_{20}}{(\omega_{10} - \omega)(\omega_{20} - 2\omega)} \right) \hat{E}_j^\omega \hat{E}_k^\omega \tag{8.42}$$

which in many cases will give a good approximation for the second harmonic polarisation. For comparison with Eq. (8.41), we can reorganise the equation into the form

$$\hat{P}_i^{2\omega} = \tfrac{1}{2} N \langle i \rangle_{20} \sum_{jk} \left(\frac{\langle j \rangle_{01} \hat{E}_j^\omega}{\hbar(\omega_{10} - \omega)} \times \frac{\langle k \rangle_{12} \hat{E}_k^\omega}{\hbar(\omega_{20} - 2\omega)} \right). \tag{8.43}$$

This shows that the most basic effect of taking perturbation to the next order has been to add a second energy ratio into the formula.

8.7.3 The second harmonic coefficient

We now need to reconcile Eq. (8.40) with Eq. (4.5) where the SHG coefficient was defined via the equation

$$\hat{P}_i^{2\omega} = \tfrac{1}{2} \varepsilon_0 \sum_{jk} \chi_{ijk}^{\text{SHG}} \hat{E}_j^\omega \hat{E}_k^\omega. \tag{8.44}$$

The notation differs slightly from that in Chapter 4, but in an obvious way. The conclusion is that

$$\chi_{ijk}^{\text{SHG}} = \frac{N}{\varepsilon_0 \hbar^2} \sum_{nn'} \left(\frac{\langle j \rangle_{0n} \langle k \rangle_{nn'} \langle i \rangle_{n'0}}{(\omega_{n0} - \omega)(\omega_{n'0} - 2\omega)} + \frac{\langle k \rangle_{0n} \langle i \rangle_{nn'} \langle j \rangle_{n'0}}{(\omega_{n0} - \omega)(\omega_{n'0} + \omega)} \right.$$

$$\left. + \frac{\langle i \rangle_{0n} \langle j \rangle_{nn'} \langle k \rangle_{n'n}}{(\omega_{n0} + 2\omega)(\omega_{n'0} + \omega)} \right). \tag{8.45}$$

We consider the special case where $i = x$ and the field contains y and z components. Then Eq. (8.44) becomes

$$\hat{P}_x^{2\omega} = \tfrac{1}{2}\varepsilon_0 \left(\chi_{xyz}^{\text{SHG}} + \chi_{xzy}^{\text{SHG}} \right) \hat{E}_y^\omega \hat{E}_z^\omega \tag{8.46}$$

while Eq. (4.10) reads

$$\hat{P}_x^{2\omega} = 2\varepsilon_0 d_{14} \hat{E}_y^\omega \hat{E}_z^\omega. \tag{8.47}$$

If we focus just on the dominant path, Eq. (8.45) gives

$$\hat{P}_x^{2\omega} = \frac{N}{2\hbar^2} \left(\frac{\langle z \rangle_{01} \langle y \rangle_{12} \langle x \rangle_{20} + \langle y \rangle_{01} \langle z \rangle_{12} \langle x \rangle_{20}}{(\omega_{10} - \omega)(\omega_{20} - 2\omega)} \right) \hat{E}_y^\omega \hat{E}_z^\omega \tag{8.48}$$

and the conclusion is that

$$\tfrac{1}{2} \left(\chi_{xyz}^{\text{SHG}} + \chi_{xzy}^{\text{SHG}} \right) = 2d_{14} = \frac{N}{2\hbar^2 \varepsilon_0} \left(\frac{\langle z \rangle_{01} \langle y \rangle_{12} \langle x \rangle_{20} + \langle y \rangle_{01} \langle z \rangle_{12} \langle x \rangle_{20}}{(\omega_{10} - \omega)(\omega_{20} - 2\omega)} \right). \tag{8.49}$$

One should not seek to associate χ_{xyz}^{SHG} and χ_{xzy}^{SHG} with the individual terms in the numerator of Eq. (8.49). Not only is there no way of distinguishing the two contributions by experiment, but the order of the fields has not been preserved in the derivation in any case.

Finally, we must address the symmetry issue. Back in Chapter 1, we used simple symmetry principles to show that SHG (and other second-order processes) are forbidden in media possessing inversion symmetry. So what basic principle ensures that the SHG coefficient of Eq. (8.44) is zero under these circumstances?

The answer is that quantum states possess definite parity in a medium with inversion symmetry. But transitions between states of the same parity are forbidden and the associated dipole matrix elements are zero. In Eq. (8.40), the sequence of states in the matrix element triple products is $0 \to n \to n' \to 0$, and there is no way of choosing even and odd parity for the three states alternately to close the loop. In general, it follows that the product of any odd number of dipole matrix elements involving a closed sequence of states is necessarily zero.

8.7.4 Sum frequency generation

The case of sum frequency generation $\omega_1 + \omega_2 = \omega_3$ can be treated by direct extension of the second-order perturbation theory techniques deployed above in the special case of SHG. The most important message is that there will be twice as many terms as for SHG; twice as many diagrams of the type shown in Fig. 8.5 can be drawn because there are now two distinct orders in which the upward arrows can appear: ω_1 first and then ω_2 or vice versa. Equation (8.42) for the dominant term now generalises to

$$\hat{P}_i(\omega_3) = \frac{N}{2\hbar^2} \sum_{jk} \left(\frac{\langle k \rangle_{01} \langle j \rangle_{12} \langle i \rangle_{20} \hat{E}_j(\omega_1)\hat{E}_k(\omega_2)}{(\omega_{10} - \omega_2)(\omega_{20} - \omega_3)} + \frac{\langle k \rangle_{01} \langle j \rangle_{12} \langle i \rangle_{20} \hat{E}_j(\omega_2)\hat{E}_k(\omega_1)}{(\omega_{10} - \omega_1)(\omega_{20} - \omega_3)} \right)$$

(8.50)

which, by exchanging j and k in the second term (which is OK because they are dummy indices under a summation) can be compacted to

$$\hat{P}_i(\omega_3) = \frac{N}{2\hbar^2} \sum_{jk} \left(\frac{\langle k \rangle_{01} \langle j \rangle_{12} \langle i \rangle_{20}}{(\omega_{10} - \omega_2)(\omega_{20} - \omega_3)} + \frac{\langle j \rangle_{01} \langle k \rangle_{12} \langle i \rangle_{20}}{(\omega_{10} - \omega_1)(\omega_{20} - \omega_3)} \right) \hat{E}_j(\omega_1)\hat{E}_k(\omega_2).$$

(8.51)

As expected, this equation contains two terms, whereas there was only one in Eq. (8.42). For the same special case that we used in Section 8.7.3, the x-component of polarisation is now

$$\hat{P}_x(\omega_3) = \frac{N}{2\hbar^2} \left(\left(\frac{\langle z \rangle_{01} \langle y \rangle_{12} \langle x \rangle_{20}}{(\omega_{10} - \omega_2)(\omega_{20} - \omega_3)} + \frac{\langle y \rangle_{01} \langle z \rangle_{12} \langle x \rangle_{20}}{(\omega_{10} - \omega_1)(\omega_{20} - \omega_3)} \right) \hat{E}_y(\omega_1)\hat{E}_z(\omega_2) \right.$$

$$\left. + \left(\frac{\langle y \rangle_{01} \langle z \rangle_{12} \langle x \rangle_{20}}{(\omega_{10} - \omega_2)(\omega_{20} - \omega_3)} + \frac{\langle z \rangle_{01} \langle y \rangle_{12} \langle x \rangle_{20}}{(\omega_{10} - \omega_1)(\omega_{20} - \omega_3)} \right) \hat{E}_y(\omega_2)\hat{E}_z(\omega_1) \right)$$ (8.52)

which parallels Eq. (8.48) for SHG. From the definition of the sum frequency generation (SFG) coefficient in Eq. (4.4), we also have

$$\hat{P}_x(\omega_3) = \varepsilon_0 \left(\chi^{\text{SFG}}_{xyz} \hat{E}_y(\omega_1)\hat{E}_z(\omega_2) + \chi^{\text{SFG}}_{xzy} \hat{E}_z(\omega_1)\hat{E}_y(\omega_2) \right).$$

(8.53)

Notice that, in contrast to SHG, χ^{SFG}_{xyz} and χ^{SFG}_{xzy} are distinct, and can be associated with the respective terms in Eq. (8.52). The first coefficient relates to the situation where the fields at ω_1 and ω_2 are, respectively, in the y- and z-directions, and the second to the case where the frequencies are reversed.

A related issue is *intrinsic permutation symmetry* (IPS), which was first discussed in Section 4.3.2. When Eq. (4.3) is written out in full for SFG, the x-component reads

$$\hat{P}_x(\omega_3) = \tfrac{1}{2}\varepsilon_0 \left(\chi^{\text{SFG}}_{xyz}(\omega_3;\, \omega_1,\omega_2) + \chi^{\text{SFG}}_{xzy}(\omega_3;\, \omega_2,\omega_1) \right) \hat{E}_y(\omega_1)\hat{E}_z(\omega_2)$$

$$+ \tfrac{1}{2}\varepsilon_0 \left(\chi^{\text{SFG}}_{xzy}(\omega_3;\, \omega_1,\omega_2) + \chi^{\text{SFG}}_{xyz}(\omega_3;\, \omega_2,\omega_1) \right) \hat{E}_y(\omega_2)\hat{E}_z(\omega_1).$$ (8.54)

Given the structural similarity between Eqs (8.54) and (8.52), it is tempting to try to match the coefficients in the latter equation with the individual terms in the former. However, as noted in Chapter 4, the effect of $\chi^{\text{SFG}}_{xyz}(\omega_3;\, \omega_1,\omega_2)$ and $\chi^{\text{SFG}}_{xzy}(\omega_3;\omega_2,\omega_1)$ cannot be distinguished, since both involve the same product $\hat{E}_y(\omega_1)\hat{E}_z(\omega_2)$, and nor can that of $\chi^{\text{SFG}}_{xzy}(\omega_3;\, \omega_1,\omega_2)$ and $\chi^{\text{SFG}}_{xyz}(\omega_3;\omega_2,\omega_1)$ in the second row of Eq. (8.54). There is in fact no justification for making the suggested identifications.

Under IPS, it is *conventional* (but not obligatory) to assume that the paired coefficients in Eq. (8.54) make equal contributions, and hence to write $\chi^{\text{SFG}}_{ijk}(\omega_3;\, \omega_1,\omega_2) =$

$\chi_{ikj}^{\mathrm{SFG}}(\omega_3; \omega_2, \omega_1)$. This enables the equation to be contracted to

$$\hat{P}_x(\omega_3) = \varepsilon_0 \left(\chi_{xyz}^{\mathrm{SFG}}(\omega_3; \ \omega_1, \omega_2)\hat{E}_y(\omega_1)\hat{E}_z(\omega_2) + \chi_{xzy}^{\mathrm{SFG}}(\omega_3; \ \omega_1, \omega_2)\hat{E}_y(\omega_2)\hat{E}_z(\omega_1) \right).$$

(8.55)

Matching Eqs (8.55) and (8.52) now yields

$$\chi_{xyz}^{\mathrm{SFG}}(\omega_3; \omega_1, \omega_2) = \frac{N}{2\hbar^2\varepsilon_0} \left(\frac{\langle z \rangle_{01}\langle y \rangle_{12}\langle x \rangle_{20}}{(\omega_{10} - \omega_2)(\omega_{20} - \omega_3)} + \frac{\langle y \rangle_{01}\langle z \rangle_{12}\langle x \rangle_{20}}{(\omega_{10} - \omega_1)(\omega_{20} - \omega_3)} \right) \quad (8.56)$$

with a closely analogous formula for $\chi_{xzy}^{\mathrm{SFG}}(\omega_3; \omega_1, \omega_2)$. Notice that the order of the frequencies is now fixed because $\chi_{xzy}^{\mathrm{SFG}}(\omega_3; \omega_2, \omega_1)$ is included within $\chi_{xyz}^{\mathrm{SFG}}(\omega_3; \ \omega_1, \omega_2)$. In conclusion, remember that IPS is merely a sensible convention, not a fundamental principle.

8.7.5 Optical rectification

The case of optical rectification (OR) can be treated if one goes back to Section 8.7.1 and extracts the zero frequency terms instead of those at 2ω. Like most perturbation theory derivations, the calculation is arduous. The result for the DC polarisation is

$$\begin{aligned}
P_i^{\mathrm{DC}} = \frac{N}{4\hbar^2} \sum_{jk}\sum_{nn'} &\left(\left(\frac{\langle i \rangle_{0n}\langle j \rangle_{nn'}\langle k \rangle_{n'0}}{\omega_{n0}(\omega_{n'0} - \omega)} + \frac{\langle i \rangle_{0n}\langle k \rangle_{nn'}\langle j \rangle_{n'0}}{\omega_{n0}(\omega_{n'0} + \omega)} \right) \right. \\
&+ \left(\frac{\langle j \rangle_{0n}\langle i \rangle_{nn'}\langle k \rangle_{n'0}}{(\omega_{n0} - \omega)(\omega_{n'0} - \omega)} + \frac{\langle k \rangle_{0n}\langle i \rangle_{nn'}\langle j \rangle_{n'0}}{(\omega_{n0} + \omega)(\omega_{n'0} + \omega)} \right) \\
&+ \left. \left(\frac{\langle j \rangle_{0n}\langle k \rangle_{nn'}\langle i \rangle_{n'0}}{(\omega_{n0} - \omega)\omega_{n'0}} + \frac{\langle k \rangle_{0n}\langle j \rangle_{nn'}\langle i \rangle_{n'0}}{(\omega_{n0} + \omega)\omega_{n0}} \right) \right) \hat{E}_j\hat{E}_k^*
\end{aligned}$$

(8.57)

and an expression for the optical rectification coefficient can be deduced from

$$P_i(0) = \tfrac{1}{2}\varepsilon_0 \sum_{jk} \chi_{ijk}^{\mathrm{OR}}(0; \ \omega, -\omega)\hat{E}_j(\omega)\hat{E}_k^*(\omega). \quad (8.58)$$

The matrix element involving the coordinate i is associated with the polarisation and those carrying j and k with the incident field. As before, the terms have been organised so that, by reading from left to right, the sequence of real and virtual states can be worked out.

An intriguing aspect of Eq. (8.57) is the need for the right-hand side to be real, since the DC polarisation is real by definition. But even if we take the matrix elements to be real, the presence of $E_j E_k^*$ makes it less than obvious that the equation satisfies the condition. In fact, detailed examination shows that the expression really is real; see Problem 8.2.

8.8 Third-order perturbation theory

8.8.1 Third harmonic generation

There is no problem in principle with taking perturbation theory to third order, but the mathematics becomes hugely tedious. Fortunately, there is no need for this labour if one

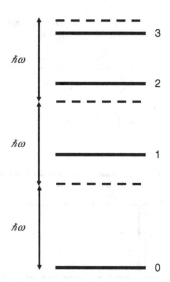

Fig. 8.6   Four-level system with virtual levels for the dominant term in third harmonic generation.

is prepared to trust the diagrams of Fig. 8.5, and make the necessary extensions. Indeed for a system of four levels 0123 of alternate parity, the contribution to the third harmonic polarisation from the dominant path shown in Fig. 8.6 can be written down by inspection, namely

$$\hat{P}_i^{3\omega} = \frac{N}{4\hbar^3} \sum_{jkl} \left(\frac{\langle j \rangle_{01} \langle k \rangle_{12} \langle l \rangle_{23} \langle i \rangle_{30}}{(\omega_{10} - \omega)(\omega_{20} - 2\omega)(\omega_{30} - 3\omega)} \right) \hat{E}_j^\omega \hat{E}_k^\omega \hat{E}_l^\omega. \qquad (8.59)$$

Notice the extra matrix element, the extra bracket, the extra field component, and the extra $2\hbar$ in the denominator. In reality, the single path $0 \to 1 \to 2 \to 3 \to 0$ would naturally be summed over multiple intermediate paths.

8.8.2 Intensity-dependent refractive index

To end this chapter, we tackle the more challenging case of intensity-dependent refractive index (IDRI). Some intriguing problems arise in the calculation of the IDRI coefficient, which provide valuable insight into the inner workings of perturbation theory, and highlight the relative advantages and disadvantages of the probability amplitude (PA) and density matrix (DM) approaches.

We simplify the discussion by restricting attention to a three-level system in which states 0 and 2 have the same parity (and hence $\mu_{02} = \mu_{20} = 0$), and we also assume that all fields are polarised in the x-direction. The terms we need to calculate in order to obtain the third-order polarisation using the two different approaches are listed in Table 8.2.

Table 8.2 Terms needed to calculate the IDRI coefficient					
	0	1	2	3	Third- order polarisation
Probability amplitude	$a_0^{(0)} = 1$	$a_1^{(0)}$	$a_0^{(2)}, a_2^{(2)}$	$a_1^{(3)}$	$a_1^{(3)} a_0^{(0)*}, a_2^{(2)} a_1^{(1)*}, +\text{c.c.}$
Density matrix	$\rho_{00}^{(0)} = 1$	$\rho_{10}^{(1)}$	$\rho_{00}^{(2)}, \rho_{11}^{(2)}, \rho_{20}^{(2)}$	$\rho_{10}^{(3)}, \rho_{21}^{(3)}$	$\rho_{10}^{(3)}, \rho_{21}^{(3)} + \text{c.c.}$

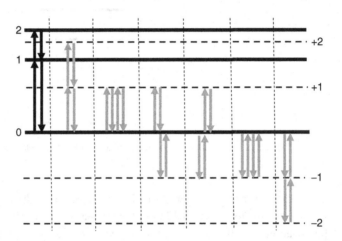

Fig. 8.7   Real level sequence (bold) for Eq. (8.60), with virtual level sequences (grey) for each of the six terms in the equation.

It can be shown quite straightforwardly (using either the PA or the DM method) that the contribution to the IDRI polarisation from the real level sequence $0 \rightarrow 1 \rightarrow 2 \rightarrow 1 \rightarrow 0$ is

$$\hat{P}_{(\text{via state2})}^{\text{IDRI}} = \frac{N \left|\mu_{01}^x\right|^2 \left|\mu_{12}^x\right|^2 \hat{E}_x^2 \hat{E}_x^*}{4\hbar^3} \times 2 \left(\frac{1}{\omega_{10}^- \omega_{20}^= \omega_{10}^-} + \frac{1}{\omega_{10}^- \omega_{20} \omega_{10}^-} \right.$$

$$\left. + \frac{1}{\omega_{10}^- \omega_{20} \omega_{10}^+} + \frac{1}{\omega_{10}^+ \omega_{20} \omega_{10}^-} + \frac{1}{\omega_{10}^+ \omega_{20} \omega_{10}^+} + \frac{1}{\omega_{10}^+ \omega_{20}^{++} \omega_{10}^+} \right) \quad (8.60)$$

where we have adopted a shorthand notation where $\omega_{n0}^\pm = \omega_{n0} \pm \omega$, $\omega_{n0}^= = \omega_{n0} - 2\omega$ and $\omega_{n0}^{++} = \omega_{n0} + 2\omega$. Diagrams representing the terms in this equation are shown in Fig. 8.7. The real state sequence $(0 \rightarrow 1 \rightarrow 2 \rightarrow 1 \rightarrow 0)$ is indicated by the black arrows at the far left, while the six sets of grey arrows show the virtual state sequences for the six terms in Eq. (8.60) in the same order. With the pre-factor of 2, these are in fact the 12 terms in Table VII of Ward [97].[8]

---

[8] Each path in Fig. 8.7 represents two terms because one downward arrow corresponds to the E_x^* in Eq. (8.60) and the other to the polarisation; these can come in either order.

We now need to incorporate the state sequence $0 \to 1 \to 0 \to 1 \to 0$ into this result, and this is where the problems begin. It turns out that the DM approach provides the easiest route to the right answer, although it is decidedly trickier than the PA method at the first step. Consider, for example, what happens when one tries to find $\rho_{11}^{(2)}$. If we assume (correctly) that $\rho_{11}^{(2)} = a_1^{(1)} a_1^{(1)*}$ in the absence of damping, the answer in that case is easily found from Eq. (8.29) to be

$$
a_1^{(1)} a_1^{(1)*} = \frac{\mu_{10}\mu_{01}}{4\hbar^2} \left(\frac{(\hat{E}_x^2 e^{i2\omega t} + \text{c.c.})}{(\omega_{10}+\omega)(\omega_{10}-\omega)} + \frac{\hat{E}_x \hat{E}_x^*}{(\omega_{10}+\omega)^2} + \frac{\hat{E}_x^* \hat{E}_x}{(\omega_{10}-\omega)^2} \right). \quad (8.61)
$$

On the other hand, if we try to derive the more general form of this result starting from Eq. (8.15), the time derivative of $\rho_{11}^{(2)}$ comes out as

$$
\dot{\rho}_{11}^{(2)} = -\frac{i}{\hbar} \sum_q \left(V_{10}\rho_{01}^{(1)} - \rho_{10}^{(1)} V_{01} \right)
$$

$$
= i\frac{\mu_{10}\mu_{01}}{4\hbar^2} \left(\left(\frac{1}{\tilde{\omega}_{10}^* - \omega} - \frac{1}{\tilde{\omega}_{10}+\omega} \right) \hat{E}_x^2 e^{i2\omega t} + \left(\frac{1}{\tilde{\omega}_{10}^* + \omega} - \frac{1}{\tilde{\omega}_{10}-\omega} \right) \hat{E}_x^{*2} e^{-i2\omega t} \right.
$$
$$
\left. + \left\{ \frac{1}{\tilde{\omega}_{10}^* + \omega} - \frac{1}{\tilde{\omega}_{10}+\omega} + \frac{1}{\tilde{\omega}_{10}^* - \omega} - \frac{1}{\tilde{\omega}_{10}-\omega} \right\} \hat{E}_x \hat{E}_x^* \right).
$$
$$
(8.62)
$$

We now face a puzzling situation. For if damping is ignored and $\tilde{\omega}_{10}$ replaced by ω_{10}, integration yields the terms in $e^{\pm i2\omega t}$ in Eq. (8.61), but the terms in $\hat{E}_x \hat{E}_x^*$ are not recovered because the curly bracket in Eq. (8.62) goes to zero.

What has gone wrong? The answer is not connected with damping. It is certainly true that, when damping is included, the cross-term in Eq. (8.62) no longer vanishes, and $\rho_{11}^{(2)}$ then diverges on integration. This feature is associated physically with the population growth of level 1 caused by absorption from level 0 in the wings of the resonance. But the PA approach on which Eq. (8.61) is based does not include damping in any case, so this cannot be the origin of the extra terms in the equation.

We can recover the missing terms if we use a device suggested by Takatsuji [98], and include a causality factor $e^{\eta t}$ in the electric fields, where η is a small positive constant that is taken to be zero at the end of the calculation. As well as attaching this factor to the fields in Eq. (8.62), it is also necessary to take account of its effect at first-order, which can be done simply by reinterpreting $\tilde{\omega}_{10}$ as $(\omega_{10} - i\eta)$. With η thus included, the curly bracket in Eq. (8.62) is no longer zero, and when the equation is integrated, the missing terms in Eq. (8.61) reappear. The moral of this story is that one should exercise caution when cancelling terms under an integral that would diverge if integrated separately.

We have in fact now surmounted the main obstacle within the DM approach to finding the contribution to the IDRI polarisation from the level sequence $0 \to 1 \to 0 \to 1 \to 0$, i.e. where the central level is the ground state. This is the same problem that arises from the appearance in the general formula for the IDRI coefficient quoted in Ward [97] of terms such as $((\omega_{n0} - \omega)\omega_{m0}(\omega_{n'0} - \omega))^{-1}$. The four central terms in Eq. (8.60) are of this type, and of course presented no difficulty when $m = 2$. But the term becomes singular if we try to set $m = 0$, and that is precisely what we are now seeking to do.

Within the PA approach, this problem manifests itself in a different but revealing way when one tries to calculate $a_0^{(2)}$ from the equation

$$\dot{a}_0^{(2)} = -\frac{i}{\hbar} V_{01} a_1^{(1)} = \cdots + i \left[\frac{|\mu_{01}|^2}{4\hbar^2} \left(\frac{1}{\omega_{10} + \omega} + \frac{1}{\omega_{10} - \omega} \right) \hat{E}_x \hat{E}_x^* \right] a_0^{(0)} \quad (8.63)$$

where only time-independent terms on the right-hand side are shown. The expression in the square brackets clearly represents a frequency change in a_0, which is associated with a shift in the energy of the ground state (0) in the presence of the excited state (1) and the coupling field. This is a real effect, known as the AC Stark shift. Since the intrinsic time dependence of a_0 is $\exp\{-iE_0t/\hbar\}$, it is clear from Eq. (8.63) that the ground state energy is lowered if $\omega < \omega_{10}$. An upward shift of the same amount will occur for level 1, which means that the two levels move apart. One would therefore expect that, for these two levels alone, IDRI causes a reduction in the polarisibility.

There are several ways of managing this problem:

- Orr and Ward [99] used a technique based on Bogoluibov and Mitropolsky's method of averages to remove the singularities and to derive a modified formula for the IDRI coefficient;
- Takatsuji [98] obtained the identical result by renormalising the zeroth-order Hamiltonian to include the AC Stark shifts;
- Boyd [1] cites Hanna, Yuratich and Cotter [100] for an algebraic device that leads to the same conclusion.

But, as noted already, we have no need to adopt any special measures because the difficulties we encountered with Eqs (8.61) and (8.62) were the manifestation within the DM approach of the same basic difficulty. Now that we have solved them, we can simply press on. The contribution to the IDRI polarisation from the sequence $0 \rightarrow 1 \rightarrow 0 \rightarrow 1 \rightarrow 0$ is included in

$$P = N \left(\rho_{01}^{(3)} \mu_{10} + \rho_{10}^{(3)} \mu_{01} \right) \quad (8.64)$$

where

$$\dot{\rho}_{10}^{(3)} + i\omega_{10}\rho_{10}^{(3)} = -\frac{i}{\hbar} \left(V_{10}\rho_{00}^{(2)} - \rho_{11}^{(2)} V_{10} \right) = -\frac{i\rho_{11}^{(2)}\mu_{10}}{\hbar} \left(\hat{E}_x e^{i\omega t} + \text{c.c.} \right). \quad (8.65)$$

A term in $V_{12}\rho_{20}^{(2)}$ has been omitted from this equation because its contribution has already been included in Eq. (8.60). The final step in Eq. (8.65) follows because $\rho_{00}^{(2)} = -\rho_{11}^{(2)}$. Of course, we already have $\rho_{11}^{(2)}$ from Eq. (8.61), so the only hurdle left to surmount is the algebra. The final result is

$$\hat{P}_{\text{via state0}}^{\omega} = N \frac{|\mu_{10}|^4}{4\hbar^3} \left(\frac{2}{\omega_{10}^-(-2\omega)\omega_{10}^-} - 2G + \frac{2}{\omega_{10}^+(2\omega)\omega_{10}^+} \right) \hat{E}_x^2 \hat{E}_x^* \quad (8.66)$$

where

$$G = \frac{1}{(\omega_{10}^+)^3} + \frac{1}{(\omega_{10}^+)^2\omega_{10}^-} + \frac{1}{\omega_{10}^+(\omega_{10}^-)^2} + \frac{1}{(\omega_{10}^-)^3}. \quad (8.67)$$

Notice that the first and last terms in Eq. (8.66) mirror the first and last terms in Eq. (8.60) and the first and last diagrams in Fig. 8.7. On the other hand, the terms in G replace the central four terms and diagrams. The total IDRI polarisation is therefore

$$\hat{P}^{\text{IDRI}} = \hat{P}^{\omega}_{\text{state2}} + \hat{P}^{\omega}_{\text{state0}} \tag{8.68}$$

where the two contributions come from Eqs (8.60) and (8.66), respectively. Notice that the contribution from level 0 is invariably negative. The contribution from level 2 could be of either sign although, for a multi-level system, the overall sign is almost invariably positive. Refractive index almost always increases with intensity.

The conclusion to draw from this section depends on one's point of view. On the one hand, perturbation theory clearly has some nasty surprises in store for the unwary; on the other, it holds some fascinating secrets for the enthusiast.

Problems

8.1 Verify Eq. (8.25) for the three-level system of Fig. 8.3 in the case where $a_0^{(0)} = 1$, and hence $a_1^{(0)} = a_2^{(0)} = 0$.

8.2 Show that the right-hand side of Eq. (8.57) is real. (This is quite easy once you know what to look for but, if you do not spot the trick quickly, you could waste a lot of time.)

8.3 For a class 11 crystal such as KDP, Eq. (4.14) reads

$$\tilde{P}_x(\omega_3) = 2\varepsilon_0 d_{14}(\tilde{E}_y(\omega_1)\tilde{E}_z(\omega_2) + \tilde{E}_z(\omega_1)\tilde{E}_y(\omega_2)).$$

Why is there only one d coefficient in this equation when there are two distinct χ coefficients in the analogous Eq. (8.39)?

Resonant effects

9.1 Introduction

Although resonant effects were largely ignored in Chapter 8, some of the machinery for handling them was included at the start of the chapter. Damping terms were, for example, introduced in Section 8.4.1, and complex frequencies were retained through much of the discussion of Section 8.6.

Moving close to a resonance does not necessarily invalidate the perturbation theory approach, as indeed the inclusion of damping in Eqs (8.30)–(8.34) indicated. We will examine the properties of those equations in the next section, and extend the discussion to include Raman resonances in Section 9.3. However, the application of sufficiently strong fields on fast time-scales certainly violates the weak perturbation assumption on which the approach of Chapter 8 was based. In Section 9.5, we will explore the properties of the density matrix equations in this 'coherent' limit and, in Sections 9.6 and 9.7, we will apply the results to two well-known coherent effects: self-induced transparency (SIT) and electromagnetically-induced transparency (EIT).

9.2 Resonance in the linear susceptibility

The linear susceptibility of a multi-level system was shown in Chapter 8 (see Eq. 8.34) to be

$$\chi^{(1)}(\omega) = \frac{N}{\varepsilon_0 \hbar} \sum_n \left(\frac{\mu_{0n}^2}{\tilde{\omega}_{n0}^* - \omega} \right) = \frac{N}{\varepsilon_0 \hbar} \sum_n \left(\frac{\mu_{0n}^2}{\omega_{n0} - \omega + i\gamma_{n0}} \right) \tag{9.1}$$

where the definition $\tilde{\omega}_{n0} = \omega_{n0} - i\gamma_{n0}$ has been used in the second step, and only the dominant terms have been included. The tensor nature of the susceptibility has been ignored, and the matrix elements μ_{0n} have been assumed to be real as before.

In the single-level case, Eq. (9.1) becomes

$$\chi^{(1)}(\Delta) = \frac{N\mu^2}{\varepsilon_0 \hbar} \left(\frac{\Delta - i\gamma}{\Delta^2 + \gamma^2} \right) = \chi'(\Delta) - i\chi''(\Delta) \tag{9.2}$$

where $\Delta = \omega_{10} - \omega$, and other subscripts have been dropped. The real and imaginary parts of $\chi(\Delta)$ are

$$\chi'(\Delta) = \frac{N\mu^2}{\varepsilon_0 \hbar} \left(\frac{\Delta}{\Delta^2 + \gamma^2} \right) \qquad \chi''(\Delta) = \frac{N\mu^2}{\varepsilon_0 \hbar} \left(\frac{\gamma}{\Delta^2 + \gamma^2} \right). \qquad (9.3)$$

Equation (2.6) can now be used to link the susceptibility to the angular wave number of the field via the sequence

$$k = \frac{\omega}{c}\sqrt{1 + \chi' - i\chi''} \cong \frac{n\omega}{c}\left(1 - \frac{i\chi''}{2n^2}\right) = \frac{n\omega}{c} - \tfrac{1}{2}i\alpha \qquad (9.4)$$

where the refractive index is $n(\Delta) = \sqrt{1 + \chi'(\Delta)}$ and the approximation is valid if χ'' is small. The negative imaginary part of k leads to absorption (via e^{-ikz}), and α is the intensity absorption coefficient given by

$$\alpha(\Delta) = \frac{\omega\chi''(\Delta)}{cn} = \frac{N\mu^2\omega\pi g(\Delta)}{c\varepsilon_0 n\hbar} \qquad (9.5)$$

where $g(\Delta) = \pi^{-1}\gamma/(\Delta^2 + \gamma^2)$ is the normalised Lorentzian lineshape function.

The linear dispersion characteristics of the medium are evidently governed by $\chi'(\Delta)$, and the absorption profile by $\chi''(\Delta)$. Graphs of these two parameters (normalised to $\chi''(0)$) are presented in Fig. 9.1. Since $n(\Delta) \cong 1 + \tfrac{1}{2}\chi'(\Delta)$, the graph of $\chi'(\Delta)/\chi''(0)$ (shown in solid black) shows how the refractive index varies in the vicinity of the resonance; notice that the abscissa scale runs from positive to negative Δ $(= \omega_0 - \omega)$ to ensure that the field frequency ω increases to the right. The refractive index falls sharply in the immediate vicinity of the resonance (anomalous dispersion), but rises with frequency (normal dispersion) further out on both wings. The Lorentzian frequency profile of the absorption coefficient is shown in grey.

Real systems are of course normally characterised by a large number of resonances, not just one. The summation in Eq. (9.1) takes care of this, and the way in which two neighbouring resonances run together is shown in Fig. 9.2.

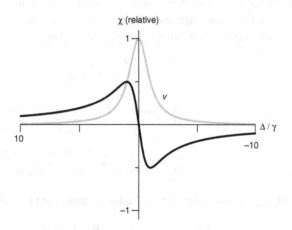

Fig. 9.1 Real part (bold) and imaginary part (grey) of the linear susceptibility near a resonance; see Eq. (9.3). Both graphs are normalised to the peak value of the imaginary part.

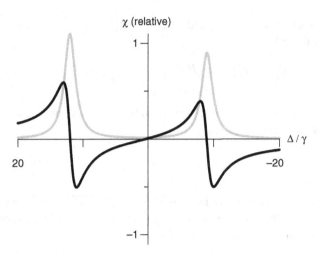

Fig. 9.2 As Fig. 9.1, but for two adjacent resonances.

9.3 Raman resonances

Several two-photon resonant effects could be examined in detail, but stimulated Raman scattering is particularly interesting and was of course discussed in Chapter 5. We therefore focus on that case. We will not engage in the same level of detail as in Chapter 8, but concentrate rather on the main features by including only the dominant path in the analysis.

A simplified energy level scheme for stimulated Raman scattering (SRS) is shown in Fig. 9.3. As explained in Chapter 5, SRS involves beams at frequencies ω_L and ω_S ($< \omega_L$), where subscripts L and S designate the 'Laser' and 'Stokes' waves, respectively. When the frequency difference ($\omega_L - \omega_S = \omega_D$) is close to the frequency separation ω_{20} of level 2 and the ground state 0 (the Raman resonance frequency), gain is recorded at ω_S, loss at ω_L, and population is transferred from level 0 to level 2 at the same time. Level 2 could be a vibrational state, a rotational state, or even an electronic state.

Under these circumstances, the methods developed in Chapter 8 lead to the hierarchy of first-, second- and third-order density matrix elements

$$\rho_{01}^{(1)} = \frac{\mu_{01} \hat{E}_L \exp\{i\omega_L t\}}{2\hbar^2 (\omega_{10} - \omega_L)} \tag{9.6}$$

$$\rho_{02}^{(2)} = \frac{\mu_{01}\mu_{12}\hat{E}_L\hat{E}_S^* \exp\{i(\omega_L - \omega_S)t\}}{4\hbar^2 (\omega_{10} - \omega_L)(\tilde{\omega}_{20}^* - \omega_D)} \tag{9.7}$$

$$\rho_{01}^{(3)} = \frac{\mu_{01}\mu_{12}\mu_{21}\hat{E}_L\hat{E}_S^*\hat{E}_S \exp\{i\omega_L t\}}{8\hbar^2 (\omega_{10} - \omega_L)(\tilde{\omega}_{20}^* - \omega_D)(\omega_{10} - \omega_L)} \tag{9.8}$$

where $\tilde{\omega}_{20} = \omega_{20} - i\gamma_2$. These lead to a term in the polarisation at ω_L of the form

$$\hat{P}_L = \frac{\mu_{01}\mu_{12}\mu_{21}\mu_{10}}{4\hbar^2 (\omega_{10} - \omega_L)(\tilde{\omega}_{20}^* - \omega_D)(\omega_{10} - \omega_L)} \left|\hat{E}_S\right|^2 \hat{E}_L. \tag{9.9}$$

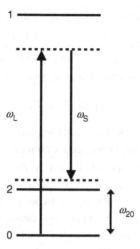

Fig. 9.3 Real and (dominant) virtual levels for stimulated Raman scattering in a three-level system.

The focus of attention here is on the middle term in the denominator. Away from resonance ($|\omega_{20} - \omega_D| \gg \gamma_2$), the polarisation in Eq. (9.9) is a manifestation of the optical Kerr effect, in which the refractive index of the wave at ω_L is modified by the presence of the wave at ω_S. On resonance, on the other hand, $\tilde{\omega}_{20}^* - \omega_D = i\gamma_2$, and it follows that $\hat{P}_L \sim -i\hat{E}_L$, a condition that we know from Chapter 2 leads to attenuation of the wave at ω_L.[1]

Clearly there must be an equation analogous to Eq. (9.9) for the Stokes polarisation $\hat{P}_S$. It turns out that this is driven by the third-order density matrix element $\rho_{21}^{(3)}$ and the associated polarisation is

$$\hat{P}_S = \frac{\mu_{01}\mu_{12}\mu_{21}\mu_{10}}{4\hbar^2(\omega_{10} - \omega_L)(\tilde{\omega}_{20} - \omega_D)(\omega_{10} - \omega_L)}\left|\hat{E}_L\right|^2\hat{E}_S. \qquad (9.10)$$

The interpretation of this equation is closely analogous to that of Eq. (9.9). Away from the Raman resonance, this equation governs the change in the refractive index of the Stokes wave in the presence of the laser field. On resonance, however, it implies that the Stokes wave will grow; notice that the middle term in the denominator of Eq. (9.10) is the conjugate of the corresponding term in Eq. (9.9), so $\hat{P}_S \sim +i\hat{E}_S$ which implies gain. Indeed, for every photon lost at ω_L, a photon is gained at ω_S. But of course the *energies* do not balance, because the photon energies are different. The energy difference is delivered to the medium, as population is transferred from level 0 to level 2.

The analysis of this section has been based on the implicit assumption that the initial (unperturbed) population of the medium is in the ground state 0. However, if the calculation were reworked with the initial population in level 2, it would emerge that an anti-Stokes wave at frequency $\omega_{AS} = \omega_L + \omega_{20}$ would receive gain, accompanied by population transfer from level 2 to level 0. But, as explained in Chapter 5, an anti-Stokes wave can also be generated without initial population in level 2 through the frequency combination

[1] See the discussion after Eq. (2.8).

$\omega_L - \omega_S + \omega_L$, a process that is mediated by the polarisation

$$\hat{P}_{AS} = \frac{\mu_{01}\mu_{12}\mu_{21}\mu_{10}}{4\hbar^2(\omega_{10} - \omega_L)(\tilde{\omega}_{20} - \omega_D)(\omega_{10} - \omega_{AS})} \hat{E}_L \hat{E}_S^* \hat{E}_L. \tag{9.11}$$

Notice that when the basic spatial dependence of the fields is made explicit, we have

$$\hat{E}_L \hat{E}_S^* \hat{E}_L = \tilde{E}_L \tilde{E}_S^* \tilde{E}_L \exp\{-i(2\mathbf{k}_L - \mathbf{k}_S).\mathbf{r}\} \tag{9.12}$$

and $\exp\{-i(2\mathbf{k}_L - \mathbf{k}_S).\mathbf{r}\}$ needs to match $\exp\{-i(\mathbf{k}_{AS}).\mathbf{r}\}$ if the interaction is to proceed efficiently. This anti-Stokes generation mechanism is therefore subject to a phase-matching condition, in contrast to the Raman generation process represented by Eqs (9.9) and (9.10).

We remark that density matrix formulae have been quoted in this section because they enable stochastic damping terms to be included. However, the same results (albeit without the damping terms) can be obtained more easily using the state vector approach, where the algebra is more straightforward. The advantages and disadvantages of the two approaches were discussed in detail in Chapter 8.

9.4 Parametric and non-parametric processes

The process described in Eqs (9.11) and (9.12) is sometimes called *parametric* anti-Stokes Raman generation to distinguish it from the anti-Stokes mechanism mentioned earlier which was based on the initial population of level 2 and which (like Stokes wave generation) counts as 'non-parametric'. Although widely used, the meaning of the word 'parametric' is obscure, and it would be as well to avoid the word altogether were it not already entrenched in the terminology of nonlinear optics.

Three questions are relevant in attempting to make the distinction between parametric and non-parametric processes:

1. Is the energy in the participating optical fields conserved?
2. Does the medium return to its initial energy level at the end of the interaction?
3. Is the process subject to a phase-matching condition?

If the answer to all three questions is Yes, the process can confidently be classed as parametric. An obvious example is second harmonic generation; two fundamental photons combine to form one harmonic photon, the state of the medium is not changed, and there is certainly a phase-matching condition to satisfy. On the other hand, if the answer to all the questions is No, the process is almost certainly non-parametric. The generation of a Stokes wave in stimulated Raman scattering seems to be a good example; a laser photon is lost, a Stokes photon of lower energy is gained, and the nonlinear medium is lifted to an excited state. Phase matching is automatically satisfied.

Unfortunately, things are not in fact so straightforward, and there are a number of complications and pitfalls in this area. How, for example, should one classify the optical Kerr effect? Far from the Raman resonance, Eqs (9.9) and (9.10) merely govern refractive index changes. On resonance, however, the same equations govern stimulated Raman scattering

(SRS), which we have just argued is 'non-parametric'. Admittedly, no phase-matching condition is involved, but energy in the EM fields is certainly conserved. However, as Butcher and Cotter [96] point out, when SRS occurs in a solid, the phonon can be treated as a wave that is subject to a phase-matching condition, as indeed has been pointed out in Chapter 5 with regard to Raman waves in general. And one could of course argue that there *is* actually a phase-matching condition in SRS, albeit one that is automatically satisfied!

Again, how should one regard a process that is subject to a phase-matching condition, but has an intermediate resonance? 'Parametric' anti-Stokes Raman scattering governed by Eq. (9.11) above is a case in point. As to question 2, which is sometimes taken to be definitive, the problem here is that different viewpoints of the same process can give different answers. After all, the matrix element sequences in Eqs (9.9) and (9.10) start and end at level 0, but that does not make SRS a parametric process.

The word 'parametric' is clearly of doubtful value and should be used with caution.

9.5 Coherent effects

9.5.1 The optical Bloch equations

We now move on to study 'coherent' effects in nonlinear optics where the interactions take place on time-scales that are short compared to the relevant damping times. Coherent processes cannot normally be handled using perturbation theory, so a new set of mathematical tools must be developed to describe them.

We start by treating a two-level atomic system excited by a near-resonant optical pulse given by $E = A(t) \cos \omega t$. To simplify the analysis, we will take the pulse envelope $A(t)$ to be real, and we will also assume that it is substantially shorter than the relaxation times of the medium so that damping effects can be ignored. Under these conditions, the off-diagonal and diagonal terms in Eq. (8.15) read

$$\dot{\rho}_{10} = -i\omega_{10}\rho_{10} + \frac{i}{\hbar}V(\rho_{11} - \rho_{00}) \tag{9.13}$$

$$\dot{\rho}_{00} = -\dot{\rho}_{11} = -\frac{iV}{\hbar}(\rho_{10} - \rho_{01}) \tag{9.14}$$

where $V = -\mu A \cos \omega t$, and we have assumed $\mu = \mu_{10} = \mu_{01}$.

These two equations have several important features. Firstly, the upper state population has been included; notice that the final term on the right-hand side involves the population difference between the two levels. Secondly, the damping terms of Eqs (8.18) and (8.19) have not been included, so Eqs (9.13) and (9.14) will provide an accurate representation of a real system only on very fast time-scales where damping plays an insignificant role.

The intrinsic time dependence of ρ_{10} governed by the first term on the right-hand side of Eq. (9.13) could, of course, be removed by using the representation based on $\sigma_{10} =$

$\rho_{10} \exp\{i\omega_{10}t\}$ from Eq. (8.16). In this case, Eqs (9.13) and (9.14) become

$$\dot{\sigma}_{10} = \frac{i}{\hbar} V'_{10}(\sigma_{11} - \sigma_{00}) \tag{9.15}$$

$$\dot{\sigma}_{00} = -\dot{\sigma}_{11} = -\frac{i}{\hbar}\left(V'_{01}\sigma_{10} - V'_{10}\sigma_{01}\right) \tag{9.16}$$

where $V'_{10} = V_{10}\exp\{i\omega_{10}t\}$ and $\sigma_{jj} = \rho_{jj}$. If we now substitute $V_{10} = -\mu_{10}A\cos\omega t$, etc., assume that $\mu = \mu_{10} = \mu_{01}$, and neglect rapidly varying terms (in what is known as the rotating wave approximation (RWA)), it follows that

$$\dot{\sigma}_{10} = -i\frac{\mu A}{2\hbar}(\sigma_{11} - \sigma_{00})\exp\{i(\omega_{10} - \omega)t\} \tag{9.17}$$

$$\dot{\sigma}_{00} = -\dot{\sigma}_{11} - \frac{i\mu A}{2\hbar}\left(\sigma_{10}\exp\{i(\omega_{10} - \omega)t\} - \sigma_{01}\exp\{-i(\omega_{10} - \omega)t\}\right). \tag{9.18}$$

Actually, it is more common to make a subtly different transformation of Eqs (9.13) and (9.14), one that is based not on the transition frequency between levels 0 and 1, but rather *on the frequency of the applied field*. If we duly define $\sigma'_{10} = \rho_{10}\exp\{i\omega t\}$, the equations read

$$\dot{\sigma}'_{10} = -i\Delta\sigma'_{10} - \frac{i\mu A(t)(\sigma'_{11} - \sigma'_{00})}{2\hbar} \tag{9.19}$$

$$\dot{\sigma}'_{00} = -\dot{\sigma}'_{11} = -\frac{i\mu A(t)}{2\hbar}\left(e^{i\omega t} + e^{-i\omega t}\right)\left(\sigma_{01}e^{i\omega t} - \sigma_{10}e^{-i\omega t}\right)$$

$$\approx \frac{i\mu A(t)}{2\hbar}\left(\sigma'_{10} - \sigma'_{01}\right) \tag{9.20}$$

where $\Delta = \omega_{10} - \omega$, and the RWA has been invoked to drop rapidly varying terms in the final form of Eq. (9.20).

It is now a simple matter to obtain the celebrated optical Bloch equations. We define

$$u = 2\text{Re}\{\sigma'_{10}\} \equiv \sigma'_{10} + \sigma'_{01}$$

$$v = -2\text{Im}\{\sigma'_{10}\} \equiv i(\sigma'_{10} - \sigma'_{01}) \tag{9.21}$$

$$w = \sigma'_{11} - \sigma'_{00} \equiv \rho_{11} - \rho_{00}$$

and Eqs (9.19) and (9.20) can now be written

$$\dot{u} = -v\Delta$$
$$\dot{v} = u\Delta + \frac{w\mu A(t)}{\hbar} \tag{9.22}$$
$$\dot{w} = -\frac{v\mu A(t)}{\hbar}.$$

The sign conventions have been chosen to correspond to those of Allen and Eberly [101].

9.5.2 The Bloch vector

A powerful geometrical interpretation of the atomic dynamics now presents itself. For if we define a vector $\mathbf{r} = (u, v, w)$, Eq. (9.22) can be written [14, 101, 102]

$$\dot{\mathbf{r}} = \gamma \times \mathbf{r} \qquad (9.23)$$

where $\gamma = (-\mu A(t)/\hbar,\ 0,\ \Delta)$. Equations (9.22)–(9.23) are known as the optical Bloch equations, and $\mathbf{r}$ as the optical Bloch vector. Since $\dot{\mathbf{r}}$ is perpendicular to $\mathbf{r}$ according to Eq. (9.23), the magnitude of the Bloch vector is clearly constant, a property that can also be seen from Eqs (9.22) where $u\dot{u} + v\dot{v} + w\dot{w}$ is evidently zero. From Eqs (9.21), it is also clear that the vector has unit magnitude, so $u^2 + v^2 + w^2 = 1$, and the vector moves on the surface of a sphere of unit radius, which is known as the Bloch sphere. However, it must be remembered that the present analysis neglects damping and, when stochastic effects are taken into account, the components of the Bloch vector are subject to relaxation.

A diagram of the Bloch sphere is shown in Fig. 9.4, where the 1, 2 and 3 axes correspond to the respective components of $\mathbf{r}$ (i.e. $u = r_1$, $v = r_2$, $w = r_3$). The first point to appreciate is that, when the Bloch vector points to the south pole of the sphere ($w = -1$), the system is in the ground state while, at the north pole ($w = +1$), it is in the excited state; this should be clear from the third of Eqs (9.21). A useful connection can also be established between the transverse components (u and v) of $\mathbf{r}$, and the polarisation of the medium. If we combine

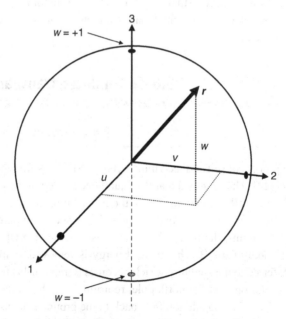

Fig. 9.4 The Bloch vector $\mathbf{r} = (u, v, w)$ showing the Bloch sphere and its north and south poles.

the definition of the polarisation in Eq. (8.21) with $\sigma'_{10} = \rho_{10}\exp\{i\omega t\}$, we obtain

$$P = \tfrac{1}{2}N\mu(u + iv)e^{i\omega t} + \text{c.c.} \tag{9.24}$$

Hence, when $w = \pm 1$ and $u = v = 0$, the polarisation is zero. However, at all other points on the sphere, the system is in a superposition state, and the polarisation is non-zero, while maximum polarisation occurs on the equator, where the magnitude of P is $N\mu$.

Stochastic damping of the off-diagonal elements of the density matrix will ultimately cause u and v, and hence the polarisation P, to relax to zero. But we are, of course, assuming the pulse envelope $A(t)$ to be much shorter than the relevant relaxation times precisely so that these effects can be ignored. However, another source of dephasing arises in the presence of inhomogeneous line broadening, when there is a distribution of detuning $\Delta\omega$ among the different two-level atoms/molecules. Under these circumstances, Eq. (9.24) generalises to

$$P = \tfrac{1}{2}N\mu(\bar{u} + i\bar{v})e^{i\omega t} + \text{c.c.} \tag{9.25}$$

where

$$\bar{u}(z,t) = \int\limits_{-\infty}^{\infty} u(\Delta, z, t)g_i(\Delta)d\Delta$$

$$\bar{v}(z,t) = \int\limits_{-\infty}^{\infty} v(\Delta, z, t)g_i(\Delta)d\Delta. \tag{9.26}$$

The parameters $\bar{u}$ and $\bar{v}$ are averages over the inhomogeneous lineshape function $g_i(\Delta)$, which is normalised so that $\int_{-\infty}^{\infty} g_i(\Delta)d\Delta = 1$. Inhomogeneous broadening plays an important role in several coherent dynamic effects including self-induced transparency, which we turn to next.

9.6 Self-induced transparency

9.6.1 Preview

The process of self-induced transparency (SIT) was first identified by McCall and Hahn in 1967 [14]. They showed that the equations derived in the previous section, in combination with Maxwell's equations, allow the propagation of a resonant pulse through a two-level medium without loss, provided the pulse is short compared to the damping times. In the simplest terms, the front of the '2π-pulse' swings the Bloch vector from the south to the north pole of the Bloch sphere. Energy is removed from the pulse in this process, and transferred to the medium, which becomes (temporarily) fully excited. But the pulse drives the vector on over the north pole, returning it to the south pole down the opposite side of the sphere. The population falls back to the ground state as the energy is returned from the medium to the field.

The detailed theory of SIT is complicated. Whole books have been written on coherent-transient phenomena, processes that include SIT itself, and related effects such as photon echoes; see [101]. We shall restrict ourselves to a highly simplified version of the theory, in which the tendency of the mathematics to obscure the physics will hopefully be avoided as far as possible.

9.6.2 Motion of the Bloch vector: the Rabi frequency

Consider first the motion of the Bloch vector in the presence of a time-independent field that is tuned to exact resonance. In this case, $\gamma = (-\mu A/\hbar,\ 0,\ 0)$, so $\mathbf{r}$ rotates about the 1-axis at an angular frequency $\mu A/\hbar$, which is known as the Rabi frequency Ω_R. If the medium is initially in the ground state, the vector starts at the south pole of the sphere and rotates in the 2–3 plane, cycling backwards and forwards between the south and north poles at the Rabi frequency, with the polarisation oscillating correspondingly. The motion of the Bloch vector in this simple case is easily shown to be described by the equations

$$u = 0$$
$$v = -\sin \Omega_R t \qquad\qquad (9.27)$$
$$w = -\cos \Omega_R t.$$

Figure 9.5 illustrates the motion, if the angle θ is interpreted as $\Omega_R t$; motion of this type is sometimes called 'Rabi flopping'. From Eq. (9.24), it is easy to show that the associated polarisation is

$$P = N\mu\ \sin \Omega_R t\ \sin \omega t. \qquad\qquad (9.28)$$

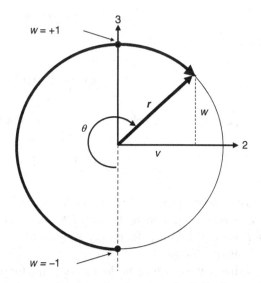

Fig. 9.5 Motion of the Bloch vector in the v-w plane for a pulse of area $\theta = 5\pi/4$. The diagram also illustrates the motion defined in Eq. (9.27) if θ is interpreted as $\Omega_R t$.

A much more elaborate calculation is needed to solve the problem when the system is detuned from resonance. If, as before, $w = -1$ initially, the solution can be shown to be [101]

$$u = ab(1 - \cos \Omega'_R t)$$
$$v = -b \sin \Omega'_R t \qquad\qquad (9.29)$$
$$w = -\left(a^2 + b^2 \cos \Omega'_R t\right)$$

where $a = \Delta / \Omega'_R$, $b = \Omega_R / \Omega'_R$ and Ω'_R is the Rabi frequency, modified to take detuning into account, and given by

$$\Omega'_R = \sqrt{\Omega_R^2 + \Delta^2}. \qquad\qquad (9.30)$$

It is easy to show (see Problem 9.4) that Eqs (9.29) describe motion in a plane that passes through the south pole of the Bloch sphere and is slanted from the 3-axis at an angle whose tangent is Δ / Ω_R.

9.6.3 Pulse area and the area theorem

We now consider the more general case of pulsed excitation, where the field is given by $E(t) = A(t)\cos \omega t$ and the pulse envelope $A(t)$ (assumed real) is short compared to the damping times. As before, we assume that the Bloch vector is initially at $w = -1$, and the detuning is zero ($\Delta = 0$); this, of course, ensures that $u = 0$ at all times, and restricts the motion to the v-w plane. It is easy to show that the solution in this case is

$$u = 0$$
$$v = -\sin \psi \qquad\qquad (9.31)$$
$$w = -\cos \psi$$

where

$$\psi(t) = \frac{\mu}{\hbar} \int_{-\infty}^{t} A(t')dt'. \qquad\qquad (9.32)$$

The final position of the Bloch vector is determined by the angle

$$\theta = \psi(\infty) = \frac{\mu}{\hbar} \int_{-\infty}^{\infty} A(t')dt' \qquad\qquad (9.33)$$

which is commonly referred to as the pulse 'area'. Notice that θ is related to the time integral of the *field* profile $A(t)$, not to the that of $A^2(t)$, which is related to the pulse energy. It is therefore possible for the pulse energy to rise while the pulse area remains unchanged, as we shall shortly discover.

The swing of the Bloch vector in the v-w plane for pulse area of $\theta = 5\pi/4$ is shown in Fig. 9.5. As noted after Eq. (2.8), when the polarisation has a $\pi/2$ phase lag with respect to the electric field (so that $\tilde{P} \sim -i\tilde{E}$), energy flows from the field to the medium. This is the case for $v < 0$ (see Eq. 9.24), which applies in the initial part of the process in Fig. 9.5 where w is rising. But as the Bloch vector passes the north pole ($\mathbf{r} = (0,\ 0,\ +1)$) of the sphere, v goes positive, the phase of the polarisation changes from a lag to a lead, the direction of energy flow reverses, and w begins to fall.

As in the case of constant field treated earlier, the path of the Bloch vector is confined to the v-w plane only for perfect tuning. When the detuning $\Delta \neq 0$, the vector traces a looped path on the surface of the sphere, starting at the south pole and returning to it, provided θ is sufficiently large. The motion is no longer in a plane, as it was for a time-independent field, except in the case of perfect tuning when the vector passes through the north pole ($w = +1$). Otherwise, the path veers away from the pole, reaching a maximum value of w ($< +1$), and then retreating south again. For very small detuning, the 'furthest north' point will be close to the north pole; for large detuning, the vector will never leave the southern hemisphere, and for very large detuning, it will never get far from the south pole.

The question that must now be addressed is: how does θ evolve as the resonant pulse propagates? The answer is contained in the celebrated 'area theorem' of McCall and Hahn, which applies in the presence of inhomogeneous line broadening, The area theorem states that, irrespective of the pulse shape, the pulse area evolves with propagation distance according to

$$\frac{d\theta}{dz} = -\left(\frac{N\omega_0\mu^2\pi g_i(0)}{2\hbar c n\varepsilon_0}\right)\sin\theta = -\frac{\alpha'}{2}\sin\theta \quad \text{(absorbing case)} \qquad (9.34)$$

where the final step defines

$$\alpha' = \frac{N\omega_0\mu^2\pi g_i(0)}{\hbar c\varepsilon_0 n}. \qquad (9.35)$$

Here, $g_i(0)$ is the inhomogeneous lineshape function at line centre, and n is the refractive index. The signs in Eqs (9.34) and (9.35) are appropriate for an absorbing medium, which is the case we have been treating throughout this section.

We now consider the implications of the area theorem. It is evident from Eq. (9.34) that $d\theta/dz = 0$ when $\theta = m\pi$ (integral m), but it is easy to see that the solution is stable for even m, and unstable for odd m. For example, if θ is close to π ($m = 1$), the sign of $d\theta/dz$ ensures that θ runs away from π, whereas, near $m = 2$, θ converges on 2π from either direction. Pulses with $\theta = 2\pi$ (known as 2π pulses) are not only stable in pulse area but, because the medium is returned to the lower state at the end of the process, the pulse energy is constant too. This suggests that 2π pulses with stationary pulse profiles might exist, a conjecture that turns out to be correct, and will be examined in a little more depth in a moment.

The graphical solutions to Eq. (9.34) shown in Fig. 9.6 provide full details of the evolution of the pulse area. The abscissa is in units of $\alpha'z$, so the characteristic length-scale is determined by $(\alpha')^{-1}$. For very weak fields, $\sin\theta \rightarrow \theta$, and the solution to Eq. (9.34) is $\theta(z) = \theta(0)e^{-\alpha'z/2}$. Now, the expressions for α' in Eq. (9.35) and α in Eq. (9.5) are identical apart from the nature of the respective lineshape functions, so it is tempting to assume that the pulse *energy* decays as $(e^{-\alpha'z/2})^2 = e^{-\alpha'z}$ in this weak-field limit. However, extreme caution is necessary in this context because, if the pulse envelope develops a negative field lobe, the area can fall to zero while energy still remains. A detailed analysis of small-area pulses has been given by Crisp [103].

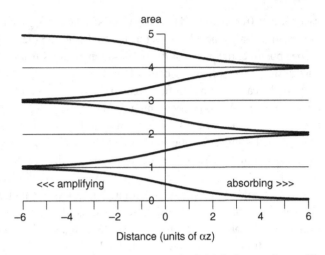

Fig. 9.6   Evolution of pulse area with propagation distance according to Eq. (9.34). In the case of an amplifier (Eq. 9.36), the evolution is from right to left.

An intriguing feature of Eq. (9.35) is the presence of $g_i(0)$ in the numerator. This implies that $\alpha' \to \infty$ in the absence of inhomogeneous broadening (IHB), because $g_i(0)$ then becomes a delta function. This in turn means that the distance scale of Fig. 9.6 shrinks to zero, and the pulse area reaches its stable limit immediately. What is the physical interpretation of this surprising feature? According to Eqs (9.24) and (9.31), a pulse of initial area θ_i leaves the medium with a polarisation $P = -\frac{1}{2}iN\mu \sin\theta_i e^{i\omega t} + \text{c.c.}$, and this optical antenna will continue to radiate until it is either washed out by IHB, or all the stored energy is used up. So, in the absence of IHB, the radiation will continue until the final area θ_f becomes $m\pi$ (m even) and $w = -1$. If $\theta_i < \pi$, the radiation adds a negative lobe to the rear of the pulse, making $\theta_f = 0$ while, if $\pi < \theta_i < 2\pi$, a positive lobe is added and $\theta_f = 2\pi$. The extension to higher values of θ_i is straightforward.

So far, the discussion has been based on the assumption that the medium is in the lower state ($w_i = -1$) before the arrival of the pulse. In the case of an amplifying medium, $w_i = +1$, and the area theorem changes to

$$\frac{d\theta}{dz} = +\frac{\alpha'}{2}\sin\theta \quad \text{(amplifying case)}. \tag{9.36}$$

The sign reversal means that the stable values of θ are now $m\pi$ with m odd; Fig. 9.6 can still be used, provided it is read from right to left since, in effect, the sign change means $z \to -z$. In the amplifying case, an initial pulse with $0 < \theta_i < 2\pi$ will gravitate towards a so-called π-pulse with $\theta = \pi$. However, a 180° swing takes the Bloch vector for perfectly tuned atoms with $\Delta\omega = 0$ from $w_i = +1$ to $w_f = -1$ so, even though the area has stabilised, the energy will continually increase as the pulse picks up energy from the medium. And the only way this combination of circumstances can occur is if the pulse continually narrows.

9.6.4 Outline proof of the area theorem

The proof of the area theorem is very involved, and only a brief sketch will be attempted here. All parameters must now be treated as functions of space and time, and $A(t)$ is replaced by $\tilde{E}(z,t)$ which is assumed real. The field equation is given by

$$\frac{\partial \tilde{E}}{dz} + \frac{n}{c}\frac{\partial \tilde{E}}{dt} = -i\frac{\omega}{2c\varepsilon_0 n}\tilde{P}^{\text{NL}} \tag{9.37}$$

which is a simplified version of Eq. (7.2). Equation (9.37) leads, with the help of Eqs (9.25) and (9.26), to

$$\frac{\partial \tilde{E}}{dz} = \frac{N\mu\omega_0}{2c\varepsilon_0 n}\int_{-\infty}^{\infty} v(\Delta,z,t)g(\Delta)d\Delta - \frac{1}{c}\frac{\partial \tilde{E}}{dt}. \tag{9.38}$$

From the definition of pulse area, and with the help of the previous equation, we have

$$\frac{d\theta}{dz} = \lim_{t\to\infty}\frac{\mu}{\hbar}\int_{-\infty}^{t} dt' \frac{\partial \tilde{E}(z,t')}{\partial z}$$

$$= \lim_{t\to\infty}\frac{\mu}{\hbar}\int_{-\infty}^{t} dt' \left(\frac{N\mu\omega_0}{2c\varepsilon_0 n}\int_{-\infty}^{\infty} v(\Delta,z,t')g(\Delta)d\Delta - \frac{1}{c}\frac{\partial \tilde{E}(z,t')}{dt'}\right). \tag{9.39}$$

The final term in this equation, when integrated, involves the factor $(\tilde{E}(z,\infty) - \tilde{E}(z,-\infty))$, which is zero for a pulse. A comparison of the remaining term with Eq. (9.34) shows that the area theorem is proved if it can be demonstrated that

$$\lim_{t\to\infty}\int_{-\infty}^{t} dt' \int_{-\infty}^{\infty} d\Delta \, v(\Delta,z,t')g(\Delta) = -\pi g(0)\sin\theta. \tag{9.40}$$

It is in this final step that most of the mathematical difficulty lies, and a lengthy analysis is necessary to verify the identity [101]. The appearance of $g(0)$ on the right-hand side underlines the fact that the double integral is dominated by the part of the lineshape in the immediate vicinity of $\Delta = 0$. In view of the second of Eqs (9.31), the result of Eq. (9.40) seems quite reasonable.

9.6.5 The steady-state solution

The fact that a 2π pulse propagates without change in either area or energy suggests that a steady-state solution might exist. This is indeed the case, although the proof leading to the solution is even more unfriendly than the one for the area theorem. The result is

$$\tilde{E}(z,t) = \frac{2\hbar}{\mu\tau}\text{sech}\left(\frac{t-z/V}{\tau}\right) \tag{9.41}$$

where τ is a measure of the pulse width, and can take any value that keeps the pulse within the coherent regime. The pulse envelope travels at speed V given by

$$\frac{1}{V} - \frac{n}{c} = \left(\frac{\alpha'\tau^2}{2\pi g_i(0)} \right) \int_{-\infty}^{\infty} \frac{g_i(\Delta')}{1 + (\tau\Delta')^2} d\Delta'. \tag{9.42}$$

This formula can be approximated when the pulse duration τ is either very short or very long compared to the inverse linewidth; see Problem 9.3.

9.7 Electromagnetically-induced transparency and slow light

To end this chapter, we apply the density matrix machinery to the study of another strong-field effect known as *electromagnetically-induced transparency* (EIT), which was first studied by Boller *et al.* in 1991 [104,15]. The treatment of EIT presented here is similar to that in Section III of Fleischhauer *et al.* [16], where research on EIT and related topics such as slow light is reviewed. The energy level scheme shown in Fig. 9.7 is the same as in [16]; much of the notation is the same too, although the numbering of the energy levels is different.

Under optimal EIT conditions, absorption at line centre of a transition from the ground state (0) to an excited state (1) is eliminated in the presence of a strong resonant coupling field between state 1 and a third state 2. At the same time, a narrow region of strong normal dispersion appears at the centre of the resonance in which the group velocity is very low; this leads to the effect known as *slow light*. Moreover, while absorption *from* the ground state is artificially removed by EIT, stimulated emission *to* the ground state turns

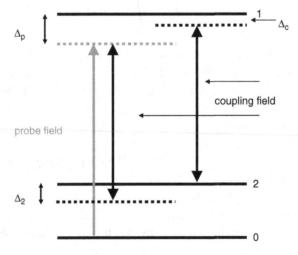

Fig. 9.7   Three-level system used for analysing electromagnetically-induced transparency, showing the three detunings defined at the beginning of Section 9.7.

out to be unaffected. This opens the way to a further phenomenon known as *lasing without inversion* [17].

The excited state 2 has the same parity as the ground state, and both are coupled to state 1, which has the opposite parity. Of the two EM waves involved in the process, the strong 'coupling' field at ω_c is tuned close to the spacing ω_{12} between state 1 and state 2, while a weak 'probe' field at ω_p scans the properties of the 0–1 transition in the presence of ω_c. We introduce detuning parameters $\Delta_p = \omega_{10} - \omega_p$ and $\Delta_c = \omega_{12} - \omega_c$ for the respective probe and coupling fields. The two-photon detuning $\Delta_2 = \Delta_p - \Delta_c = \omega_{20} - (\omega_p - \omega_c)$ also turns out to be useful. The coupling field is shown twice (by double arrows) in Fig. 9.7, in order to highlight all three detuning parameters.

Nine density matrix equations can be written down for this three-level system, three for the diagonal elements and six for the off-diagonal elements (of which three are conjugates of the other three). It turns out, however, that the key features of EIT can be understood by considering the equations for just two off-diagonal elements, namely[2]

$$\dot{\sigma}_{01} = -\gamma_{10}\sigma_{01} - \tfrac{1}{2}i\left(\Omega_{Rc}\sigma_{02}\exp\{-i\Delta_c t\} + \frac{\mu_{01}\hat{E}_p}{\hbar}(\sigma_{00} - \sigma_{11})\exp\{-i\Delta_p t\}\right) \quad (9.43)$$

$$\dot{\sigma}_{02} = -\gamma_{20}\sigma_{02} - \tfrac{1}{2}i\Omega_{Rc}\sigma_{01}\exp\{i\Delta_c t\}. \quad (9.44)$$

In these equations, $\Omega_{Rc} = \mu_{12}A_c/\hbar$ is the Rabi frequency associated with the coupling field, and $\hat{E}_p$ is the probe field; damping terms have been included. An equation for σ_{12} can easily be written down too, but this element plays no direct part in the action. As for the diagonal elements, it turns out to be sufficient simply to assume that $\sigma_{00} \cong 1$ and $\sigma_{11} \cong \sigma_{22} \cong 0$.

The steady-state solutions of the equations are readily obtained once one spots that σ_{01} varies as $\exp\{-i\Delta_p t\}$ and σ_{02} as $\exp\{-i\Delta_2 t\}$. This enables the replacements $\dot{\sigma}_{01} \rightarrow -i\Delta_p\sigma_{01}$ and $\dot{\sigma}_{02} \rightarrow -i\Delta_2\sigma_{02}$ to be made, and one then quickly obtains the solutions

$$\sigma_{02} = \frac{\Omega_{Rc}\exp\{i\Delta_c t\}}{2(\Delta_2 + i\gamma_{20})}\sigma_{01} \quad (9.45)$$

$$\sigma_{01} = \frac{\Omega_{Rc}\exp\{-i\Delta_c t\}}{2(\Delta_p + i\gamma_{10})}\sigma_{02} + \frac{\mu_{01}\hat{E}_p\exp\{-i\Delta_p t\}}{2\hbar(\Delta_p + i\gamma_{10})}. \quad (9.46)$$

Combining these equations yields

$$\sigma_{01} = -\exp\{-i\Delta_p t\}\frac{\mu_{01}\hat{E}_p}{\hbar}\frac{(\Delta_2 + i\gamma_{20})}{2D} \quad (9.47)$$

where

$$D = \left(\tfrac{1}{2}\Omega_{Rc}\right)^2 - (\Delta_p + i\gamma_{10})(\Delta_2 + i\gamma_{20}). \quad (9.48)$$

The structure of the 01 transition can now be explored by studying the susceptibility at the probe frequency. From Eq. (8.21), it can be shown that the dominant term in the

---

[2] The damping constants defined in Eqs (9.43) and (9.44) are half the corresponding factors in [16].

polarisation is

$$\hat{P}_p = N\sigma_{01}\mu_{10}\exp\{i(\omega_{10} - \omega_p)t\} = -\frac{N|\mu_{01}|^2}{\hbar}\frac{(\Delta_p + i\gamma_{20})}{D}\hat{E}_p. \tag{9.49}$$

Writing the susceptibility in the form $\chi = \chi' - i\chi''$ as in Section 9.2, we obtain

$$\chi' = \frac{N|\mu_{01}|^2}{\hbar\varepsilon_0}\frac{\Delta_2(\Delta_2\Delta_p - (\frac{1}{2}\Omega_{Rc})^2) + (\gamma_{20})^2\Delta_p}{|D|^2} \tag{9.50}$$

$$\chi'' = \frac{N|\mu_{01}|^2}{\hbar\varepsilon_0}\frac{\Delta_2^2\gamma_{10} + \gamma_{20}((\frac{1}{2}\Omega_{Rc})^2 + \gamma_{10}\gamma_{20})}{|D|^2}. \tag{9.51}$$

Despite the relative simplicity of Eqs (9.43) and (9.44) with which we started, the structure of the susceptibility is evidently quite complicated. However, careful examination of the formulae yields a rich reward, because EIT, slow light, and indeed the Autler–Townes splitting too, are all hidden inside Eqs (9.50) and (9.51).

We will look at the special case where γ_{20} is negligible, and the coupling field is perfectly tuned to the 12 transition, so that $\Delta_c = 0$ and $\Delta_2 = \Delta_p$. Equations (9.50) and (9.51) then reduce to

$$\chi' = \frac{N|\mu_{01}|^2}{\hbar\varepsilon_0\gamma_{10}}\left[\frac{\gamma_{10}\Delta_p(\Delta_p^2 - (\frac{1}{2}\Omega_{Rc})^2)}{|D|^2}\right] \tag{9.52}$$

$$\chi'' = \frac{N|\mu_{01}|^2}{\hbar\varepsilon_0\gamma_{10}}\left[\frac{\Delta_p^2\gamma_{10}^2}{|D|^2}\right] \tag{9.53}$$

where $|D|^2 = ((\frac{1}{2}\Omega_{Rc})^2 - \Delta_p^2)^2 + \Delta_p^2\gamma_{10}^2$.

In Figs 9.8 and 9.9, the square-bracketed terms in Eqs (9.52) and (9.53) are plotted as a function of the normalised detuning Δ_p/γ_{10} for two different values of Ω_{Rc}. Notice that the abscissa scale decreases to the right, which ensures that the probe frequency $\omega_p = \omega_{10} - \Delta_p$ increases in that direction. Positive values of χ'' represent absorption.

Figure 9.8 shows the behaviour when $\Omega_{Rc} = 4.0\gamma_{10}$, with graphs for $\Omega_{Rc} = 0$ (i.e. zero coupling field) shown dotted for comparison. Several important features are apparent from the figure. Firstly, the absorption profile, which consists of a single maximum when $\Omega_{Rc} = 0$ (dotted line), is split into two in the presence of the coupling field. This effect is known at the *Autler–Townes splitting*. Inspection of Eqs (9.52) and (9.53) reveals that the peak separation corresponds to the Rabi frequency, and this is confirmed by the figure.

Secondly, the absorption near exact resonance is very small, and careful inspection reveals that it is actually zero at $\Delta_p = 0$; see Eq. (9.53). This is the basis of *electromagnetically-induced transparency* (EIT). Exactly on resonance, the medium becomes perfectly transparent!

Thirdly, there are now two regions of anomalous dispersion, one associated with each component of the double peak. But between them lies a region of normal dispersion so, at the point of exact resonance ($\Delta_p = 0$), the type of dispersion has changed.

What happens when the strength of the coupling field is reduced is no less remarkable. In Fig. 9.9, where $\Omega_{Rc} = 0.6\gamma_{10}$, the Autler–Townes splitting has narrowed, but the feature at

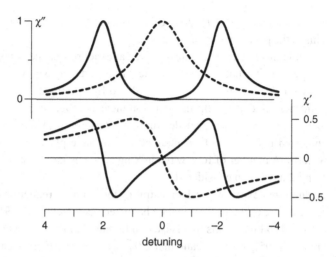

Fig. 9.8 Real part (top) and imaginary part (bottom) of the susceptibility of Eqs (9.52) and (9.53) as a function of normalised probe detuning for $\Delta_c = 0$ and $\Omega_{Rc} = 4\gamma_{10}$, showing the Autler–Townes splitting, strong normal dispersion near the centre of the resonance, and transparency at exact resonance. The results for zero coupling field are shown dotted.

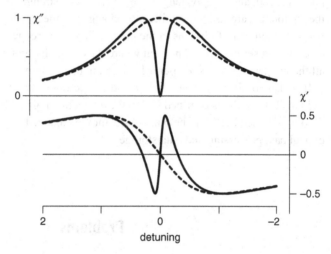

Fig. 9.9 As Fig. 9.8 but for $\Omega_{Rc} = 0.6\gamma_{10}$.

exact resonance still exists, and has become notably sharper. And it continues to sharpen as the coupling field is further reduced. Indeed, under these idealised conditions, the frequency dependence of the real part at $\Delta_p = 0$ increases without limit, tending to infinity as $\Omega_{Rc} \to 0$ at which point the EIT features naturally vanish. From Section 6.2,

$$v_{\text{group}} = \frac{c}{n + \omega \dfrac{dn}{d\omega}} \tag{9.54}$$

and, since $dn/d\omega \cong \frac{1}{2}d\chi'/d\omega$, EIT is associated with very low values of the group velocity. This is the phenomenon of *slow light*.

While the existence of transparency at exact resonance is fascinating from a theoretical perspective, it is not *in itself* of particular importance experimentally. After all, if one wants to avoid absorption, one can easily do so by moving away from line centre. The crucial point here is that the third-order nonlinear coefficient $\chi^{(3)}$ can be shown to be subject to *constructive* rather than destructive interference at the resonance, and so resonantly enhanced nonlinear optical interactions can take place at the centre of the transparency window. A number of four-wave mixing schemes involving the coherent linkage of states 1 and 2 have been considered.

Numerous experiments demonstrating the dramatic reduction in the optical group velocity at the centre of the EIT resonance have been conducted [18,19]. In one example, based on ultracold sodium atoms in a Bose condensate, light was slowed to 17 ms^{-1}, or roughly 60 kmhr^{-1} [105], which means that $c/v_{group} \cong 2 \times 10^7$. It is easy to see that this massive speed reduction will be accompanied by a corresponding spatial compression of the pulse envelope by the same factor. After all, a 1 μs pulse is spread over 300 m at the speed of light in vacuum, but at 17 ms^{-1} its extent will be only 17 μm. However, it is only the envelope that is slowed down and compressed; the optical carrier wave continues to travel at the normal speed, and its wavelength is unaltered. The strength of the electric field and hence the photon flux are also not significantly changed by the slowing down, and the consequence is that the number of photons in the pulse will also be reduced by the compression factor. This raises a serious question about where the energy has gone, and it turns out that almost all the incident energy is temporarily stored in the medium.

It is also possible to bring a light pulse to a complete halt by switching off the coupling field while the pulse is confined within the interaction region [106]. And, when the coupling field is switched back on, the pulse starts moving again. This has potential applications in optical data processing and data storage.

Problems

9.1 Convince yourself that the approximation in Eq. (9.4) is correct.

9.2 Use Eq. (2.8) in conjunction with Eqs (9.9)–(9.10) to verify that the energies gained at ω_S and lost at ω_L in stimulated Raman scattering are in the ratio of the respective photon energies.

9.3 When the steady-state pulse duration τ in self-induced transparency is very small compared to the inverse inhomogeneous linewidth $g_i(0)^{-1}$, show that Eq. (9.42) can be approximated to $\dfrac{1}{V} - \dfrac{n}{c} \cong \dfrac{\alpha'\tau^2}{2\pi g_i(0)}$. Show similarly that $\dfrac{1}{V} - \dfrac{n}{c} \cong \frac{1}{2}\alpha'\tau$ in the opposite limit $(\tau \gg g_i(0)^{-1})$.

9.4 Show that Eq. (9.29) represents motion in a plane that passes through $w = -1$ and is slanted away from the 3-axis at an angle of $\tan^{-1}\{\Delta/\Omega_R\}$.

9.5 Show that the refractive index gradient at the centre of the EIT resonance in Fig. 9.9 is given by

$$\frac{dn}{d\omega} = \frac{2N\,|\mu_{01}|^2}{\varepsilon_0\,\hbar(\Omega_{Rc})^2}.$$

9.6 Use a PC to explore the properties of Eqs (9.52) and (9.53) in a more general case than that considered in this chapter. For example, how are Figs 9.8 and 9.9 modified when $\gamma_{20} \neq 0$? See if you can discover the Raman resonance, when $\Delta_c \neq 0$ and Δ_2 is scanned through zero.

10 High harmonic generation

10.1 Introduction to high harmonic generation

The treatment of frequency mixing in the early chapters of this book was based on the assumption that the applied EM fields were weak compared to the internal fields within the nonlinear media. This enabled the polarisation to be expanded as a power series in the field, and perturbation theory to be used to calculate the nonlinear coefficients.

In this chapter, we consider the generation of optical harmonics in the strong-field regime, where the fields are sufficiently intense to ionise the atoms in a gaseous medium, and the electrons released then move freely in the field, at least to a good approximation. As we shall discover, an enormous number of harmonics can be generated in these circumstances, extending across a broad spectral plateau that extends deep into the soft X-ray region. High harmonic generation (HHG) therefore creates a table-top source of coherent X-rays, which has already been applied in lensless diffraction imaging [107]. Moreover, if the huge spectral bandwidth is suitably organised, HHG enables pulses as short as 100 attoseconds[1] or even less to be generated [108]. These can then serve as diagnostic tools on unprecedented time-scales. They have already been used to probe proton dynamics in molecules with a 100-as time resolution [109], and to measure the time delay of electron emission in the photoelectric effect for the first time [110]. Other applications are described in [111].

Experiments on third harmonic generation in the noble gases in the late 1960s involved laser intensities of around 1 GW cm^{-2}, far below those employed in current HHG experiments, which are typically five or six orders of magnitude higher. The early experiments [22] were definitely in the perturbative regime, and they were performed under non-phase-matched conditions in an extended nonlinear medium. Conversion efficiencies were minuscule, at around 1 in 10^{14}.

Experimental results on harmonic generation in gases and metal vapours in the 1970s and early 1980s continued to exhibit the typical characteristics of the perturbative regime, namely that the conversion efficiency for successive harmonics dropped off rapidly.[2] However, by the late 1980s, much higher laser intensities were being deployed, and some remarkable experimental results on the noble gases [113,114] suggested that a new strong-field regime was being entered. For lower harmonics, the conversion efficiency still fell away as before but, beyond a certain point (typically around the seventh to the ninth

---

[1] 1 attosecond (1 as) = 10^{-18}s.
[2] See [112] for citations to this work.

harmonic), the development of the spectral plateau mentioned earlier was clearly seen. Once on the plateau, the efficiency remained essentially constant up to a fairly well-defined high-frequency cut-off. The plateau could be extended and the cut-off pushed further into the UV by raising the laser intensity, although a saturation intensity existed beyond which further extension was not possible.

By the early 1990s, harmonic orders well into the hundreds had been recorded in Ne [115,116] and, by the end of the decade, orders around 300 were reported [117]. At the time of writing, the figure is over 1000 [118] which, with a fundamental wavelength of 800 nm, corresponds to a harmonic wavelength of around 0.8 nm at the spectral extremity, which is deep in the soft X-ray part of the spectrum beyond the water window.

Thin gas jets are normally used in HHG work, as this gets round at least some of the problems associated with focused beams in normally dispersive media (see Section 2.5). Conversion efficiencies are many orders of magnitude higher than in the very early work, but are still small at around 1 in 10^6. Excitation intensities are typically around 10^{14}–10^{15} W cm^{-2} for Nd:YAG and Ti:sapphire lasers or perhaps a little higher for KrF excitation at 249 nm.[3] These are the levels at which ionisation of the atoms becomes significant, and this means that an entirely new set of concepts is needed to model the HHG process.

Highly sophisticated theoretical techniques are needed to do full justice to the HHG problem, and some aspects lie at the current frontier of knowledge. However, a general appreciation of how high harmonics are produced in the strong-field regime can be gained by using the 'three-step model', the foundations of which were laid in [119,120]. The process is divided into the following stages:

(1) Tunnel-ionisation in which an electron is removed from an atom by the field.
(2) The subsequent motion of the ionised electron in the oscillating EM field, in the course of which the electron acquires significant kinetic energy.
(3) The 'recollision' of the energised electron with the parent atom, in which the accumulated energy is emitted in the form of radiation. The frequency of the emission corresponds to the kinetic energy of the electron, plus the energy of ionisation.

The schematic diagram of Fig. 10.1 illustrates the concept behind stage 1. It shows a one-dimensional Coulomb potential (frame b) tilted by a strong electric field, which is positive on the left (frame a) and negative on the right (frame c). In an oscillating field, the slope naturally changes with time, so a movie would show a see-saw motion with the potential sloping to the left and to the right on successive half-cycles. The key point is that the potential barrier is lowered on each side in turn, increasing the probability of ionisation in the favoured direction, the more so as the amplitude of the field is increased.

Many of the main features of the second stage (the motion of the ionised electron in the field) can be understood within the simple classical model outlined in the next section. It is fairly easy to understand that linearly polarised light is a prerequisite if the electron is to be returned accurately to its starting point in the parent atom. We already know from Chapter 5 that the effective *third harmonic generation* coefficient is zero for circularly polarised light, so this feature of HHG should not come as a complete surprise. However, the criterion on

---

[3] Intensities are traditionally quoted in W cm^{-2} in the research literature.

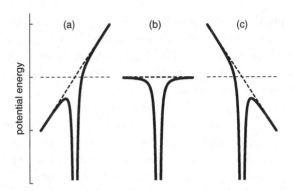

Fig. 10.1 Schematic diagram showing a Coulomb potential (frame b) tilted by a positive electric field (frame a) and a negative electric field (frame c). Notice the lowering of the potential barrier on the left in frame a, and the right in frame c.

the polarisation is more stringent in the HHG case; indeed it turns out that any significant ellipticity will cause the electron to move away from the atom along a curved path which is highly unlikely ever to return to the starting point.[4] The need to employ linearly polarised light has the benefit that a meaningful quantum mechanical treatment based on the time-dependent Schrödinger equation *in one dimension* can be used to trace the evolution of the electron wave packet in the interval between ionisation and recollision. We shall, however, confine ourselves to a classical approach in this chapter.

The simplest statement that can be made about the recollision event is that the electron returns to its bound state, and the energy of the emitted photon corresponds to the kinetic energy acquired by the electron during its journey in the continuum, plus the energy of ionisation. Quantum mechanically, the optical antenna driving the emission is formed in the interference between the electron wave packet and the bound-state wave function of the atom. The cut-off in the spectrum is determined by the maximum kinetic energy that the electron can acquire, and a simple formula for this is easily obtained from the classical model that we now describe.

10.2 Classical model

10.2.1 Electron trajectories

The treatment of HHG that follows is essentially classical, although Planck's constant is used to convert recollision energy to emission frequency, and a formula derived from a more sophisticated model is used for a weighting factor (see Eq. 10.11 below). It is remarkable how far one can progress with the HHG problem on this very simplistic basis. The reason is

---

[4] If the electron and the ion are treated classically as point particles, even the slightest ellipticity in the field polarisation precludes a recollision; see Problem 10.2. Quantum mechanically, however, the electron wave packet and the ion have finite size, which reduces the strength of the recollision while making it more likely to occur.

that classical physics does a relatively good job at handling the motion of the ionised electron in the continuum, which turns out to be the critical stage of the process. The conclusion is confirmed by quantum mechanical results in the strong-field approximation [121]. The excursions of the electron are also quite large (see Problem 10.4), so it is an acceptable approximation to neglect the influence of the parent ion.

We assume that the HHG process is excited by an optical field, linearly polarised in the x-direction, and given by $E(t) = A(t) \cos \omega t$, where $A(t)$ is in general the time-dependent envelope of an optical pulse. The equation of motion of the electron is

$$m\ddot{x} = -eA(t) \cos \omega t \tag{10.1}$$

where e is the elementary charge and m the electron mass. If we ignore the time variation of A for the moment and assume the field is monochromatic, simple integration yields the electron velocity as

$$\dot{x} = \dot{x}_{\mathrm{drift}} - \frac{eA \sin \omega t}{m\omega} \tag{10.2}$$

where the constant of integration is the cycle-averaged velocity. Setting $\dot{x}_{\mathrm{drift}} = 0$ yields a formula for the 'quiver energy' (or *ponderomotive energy*) of the electron in the field, namely

$$U_{\mathrm{P}} = \tfrac{1}{2}m\overline{(\dot{x}^2)} = \frac{(eA)^2}{2m\omega^2}\overline{\sin^2 \omega t} = \frac{(eA)^2}{4m\omega^2}. \tag{10.3}$$

Consider now the situation of an electron immediately after ionisation. If the electron is released from the atom at time t_0, and its initial velocity is assumed to be zero, the solution to Eq. (10.1) is

$$\dot{x} = V\sqrt{\frac{4U_{\mathrm{P}}}{m}} \tag{10.4}$$

where

$$V(t) = (\sin \omega t_0 - \sin \omega t). \tag{10.5}$$

The nature of this solution depends on the point in the cycle where ionisation occurred (i.e. on ωt_0). The electron is initially accelerated away from the atom by the field but, sooner or later, the field changes sign and the electron is slowed to a standstill ($V(t) = 0$). The velocity then changes sign, and the electron starts moving back toward the parent ion, but whether it actually reaches it before the velocity changes sign again depends on ωt_0.

If ionisation occurs while the field magnitude is rising (i.e. in a quarter-cycle preceding an extremum), it can be shown (see Problem 10.3) that the electron will never return to the parent atom. On the other hand, in quarter-cycles where the field magnitude is falling, the time of recollision can be found by integrating Eq. (10.4) and setting the net excursion to zero, i.e.

$$x = \int_{t_0}^{t_{\mathrm{R}}} \dot{x}\, dt' = \frac{eAX}{m\omega^2} = 0 \tag{10.6}$$

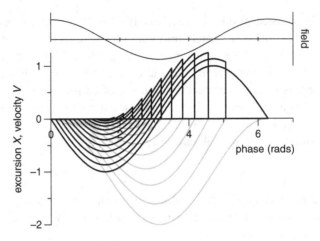

Fig. 10.2   Set of one-dimensional electron trajectories showing the displacements X (in grey) from Eq. (10.7), and the velocities V (in black) from Eq. (10.5), both as a function of ωt (in radians). The (co)sinusoidal field profile is shown at the top.

where

$$X(t) = (\cos \omega t - \cos \omega t_0 + \omega(t - t_0) \sin \omega t_0).$$ (10.7)

The transcendental equation $X(t_R) = 0$ can now be solved for the recollision time t_R, and the result substituted into Eq. (10.5) to obtain the velocity and kinetic energy at the point of recollision. The equation is in fact represented by an interesting geometrical construction that links the times of ionisation and recollision (see Problem 10.5). If the tangent to $E(t) = A \cos \omega t$ at $t = t_0$ intersects $E(t)$ at some subsequent time, this is the point of recollision. As noted earlier, the light must be plane-polarised if the electron is to remain on the correct line; otherwise the electron trajectory will be curved and, classically at least, the electron will never return to its starting point (see Problem 10.2).

Some results are displayed in Fig. 10.2 in which the solid grey lines are a set of one-dimensional electron trajectories $X(t)$ from Eq. (10.7) for departure times from the parent atom at 12 equally-spaced values of ωt_0 between 0 and $\pi/2$ radians. The electric field is shown at the top for reference. The black lines trace the corresponding variations of $V(t) = (\sin \omega t_0 - \sin \omega t)$ from Eq. (10.5), and represent the electron velocities as a function of time. Both sets of lines terminate at the point of recollision where $X(t_R) = 0$. On the graphs of $V(t)$, the length of the verticals to the axis where the lines end represents the recollision velocity V_R. The kinetic energy at recollision is therefore given by $U_{KE} = \frac{1}{2}m\dot{x}^2 = 2U_P V_R^2$ from Eqs (10.3) and (10.5).

All trajectories shown in Fig. 10.2 involve a negative X excursion from the phase (or time) axis, and end at recollision.[5] Electrons leaving the atom at exactly the peak of the field ($\omega t_0 = 0$) have the longest trajectories in terms of both distance and duration, but these

---

[5] The excursion is negative because negatively charged electrons move in a field that is initially positive.

electrons only just manage to complete the return journey, arriving back at the starting point at $\omega t = 2\pi$ with zero velocity and zero kinetic energy. For electrons ionised after the peak, the trajectories do not extend quite so far from the parent ion, although they are still counted as 'long trajectories' at this stage. The recollision velocities rise quite rapidly as the ionisation time increases, reaching a maximum of $|V| \cong 1.26$ when $\omega t_0 \simeq 0.314$, which corresponds to a recollision kinetic energy of $2 \times 1.26^2 U_P$. This is the origin of the equation, frequently repeated in the literature, that the maximum energy available at recollision is [120]

$$U_R^{max} = 3.17 U_P + U_I. \tag{10.8}$$

The ionisation energy U_I is included here because the electron returns to a bound state in the parent atom.

As ωt_0 increases further, the trajectories continue to shorten, and the recollision velocities for these 'short trajectories' fall away too, until $\omega t_0 = \pi/2$ where the field is zero. Throughout the next quarter-cycle of the field when the field magnitude is rising, no trajectories return to the starting point, and all electrons escape. The quarter-cycle ends at $\omega t_0 = \pi$, after which the entire sequence repeats with the opposite polarity.

It is worth emphasising that trajectories are classified as 'long' or 'short' according to the time interval between ionisation and recollision. From Fig. 10.2, it is clear that the earlier the time of ionisation, the later the time of recollision, so the long trajectories correspond to early departures and late arrivals, and short trajectories to late departures and early arrivals. The boundary between the long and short trajectories is taken to be the point where the recollision energy is maximised according to Eq. (10.8).

10.2.2 Recollision energy

The kinetic energy component U_{KE} of the recollision energy is plotted (in units of U_P) in Fig. 10.3. The solid black line (on the left) shows U_{KE} plotted as a function of the phase at ionisation. The graph shows a sharp initial rise in the energy, the peak at 3.17, and the subsequent fall to zero. The solid grey curve to the right is exactly the same data, but now plotted as a function of the phase *at recollision*, rather than at ionisation. Since electrons ionised early follow long trajectories and recollide late, the left-hand side of the black curve corresponds to the right-hand side of the grey curve, which is labelled 'long' for that reason. Similarly, the right-hand wing of the black curve and the left-hand wing of the grey curve are associated with the short trajectories. These two wings meet where the phase is $\pi/2$ (arrowed), which corresponds to the first zero of the field. Notice that ionisation times are spread over a quarter-period of the field, whereas the recollisions are spread over the remaining three-quarters of the cycle.

As explained in Section 10.1, when the electron returns to the parent atom at recollision, the available energy, made up of the acquired kinetic energy plus the ionisation energy, is emitted in the form of radiation at frequency $\nu = (U_{KE} + U_I)/h$. Written in terms of laser intensity, the kinetic energy at recollision is given by

$$U_{KE} = 3.17 \times \left(\frac{e^2}{8\pi^2 \varepsilon_0 c^3 m} \right) I\lambda^2 \tag{10.9}$$

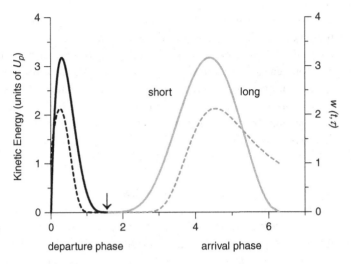

Fig. 10.3   The kinetic energy at recollision (in units of U_P) plotted as a function of the phase at ionisation ωt_0 (on the left) and the phase at recollision ωt_R (on the right). The dotted lines show the corresponding variation of the weighting function of Eq. (10.11).

where Eqs (10.3) and (10.8) have been used. In terms of intensity and wavelength, the corresponding number of harmonics is

$$H_{\text{KE}} = \frac{U_{\text{KE}}}{h\nu} = 2.39 \times 10^{-13}(\lambda(\mu\text{m}))^3 I(\text{W cm}^{-2}).   \tag{10.10}$$

An ionisation component (see Eq. 10.8) must of course be added to both these equations to obtain the corresponding values at recollision.

   Notice especially that the kinetic energy element of the recollision energy is proportional to the laser intensity, and so varying the intensity has a direct effect on the cut-off point. However, the range of viable intensities is quite limited. At the lower end, Eq. (10.10) indicates that, for $\lambda = 1$ μm, little of interest is likely to happen at intensities under about 5×10^{13} W cm^{-2}. On the other hand, an upper limit on the intensity is set by the point at which the atoms become fully ionised, and the efficiency of the recollision process is consequently seriously impaired. In practice, this limit is maybe 3×10^{15}W cm^{-2}. Incidentally, at even higher intensities, the Lorentz force associated with the magnetic component of the EM field would become significant. This would drive the electron along a curved path and, as for elliptically-polarised light, recollisions would be prevented. But this regime is well above the ionisation limit.

   Equations (10.9) and (10.10) also suggest that higher recollision energies can be realised by moving to higher wavelengths. This is a controversial issue at the time of writing because it seems that the advantage may be offset by a severe reduction in conversion efficiency as the wavelength is increased. At a lower fundamental frequency, the time interval between ionisation and recollision is lengthened, and this gives the electron wave packet more time

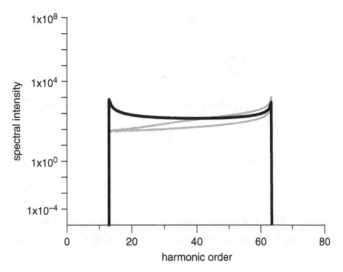

Fig. 10.4 Simulated emission spectra for a neon atom excited by 750 nm radiation at 5×10^{14} W cm^{-2}, under three different conditions: no weighting (black line), full weighting factor of Eq. (10.11) (upper grey line), ionisation weighting term only (lower grey line).

to disperse, reducing the strength of the recollision. Some degree of trade-off between the two effects may be possible but, as noted earlier, conversion efficiencies in HHG are already small, with best results in the region of 1 in 10^5.

10.2.3 High harmonic spectra

A rough idea of the overall emission spectrum can be gained by computing the emission frequency for a large number of electron trajectories spread over a quarter-period, and plotting the distribution of values. The result of this exercise for a neon atom excited by a 5×10^{14} W cm^{-2} beam at 750 nm is plotted as the black line in Fig. 10.4. At this frequency, the 21.56 eV ionisation potential of neon lies close to the thirteenth harmonic, while the kinetic energy gained in the continuum ($U_{\mathrm{KE}} = 1.33 \times 10^{-17}$ J) corresponds to roughly 50 harmonic orders from Eq. (10.10). The figure duly shows that the spectrum exhibits a broad plateau, starting at the thirteenth harmonic and running up to a sharp cut-off close to the sixty-third harmonic, which is at a wavelength of approximately 12 nm, in the XUV region of the EM spectrum.

There are several weaknesses in the argument as it stands, apart from the obvious one that it is essentially classical. First of all, it is based on the false assumption that the atom is equally likely to be ionised at all points within the quarter-cycle. In fact, the ionisation rate will inevitably drop as the magnitude of the field falls, so the contribution of the late departing electrons to the low-frequency side of the spectrum will certainly have been exaggerated in our analysis. However, this effect is offset to some extent in more sophisticated treatments by the fact that the quantum mechanical wave packet representing the electron in the continuum has more time to spread out on the long trajectories, and this weakens the

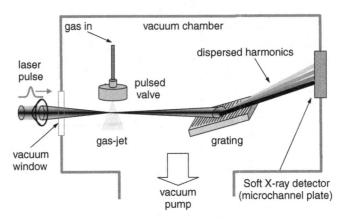

Fig. 10.5 Schematic diagram of the experimental set-up used to generate the data of Fig. 10.6, showing the grazing incidence flat-field diffraction grating and the soft X-ray sensitised microchannel plate detector. (Courtesy of J.W.G. Tisch, Imperial College Attosecond Laboratory.)

emission at recollision. The precise weighting functions that should be used to account for these effects is hotly debated, but one possible expression that takes both into account is [122,123]

$$w(t,\tau) = \underbrace{\left(\frac{T}{\tau}\right)^3}_{\text{spreading}} \underbrace{\sqrt{\frac{E(0)}{E(t)}} \exp\left\{-\frac{2(2U_I')^{3/2}}{3E'(t)}\right\}}_{\text{ionisation}}. \qquad (10.11)$$

In the first factor, which governs wave packet spreading, τ is the time the electron spends in the continuum and T is the optical period. Primes on the ionisation potential and the optical field in the final factor indicate that values in atomic units are to be used.[6]

The dotted lines in Fig. 10.3 show plots of Eq. (10.11) for the parameter values used for Fig. 10.4, while the grey lines in Fig. 10.4 show the effect of the weighting factor on the spectrum. The lower of the two shows the result when the spreading factor in the equation is omitted; as expected, weighting the distribution in this way has a stronger effect at lower frequencies.

A schematic diagram of a typical HHG experimental set-up is shown in Fig. 10.5 [72]. The fundamental laser beam is focused into a pulsed gas jet, and the resulting harmonics are dispersed by a grating onto a soft X-ray detector. All the equipment must be enclosed in a vacuum chamber in order for the gas jet to function properly, and of course to ensure that the harmonics are transmitted to the detector.

An experimental spectrum obtained under broadly similar conditions to those of Fig. 10.4 is shown in Fig. 10.6 [72]. This confirms the presence of a broad plateau of harmonics, extending to around the fifty-ninth harmonic in this case, after which there is a sharp cut-off. The main qualitative difference between the two figures is the appearance of discrete odd harmonics in Fig. 10.6, which arise from the periodic nature of the driving

---

[6] 1 a.u. of potential = 27.212 V, and 1 a.u. of field = $5.142 \times 10^{11}\,\mathrm{V\,m^{-1}}$.

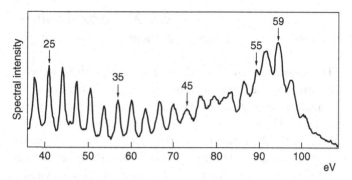

Fig. 10.6   Experimental HHG spectrum in neon excited by a 7 fs 250 μJ CEP stabilised Ti:sapphire laser pulse centred at 780 nm (average over 25 shots). See the caption to Fig. 10.5 for more details. (Courtesy of J.W.G. Tisch, Imperial College Attosecond Laboratory.)

field. By contrast, Fig. 10.4 was based on ionisations within a single quarter-cycle of the field. This point is discussed further in Section 10.4 below.

10.2.4 Phase matching

A further weakness in the model outlined so far is that it is based on the behaviour of a single atom, rather than on a spatial distribution of atoms. The propagation of harmonic radiation from individual source atoms to the detector, through the gas, and quite possibly through the focal region of the driving field too, needs to be taken into account in a fully-fledged analysis.

The phase-matching problem for second and third harmonic generation in a focused beam was treated in Section 2.5. There are of course now large numbers of harmonics in the analysis, rather than just one, although, as indicated in Table 2.1, the diffraction integrals are zero under normal dispersion for all harmonics higher than the second. Thin gas jets are used in HHG work for this reason. But there are two major complications in the HHG case. For one thing, the presence of ionised free electrons in the interaction region changes the local refractive index. But a more serious problem is that whereas, for harmonic generation in crystals, the phase of the polarisation antenna is linked directly to the phase of the relevant power of E, now the antenna phase is dependent on intensity, and it therefore varies *both spatially and temporally* within the interaction region. It may turn out to be possible to balance this effect against the variation of the Gouy phase, but this is certainly a difficulty that one could well do without.

A full discussion of these issues is beyond the scope of the present chapter, although an overall feature that emerges is that phase matching is harder to achieve for radiation from the long trajectories than from the short trajectories. There is therefore this further aspect to include in the balance between long- and short-trajectory emissions, in addition to ionisation weighting and wave-packet spreading.

For a wider discussion of HHG and related issues, readers are referred to [112,124].

10.3 Attosecond pulses

The sequence shown in Fig. 10.2 repeats every half-cycle of the field with alternating polarity, but the same emission spectrum. Two bursts of harmonic radiation are therefore emitted in each cycle of the field, one on the positive field excursion, and one on the negative. These are commonly known as 'half-cycle bursts'. The overall behaviour is shown in Fig. 10.7 where the emission frequency (in units of harmonic order) is plotted as a function of time (in cycles). The driving waveform is shown at the bottom for reference. The section of each burst where $w(t, \tau) > 0.1$ (see Eq. 10.11) is drawn as a solid line. The tails where the weighting factor is smaller are drawn dotted. The leftmost burst in the figure derives from electrons ionised in the first quarter-cycle of the field. A burst further to the left, generated by electrons ionised in the quarter-cycle between $-\frac{1}{2}$ and $-\frac{1}{4}$ is not shown.

The key question that now arises is: what kind of intensity profile is associated with the time-dependent spectra of Fig. 10.7? Clearly there will be some kind of high harmonic pulse train, but what will be the form of its individual components? The overall bandwidth, covering a range up to and beyond the sixtieth harmonic, is certainly sufficient to support pulses as short as 30 as (1 as $= 10^{-18}$s), provided the phases can be properly organised, but how can that be done? While more sophisticated theoretical models are needed to answer this question definitively, some aspects of the classical picture are preserved in the various quantum mechanical treatments. For example, Fig. 10.7 implies (correctly) that, if attention is focused on a band of harmonics, between (say) order 40 and 50, this will receive two distinct contributions in each half-cycle, an early burst from the short trajectories to the left of each peak, and a later burst from the long trajectories to the right. The figure also

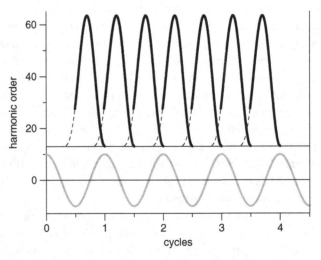

Fig. 10.7   Emission frequency as a function of time for neon excited by 750 nm CW laser radiation at an intensity of 5×10^{14} W cm^{-2} showing the 'half-cycle bursts'. Frequency is plotted in units of harmonic order. Sections of the spectral profile where the weighting factor of Eq. (10.11) is less than 0.1 are shown dotted.

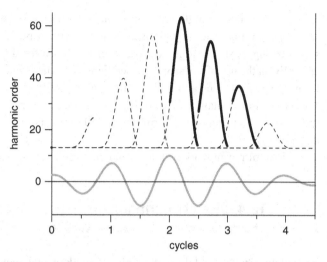

As Fig. 10.7 but for a pulse containing only two optical cycles (full width at half maximum intensity). The peak
intensity is 5×10^{14} W cm^{-2}.

suggests that the pulses are likely to be chirped, since the frequency obviously rises with
time before each peak, and falls with time subsequently.

Once a train of attosecond pulses (i.e. pulses significantly shorter than 1 fs) has been
created, the next question is how to isolate a single pulse from the train. A number of
strategies have been devised for this purpose, the obvious first step being to use a pulsed
rather than a monochromatic fundamental field to drive the HHG process. Figure 10.8
shows a typical prediction from the classical model for a fundamental pulse with an intensity
FWHM of two optical cycles. The first thing to notice about Fig. 10.8 is its asymmetry. As
in Fig. 10.7, only those parts of the profiles where $w(t, \tau) > 0.1$ are drawn as solid lines
and, whereas the first three bursts can be ignored on this criterion, the corresponding bursts
on the right contain solid sections. The main reason for the skewing is that the weighting
factor depends on the field *at the time of ionisation*, whereas the spectra are based on the
electron kinetic energy at the time of recollision which may be up to one period later.[7]

The second noteworthy aspect of Fig. 10.8 is that significant harmonic content is confined
to just three half-cycles; the spectrum extends beyond the fortieth harmonic on only two
of these, and beyond the sixtieth on only one. This feature is quite easy to understand
from Eq. (10.8) once U_P (defined in Eq. 10.3) is recognised as time dependent. It certainly
suggests that using a few-cycle fundamental pulse is a valuable first step.

Further refinement is possible by tuning the carrier-envelope phase (CEP), although a
potential complication is that the CEP changes as a few-cycle pulse passes through a focus
as a result of the Gouy phase; see Section 2.5. Various other tactics for creating single
attosecond pulses have been explored. For example, a high-pass spectral filter with a cut-
off around the fiftieth harmonic would deliver the highest harmonics in the central burst

---

[7] Another consequence of pulsed excitation is that on the leading edge where the pulse amplitude is rising, the
bursts are incomplete, even when the weighting factor is ignored.

of Fig. 10.8, while strongly attenuating the other bursts in the sequence. Another effective strategy is to modulate the polarisation of the driving pulse in a technique called polarisation gating. A simple argument has already been offered to suggest that HHG depends critically on the use of linearly polarised light. So if the polarisation can be changed rapidly from elliptical to plane and back to elliptical, a powerful switch can be created. Pulses of 130-as duration containing 1.2 optical cycles, and with spectra extending to 36 eV (corresponding to the twenty-third harmonic of 750 nm radiation) have been achieved using this technique [125].

Reviews of recent work on attosecond pulse generation can be found in [13, 111, 124].

10.4 Discrete harmonics in high harmonic generation

We end this chapter, and the book, with two fundamental questions about HHG. The first is this. Why does Fig. 10.4 show a broad continuous spectrum with no evidence of individual harmonics, while Fig. 10.6 clearly indicates the presence of discrete odd harmonics? Given the way in which Fig. 10.4 was created, there is of course no way it could have been been any different. A continuous range of electron trajectories was used to generate a continuous spread of recollision energies, which were divided by Planck's constant to create a continuous spectrum.

Where then does the discreteness in Fig. 10.6 come from? The answer lies in Fourier analysis. A spectrum containing regularly spaced discrete components is necessarily associated with some degree of temporal periodicity, so a single half-cycle burst can never deliver a spectrum in which individual harmonic components are identifiable. Conversely, the appearance of discrete spectral components in Fig. 10.6 indicates that two or more half-cycle bursts are involved [111].

A second issue now arises. The fact that the discrete components in Fig. 10.6 correspond to *odd* harmonics is of course associated with the centrosymmetric nature of the gaseous medium. However, the figure is not particularly convincing on this score. The harmonics are reasonably well separated up to about the forty-fifth harmonic but, from there on, the spectrum becomes more continuous in nature.

One possible reason for this is simply that the spectral resolution is limited. However, only a single half-cycle burst in Fig. 10.8 extends to the sixtieth harmonic, so one should not expect to see significant discreteness at the high end of the spectrum anyway.[8] But that argument also suggests that even harmonics really can be generated in HHG experiment, which raises the further question of how this can be so without violating basic symmetry principles.

The answer is simply that few-cycle driving pulses possess their own inherent 'one-wayness' and hence themselves lack inversion symmetry. For example, the two-cycle pulse used to create Fig. 10.8, and shown at the base of that figure, is certainly different in the

---

[8] The fact that there is some discreteness in Fig. 10.6 up to the highest harmonic is probably because the driving pulse was slightly longer than two optical cycles and/or the CEP was different from that in Fig. 10.8.

positive and negative directions; the central maximum is at $+10$ (with respect to the left-hand scale), while the two neighbouring negative peaks are at -9.1. In fact, it is this very feature that has created only one burst in Fig. 10.8 with spectral content beyond the sixtieth harmonic.

With regard to this aspect of nonlinear optics, we have therefore come full circle in this book. At the beginning of Chapter 1, it was stressed that a non-centrosymmetric nonlinear medium was essential for generating even harmonics, and that amorphous solids, liquids and gases were ruled out for symmetry reasons. Now, at the end of Chapter 10, we have discovered an exception to the rule. But there is, of course, no violation of fundamental symmetry principles, merely a new source of asymmetry. It turns out that even harmonics can be generated in a centrosymmetric medium provided a 'non-centrosymmetric' optical pulse is used to excite the nonlinearity.

Problems

10.1 Show that when a one-dimensional Coulomb potential is modified in the presence of an optical field (see Fig. 10.1), the potential barrier is lowered by an amount that is proportional to the inverse square root of the field. (But note that a proper treatment of ionisation suggests a far stronger dependence on field; see Eq. 10.11.)

10.2 Show that if the electron and the ion are treated as point particles, even the slightest degree of ellipticity in the polarisation will prevent a recollision

10.3 If ionisation occurs while the field magnitude is rising (i.e. in a quarter-cycle preceding an extremum), show that the electron will never return to the parent atom.

10.4 On the basis of the classical model, estimate the maximum excursion of the electron from the parent ion if the fundamental wavelength is 800 nm and the laser intensity is 10^{15} W cm^{-2}.

10.5 In Fig. 10.2, show that the tangent to the electric field curve $E(t) = A \cos \omega t$ at the time of ionisation intersects $E(t)$ at the time of recollision.

10.6 In an HHG experiment, He atoms are excited by 800 nm Ti:sapphire laser pulses with a peak intensity of 2×10^{15} W cm^2. On the basis of the classical model developed in this chapter, what values would you predict for (a) the highest harmonic and (b) the minimum wavelength. [Ionisation potential of helium $= 24.58$ eV.]

Appendix A Conventions and units

A1 Introduction

The formulae governing nonlinear frequency mixing appear in a bewildering variety of forms in the literature. This is caused by the use of different unit systems, different conventions for defining complex field amplitudes, and different sign conventions in wave propagation. To make matters worse, apparent discrepancies (often by factors of 2) have sometimes turned into real discrepancies (errors) as formulae have been copied without allowing for the conventions in use when they were originally written. The purpose of this appendix is therefore to restate the conventions used in this book, and to attempt to reconcile standard results quoted in different sources.

Many of the issues can be traced back to the different ways of defining the complex envelopes of real variables. We start by defining electric fields in the general form[1]

$$E(z,t) = p \sum_n \left[\hat{E}_n(z,t) \exp\{i\sigma_e \omega_n t\} + \text{c.c.} \right]$$

$$= p \sum_n \left[\tilde{E}_n(z,t) \exp\{i\sigma_e(\omega_n t - k_n z)\} + \text{c.c.} \right] \tag{A1.1}$$

where $\hat{E}_n = \tilde{E}_n \exp\{-ik_n z\}$, and the envelope function $\tilde{E}_n$ is slowly varying in both time and space. Analogous formulae apply for the polarisation (see Eq. 2.5) and for the associated magnetic field variables.

The parameter $\sigma_e \ (= \pm 1)$ in Eq. (A1.1) allows for a choice of sign in the space-time coefficient of the exponential, while the pre-factor p can be either $\frac{1}{2}$ or 1. The rationale for $p = \frac{1}{2}$ is that the moduli of the complex fields are then the same as the amplitudes of the real signal. On the other hand, setting $p = 1$ makes Eq. (A1.1) marginally neater, and simplifies the pre-factors in several common formulae in nonlinear optics at the same time.

The values of p and σ_e work their way through into all subsequent equations. The choices used by different authors are listed in columns 2 and 5 of Table A1. The table also includes further parameters defined in the next section.

---

[1] Equation (2.4) is the special case where $p = \frac{1}{2}$ and $\sigma_e = +1$.

Table A1 Conventions used in different sources.								
	p	q	γ	σ_e	σ_m	$\sigma \equiv \sigma_e \sigma_m$	$f = 2p\gamma q\sigma_e$	
Boyd [1] (2nd edn)	1	1	4π	-1	-1	1	-8π	
Boyd [1] (3rd edn)	1	ε_0	ε_0^{-1}	-1	-1	1	-2	
Sutherland [126]	1	$K = q\gamma$		-1	-1	1	$-2K$	note 1
Yariv [20] (3rd edn)	$\frac{1}{2}$	1	ε_0^{-1}	1	1	1	ε_0^{-1}	note 2
This book	$\frac{1}{2}$	ε_0	ε_0^{-1}	1	1	1	1	

Note 1: Sutherland, who uses the symbol K for our $q\gamma$, quotes results in both SI ($K = 1$) and Gaussian units ($K = 4\pi$).

Note 2: Yariv uses the symbol d_{eff} for our $\varepsilon_0 d_{\text{eff}}$. This means that an extra ε_0 appears in the denominator of his coupled-wave equations, although the equations are actually identical to ours.

A2 The nonlinear polarisation

We take the case of sum and difference frequency generation as our basic example. As explained in Chapter 2, the frequencies of the three interacting fields are linked by the equation

$$\omega_1 + \omega_2 \rightleftharpoons \omega_3. \tag{A2.1}$$

In the case of parametric amplification, which is governed by the same set of differential equations, the wave at ω_3 is the pump, and those at ω_2 and ω_1 are the signal and the idler. The signal frequency is normally the higher of the two, although this is not prescriptive.

In terms of the effective nonlinear coefficient d_{eff} (see Section 4.5), the equations for the polarisations are

$$\hat{P}(\omega_3) = 4pq d_{\text{eff}} \hat{E}(\omega_2) \hat{E}(\omega_1);$$

$$\hat{P}(\omega_2) = 4pq d_{\text{eff}} \hat{E}(\omega_3) \hat{E}^*(\omega_1); \quad \hat{P}(\omega_1) = 4pq d_{\text{eff}} \hat{E}(\omega_3) \hat{E}^*(\omega_2). \tag{A2.2}$$

In Gaussian units, $q = 1$, P and E are both in statvolts per cm, and hence d_{eff} is in cm per statvolt.[2] In SI units, it is common to write $q = \varepsilon_0$, in which case d_{eff} is in metres per volt (or more often pm V^{-1}). However, some authors (e.g. Yariv [20]) use SI units, but set $q = 1$ and absorb ε_0 into d_{eff}. Equation (A2.2) then becomes

$$\hat{P}(\omega_3) = 4p\bar{d}_{\text{eff}} \hat{E}(\omega_2) \hat{E}(\omega_1), \text{etc.} \tag{A2.3}$$

where the modified coefficient $\bar{d}_{\text{eff}} = \varepsilon_0 d_{\text{eff}}$ is in coulombs per volt (C V^{-1}).[3] Conversion factors between the various unit systems are listed in Table A2.

For a type I interaction in KDP, for example, d_{eff} is given by

$$d_{\text{eff}} = d_{36} \sin \theta \sin 2\phi \tag{A2.4}$$

[2] 1 statvolt = 300 V.
[3] Yariv writes $\bar{d}_{\text{eff}}$ as d_{eff}, so his equations appear superficially to be different from ours.

Table A2	Conversion factors linking Gaussian and SI Units.	
Multiply value in	**by**	**to get value in**
cm/statvolt (esu)	$(4\pi\varepsilon_0/3) \times 10^{-4} = 3.709 \times 10^{-15}$	C/V
cm/statvolt (esu)	$(4\pi/3) \times 10^8 = 4.189 \times 10^8$	pm/V
C/V	$(3/4\pi\varepsilon_0) \times 10^4 = 2.696 \times 10^{14}$	cm/statvol (esu)
C/V	$10^{12}/\varepsilon_0 = 1.129 \times 10^{23}$	pm/V
pm/V	$(3/4\pi) \times 10^{-8} = 2.387 \times 10^{-9}$	cm/statvolt (esu)
pm/V	$\varepsilon_0 \times 10^{-12} = 8.854 \times 10^{-24}$	C/V

where θ is the phase-matching angle, ϕ is the azimuthal angle, and $d_{36} \cong 0.43$ pm V^{-1} in the SI system; see Eq. (4.18).

A3 Coupled-wave equations

The general form of the coupled-wave equation of Eq. (2.8) is

$$\frac{\partial \tilde{E}_n}{\partial z} = -i(\gamma\sigma_e)\frac{\omega_n}{2cn_n}\tilde{P}_n^{\mathrm{NL}} \tag{A3.1}$$

where $\gamma = 4\pi$ in Gaussian units or ε_0^{-1} in SI. Using Eq. (A2.2) for the nonlinear polarisation, and recalling that $\hat{E}_n = \tilde{E}_n \exp\{-ik_nz\}$ and $\hat{P}_n = \tilde{P}_n \exp\{-ik_nz\}$, we obtain the coupled-wave equations in the general form

$$\frac{\partial \tilde{E}_3}{\partial z} = -if\left(\frac{\omega_3}{n_3 c}\right) d_{\mathrm{eff}} \tilde{E}_2 \tilde{E}_1 \exp\{i\sigma \Delta kz\} \tag{A3.2a}$$

$$\frac{\partial \tilde{E}_2}{\partial z} = -if\left(\frac{\omega_2}{n_2 c}\right) d_{\mathrm{eff}} \tilde{E}_3 \tilde{E}_1^* \exp\{-i\sigma \Delta kz\} \tag{A3.2b}$$

$$\frac{\partial \tilde{E}_1}{\partial z} = -if\left(\frac{\omega_1}{n_1 c}\right) d_{\mathrm{eff}} \tilde{E}_3 \tilde{E}_2^* \exp\{-i\sigma \Delta kz\} \tag{A3.2c}$$

where $f = 2p\gamma q\sigma_e$, and n_3, n_2 and n_1 are the respective refractive indices. The parameter σ in the exponential coefficients is the product $\sigma_e\sigma_{\mathrm{m}}$, where σ_{m} $(= \pm1)$ determines the sign convention used in defining the mismatch parameter

$$\Delta k = \sigma_{\mathrm{m}}(k_3 - k_2 - k_1) \tag{A3.3}$$

in which $k_3 = n_3\omega_3/c$, etc.

The equation for intensity (power per unit area) is

$$I = \frac{2p^2 nc |E|^2}{\gamma} \tag{A3.4}$$

where n is the refractive index. Notice the presence of p in this equation.

The conventions used by different authors for the further parameters introduced in this section are included in Table A1.

Appendix B Linear and nonlinear susceptibilities in the time and frequency domains

B1 Non-instantaneous response: the time and frequency domain equations

In this appendix, the structure of the polarisation expansion that has appeared at various levels of sophistication in this book will be examined in more detail. The introductory treatment of Chapter 1 was based on the simple statement of Eq. (1.24), namely

$$P = \varepsilon_0(\chi^{(1)}E + \chi^{(2)}E^2 + \chi^{(3)}E^3 + \cdots). \tag{B1.1}$$

The tensor nature of the nonlinear coefficients has been ignored. The polarisation, P, and electric field, E, are real variables in $C\,m^{-2}$ and $V\,m^{-1}$, respectively, which implies that the nth-order susceptibilities $\chi^{(n)}$ are in $(m\,V^{-1})^{n-1}$, or more conveniently in $(pm\,V^{-1})^{n-1}$.

In Chapter 1, the frequency dependence of the nonlinear coefficients was ignored. The coefficients in Eq. (B1.1) are therefore simple constants, which means that the polarisation responds instantaneously to changes in the field. This property is never realised in actual physical systems, and more sophisticated mathematical machinery is needed to describe what happens in practice.

The following sections provide a brief review of what this involves. The discussion will also address an issue surrounding equations such as Eq. (4.5), namely

$$\hat{P}_i(2\omega) = \tfrac{1}{2}\varepsilon_0 \sum_{jk} \chi_{ijk}^{\mathrm{SHG}}(2\omega;\; \omega, \omega)\hat{E}_j(\omega)\hat{E}_k(\omega). \tag{B1.2}$$

In this, the defining equation for second harmonic generation, the polarisation, the fields and the coefficient are all written as functions of frequency, which immediately raises the question of whether this is a time-domain or a frequency-domain equation. As we will show below, the equation is in reality a hybrid form. On the one hand, $\hat{E}_j(\omega)$ retains its normal units of $V\,m^{-1}$, and represents the electric field in a narrow frequency band around ω. (If $\hat{E}_j(\omega)$ were the Fourier transform of the time-dependent field, it would of course be in $V\,m^{-1}\,Hz^{-1}$.) But if Eq. (B1.2) is not a frequency-domain equation, it is not a true time-domain equation either. Equation (B3.1) below shows what that would look like.

B2 The linear term

We focus on the linear term, because it is the easiest to handle mathematically and the principles extend to the higher-order terms. To take account of non-instantaneous response, the first term on the right-hand side of Eq. (B1.1) should be replaced by

$$P^{(1)}(t) = \varepsilon_0 \int_{-\infty}^{\infty} \rho_1(\tau) E(t - \tau) d\tau. \tag{B2.1}$$

The equation specifies the polarisation of the medium at time t in terms of the electric field at all previous times $t-\tau$, etc. through the response function ρ_1, which contains all the information about the material properties, and must be zero for $\tau < 0$ to preserve causality. The response function is properly a tensor quantity, but we ignore this feature here for the sake of simplicity.[1]

The connection between non-instantaneous response and the frequency dependence of the linear susceptibility can be established by introducing $\bar{E}(\omega)$, which is the Fourier transform of $E(t)$, given by $E(t) = \int_{-\infty}^{\infty} \bar{E}(\omega) e^{i\omega t} d\omega$. Equation (B2.1) can now be written

$$P^{(1)}(t) = \varepsilon_0 \int_{-\infty}^{\infty} d\omega \left(\int_{-\infty}^{\infty} d\tau \rho_1(\tau) e^{-i\omega\tau} \right) \bar{E}(\omega) e^{i\omega t} = \varepsilon_0 \int_{-\infty}^{\infty} d\omega \chi^{(1)}(\omega) \bar{E}(\omega) e^{i\omega t} \tag{B2.2}$$

where the final step defines the frequency-dependent coefficient

$$\chi^{(1)}(\omega) = \int_{-\infty}^{\infty} d\tau \rho_1(\tau) e^{-i\omega\tau}. \tag{B2.3}$$

If we also introduce the Fourier transform of $P^{(1)}(t)$ and match terms, we obtain

$$\bar{P}(\omega) = \varepsilon_0 \chi^{(1)}(\omega) \bar{E}(\omega) \tag{B2.4}$$

which is the frequency-domain analogue of Eq. (B2.1). The form of this equation is similar to that of Eq. (4.1), which reads

$$\hat{P}(\omega) = \varepsilon_0 \chi^{(1)}(\omega) \hat{E}(\omega). \tag{B2.5}$$

However, the key difference between Eqs (B2.4) and (B2.5) is that $\bar{P}(\omega)$ and $\bar{E}(\omega)$ are the Fourier transforms of $P(t)$ and $E(t)$, whereas $\hat{P}(\omega)$ and $\hat{E}(\omega)$ are the amplitudes of quasi-monochromatic components of the polarisation and electric field; see Eqs (2.4) and (2.5) in Chapter 2. As we shall now show, these are different things; indeed the corresponding parameters do not even have the same dimensions!

---

[1] Refer to Chapter 4 and Appendix C for detailed discussion on the tensor nature of the coefficients.

In Eq. (2.4), the electric field was written

$$E(t) = \tfrac{1}{2} \sum_n \hat{E}(\omega_n) e^{i\omega_n t} + \text{c.c.} \tag{B2.6}$$

where $\hat{E}(-\omega_n) = \hat{E}^*(\omega_n)$ since $E(t)$ is real.[2] The Fourier transform of $E(t)$ is

$$\begin{aligned}
\bar{E}(\omega) &= \frac{1}{2\pi} \int\limits_{-\infty}^{\infty} dt \, \tfrac{1}{2} \sum_n \Big(\hat{E}(\omega_n) e^{i\omega_n t} + \text{c.c} \Big) e^{-i\omega t} \\
&= \tfrac{1}{2} \sum_n \Big(\hat{E}(\omega_n) \delta(\omega - \omega_n) + \hat{E}(-\omega_n) \delta(\omega + \omega_n) \Big)
\end{aligned} \tag{B2.7}$$

which is a set of delta functions, as expected. Substituting Eq. (B2.7) into Eq. (B2.2) leads to

$$P^{(1)}(t) = \varepsilon_0 \int\limits_{-\infty}^{\infty} d\omega \, \chi^{(1)}(\omega) \bar{E}(\omega) e^{i\omega t} = \tfrac{1}{2} \varepsilon_0 \sum_n \chi^{(1)}(\omega_n) \hat{E}(\omega_n) e^{i\omega_n t} + \text{c.c.} \tag{B2.8}$$

If we now write the polarisation as in Eq. (2.5), we have

$$P^{(1)}(t) = \tfrac{1}{2} \sum_n \hat{P}(\omega_n) e^{i\omega_n t} + \text{c.c.} \tag{B2.9}$$

Matching terms with Eq. (B2.8), we recover Eq. (B2.5).

Notice that $\hat{E}(\omega_n)$ is in $\mathrm{V\,m^{-1}}$, like $E^{(1)}(t)$, whereas $\bar{E}(\omega)$ is in $\mathrm{V\,m^{-1}\,Hz^{-1}}$ because it is linked to $E^{(1)}(t)$ via a Fourier integral. Similarly, $\hat{P}(\omega_n)$ is in $\mathrm{C\,m^{-2}}$, whereas $\bar{P}(\omega)$ is in $\mathrm{C\,m^{-2}\,Hz^{-1}}$. The parameters $E(t)$, $P^{(1)}(t)$ and $\rho^{(1)}(\tau)$ in Eqs (B2.1)–(B2.3) are real quantities, while $\bar{E}(\omega)$, $\bar{P}^{(1)}(\omega)$ and (potentially) $\chi^{(1)}(\omega)$ are complex.

In a medium with no memory (the case of instantaneous response), $\rho_1(\tau) = \chi^{(1)} \delta(\tau)$. The linear term then becomes $P^{(1)}(t) = \varepsilon_0 \chi^{(1)} E(t)$ in the time domain, or $\bar{P}^{(1)}(\omega) = \varepsilon_0 \chi^{(1)} \bar{E}(\omega)$ in the frequency domain, where $\chi^{(1)}$ is now a real constant that is independent of both time and frequency.

B3   The second-order nonlinear term

A similar procedure can be applied to the second-order nonlinear term. Written in an analogous form to Eqs (B2.1) and (B2.2), the results are

$$\begin{aligned}
P^{(2)}(t) &= \varepsilon_0 \int\limits_{-\infty}^{\infty} d\tau \int\limits_{-\infty}^{\infty} d\tau' \rho_2(\tau, \tau') E(t - \tau) E(t - \tau') \\
&= \varepsilon_0 \int\limits_{-\infty}^{\infty} d\omega \int\limits_{-\infty}^{\infty} d\omega' \chi^{(2)}(\omega, \omega') \bar{E}(\omega) \bar{E}(\omega') \exp\{i(\omega + \omega')t\}
\end{aligned} \tag{B3.1}$$

---

[2] $\hat{E}(\omega_n)$ was written $\hat{E}_n$ in Eq. (2.4).

where $\rho_2(\tau, \tau') = 0$ for $\tau, \tau' < 0$ (to ensure causality) and

$$\chi^{(2)}(\omega, \omega') = \int_{-\infty}^{\infty} d\tau \int_{-\infty}^{\infty} d\tau' \rho_2(\tau, \tau') e^{-i(\omega\tau + \omega'\tau')}. \tag{B3.2}$$

Once again, if the field spectrum is made up of delta functions (as in Eq. B3.2), the frequency integrals in Eq. (B3.1) reduce to summations, and we obtain

$$P^{(2)}(t) = \tfrac{1}{4}\varepsilon_0 \sum_{n,n'} \chi^{(2)}(\omega_n, \omega_{n'}) \hat{E}(\omega_n) \hat{E}(\omega_{n'}) \exp\{i(\omega_n + \omega_{n'})t\}. \tag{B3.3}$$

If $P^{(2)}(t)$ is now written in the form of Eq. (B2.5), we obtain terms of the type

$$\hat{P}^{(2)}(\omega_n + \omega_{n'}) = \tfrac{1}{2}\varepsilon_0 \chi^{(2)}(\omega_n, \omega_{n'}) \hat{E}(\omega_n) \hat{E}(\omega_{n'}). \tag{B3.4}$$

This result is the basis of the sum frequency generation formula of Eq. (4.3), which reads

$$\hat{P}_i(\omega_3) = \tfrac{1}{2}\varepsilon_0 \sum_{jk} \left[\chi^{(2)}_{ijk}(\omega_3; \, \omega_1, \omega_2) \hat{E}_j(\omega_1) \hat{E}_k(\omega_2) + \chi^{(2)}_{ijk}(\omega_3; \omega_2, \omega_1) \hat{E}_j(\omega_2) \hat{E}_k(\omega_1) \right].$$

$$\tag{B3.5}$$

The notation in these last two formulae is slightly different, mainly because the second includes the tensor nature of the coefficients and the first does not. Also, in line with convention, the susceptibilities in Eq. (B3.5) carry three frequency arguments rather than two.

Appendix C Definition of the nonlinear coefficients

C1 Introduction

The defining equations for nonlinear optical coefficients are inherently intricate and the complexity is worsened by the use of different conventions by different authors. A perennial problem is that the equations are huge when written out in full, but opaque when shortened by the use of summations and other notational devices.

This appendix contains a blow-by-blow discussion of the most important definitions, with a selection of examples and special cases that will hopefully serve to clarify the general equations.

C2 Second-order processes

C2.1 Sum frequency generation and intrinsic permutation symmetry

The nonlinear coefficient for the basic second-order process of sum frequency generation (SFG) is defined by[1]

$$\hat{P}_i(\omega_3) = \tfrac{1}{2}\varepsilon_0 \sum_p \sum_{jk} \chi_{ijk}^{(2)}(\omega_3; \omega_1, \omega_2)\hat{E}_j(\omega_1)\hat{E}_k(\omega_2) \tag{C2.1}$$

where $\omega_1 + \omega_2 = \omega_3$, and the indices ijk can each be x, y or z. The symbol $\sum_p$ indicates that the right-hand side should be summed over all distinct permutations of ω_1 and ω_2. So, provided $\omega_1 \neq \omega_2$, terms with the two frequencies in reverse order should be included. This feature has its origin in the trivial fact that when one works out $(E(\omega_1) + E(\omega_2))^2$, the cross-term is $E(\omega_1)E(\omega_2) + E(\omega_2)E(\omega_1) = 2E(\omega_1)E(\omega_2)$.

When the sum over permutations is carried out, Eq. (C2.1) becomes

$$\hat{P}_i(\omega_3) = \tfrac{1}{2}\varepsilon_0 \sum_{jk} \left(\chi_{ijk}^{(2)}(\omega_3; \omega_1, \omega_2)\hat{E}_j(\omega_1)\hat{E}_k(\omega_2) + \chi_{ijk}^{(2)}(\omega_3; \omega_2, \omega_1)\hat{E}_j(\omega_2)\hat{E}_k(\omega_1) \right)$$

$$= \tfrac{1}{2}\varepsilon_0 \sum_{jk} \left(\chi_{ijk}^{(2)}(\omega_3; \omega_1, \omega_2) + \chi_{ikj}^{(2)}(\omega_3; \omega_2, \omega_1) \right)\hat{E}_j(\omega_1)\hat{E}_k(\omega_2) \tag{C2.2}$$

---

[1] The pre-factor of $\tfrac{1}{2}$ arises from the field definitions of Eq. (2.4).

where the second line follows by exchanging the dummy indices j and k in the second term of the first line. Since $\chi_{ijk}(\omega_3;\omega_1,\omega_2)$ and $\chi_{ikj}(\omega_3;\omega_2,\omega_1)$ are associated with the same pair of fields, there is no way of distinguishing their effect, and any measurement will record the sum of the two. It therefore makes sense to contract Eq. (C2.2), which can be done in two slightly different ways. One possibility is to write

$$\hat{P}_i(\omega_3) = \varepsilon_0 \sum_{jk} \bar{\chi}_{ijk}^{(2)}(\omega_3;\omega_1,\omega_2)\hat{E}_j(\omega_1)\hat{E}_k(\omega_2) \tag{C2.3}$$

where

$$\bar{\chi}_{ijk}^{(2)}(\omega_3;\omega_1,\omega_2) = \tfrac{1}{2}\left(\chi_{ijk}^{(2)}(\omega_3;\omega_1,\omega_2) + \chi_{ikj}^{(2)}(\omega_3;\omega_2,\omega_1)\right) \tag{C2.4}$$

and the overbar indicates that an average has been taken. The other possibility is to invoke the convention known as *intrinsic permutation symmetry* (IPS) and assume that

$$\chi_{ijk}^{(2)}(\omega_3;\omega_1,\omega_2) = \chi_{ikj}^{(2)}(\omega_3;\omega_2,\omega_1). \tag{C2.5}$$

In this case Eq. (C2.2) becomes

$$\hat{P}_i(\omega_3) = \varepsilon_0 \sum_{jk} \chi_{ijk}^{(2)}(\omega_3;\omega_1,\omega_2)\hat{E}_j(\omega_1)\hat{E}_k(\omega_2) \tag{C2.6}$$

which is identical to Eq. (C2.3) apart from the overbar. The key point about both Eq. (C2.3) and Eq. (C2.6) is that, one way or another, coefficients with ω_1 and ω_2 in reverse order have been eliminated, and the ordering is now fixed by definition.

In some ways, the averaging approach is preferable to the imposition of IPS. An arbitrary assumption is avoided, and the overbar on the coefficient in Eqs (C2.3) and (C2.4) serves as a reminder that the frequency ordering is fixed. However, it is common practice to use IPS, and so we will do so here. Reminders that the frequency ordering in Eqs (C2.3), (C2.6) and similar equations is fixed by definition are included where appropriate.

As an example, consider the case where $i = x$ and $jk = yz, zy$. Equation (C2.2) then reads

$$\hat{P}_x(\omega_3) = \tfrac{1}{2}\left(\chi_{xyz}^{(2)}(\omega_3;\omega_1,\omega_2)\hat{E}_y(\omega_1)\hat{E}_z(\omega_2) + \chi_{xyz}^{(2)}(\omega_3;\omega_2,\omega_1)\hat{E}_y(\omega_2)\hat{E}_z(\omega_1)\right)$$

$$+ \tfrac{1}{2}\left(\chi_{xzy}^{(2)}(\omega_3;\omega_1,\omega_2)\hat{E}_z(\omega_1)\hat{E}_y(\omega_2) + \chi_{xzy}^{(2)}(\omega_3;\omega_2,\omega_1)\hat{E}_z(\omega_2)\hat{E}_y(\omega_1)\right) \tag{C2.7}$$

which can be rearranged as

$$\hat{P}_x(\omega_3) = \tfrac{1}{2}\left(\chi_{xyz}^{(2)}(\omega_3;\omega_1,\omega_2) + \chi_{xzy}^{(2)}(\omega_3;\omega_2,\omega_1)\right)\hat{E}_y(\omega_1)\hat{E}_z(\omega_2)$$

$$+ \tfrac{1}{2}\left(\chi_{xzy}^{(2)}(\omega_3;\omega_1,\omega_2) + \chi_{xyz}^{(2)}(\omega_3;\omega_2,\omega_1)\right)\hat{E}_z(\omega_1)\hat{E}_y(\omega_2). \tag{C2.8}$$

By assuming that $\chi_{xyz}^{(2)}(\omega_3;\omega_1,\omega_2) = \chi_{xzy}^{(2)}(\omega_3;\omega_2,\omega_1)$ and $\chi_{xzy}^{(2)}(\omega_3;\omega_1,\omega_2) = \chi_{xyz}^{(2)}(\omega_3;$ $\omega_2,\omega_1)$ under IPS, Eq. (C2.8) reduces to

$$\hat{P}_x(\omega_3) = \varepsilon_0 \left(\chi_{xyz}^{(2)}(\omega_3;\omega_1,\omega_2)\hat{E}_y(\omega_1)\hat{E}_z(\omega_2) + \chi_{xzy}^{(2)}(\omega_3;\omega_1,\omega_2)\hat{E}_z(\omega_1)\hat{E}_y(\omega_2) \right).$$

(C2.9)

The corresponding result for $i = x$ and $jk = yy$ is

$$\hat{P}_x(\omega_3) = \varepsilon_0 \chi_{xyy}^{(2)}(\omega_3;\omega_1,\omega_2)\hat{E}_y(\omega_1)\hat{E}_y(\omega_2)$$

(C2.10)

where $\chi_{xyy}^{(2)}(\omega_3;\omega_2,\omega_1)$ has been identified with $\chi_{xyy}^{(2)}(\omega_3;\omega_1,\omega_2)$.

C2.2 Second harmonic generation

The results of Section C2.1 have to be adjusted in the case of second harmonic generation. Provided $\omega_1 = \omega_2 = \omega$, and only a single beam at ω is involved (see Section C2.3 below), the summation $\sum_p$ is irrelevant and can be removed. Equation (C2.1) now reads

$$\hat{P}_i(2\omega) = \tfrac{1}{2}\varepsilon_0 \sum_{jk} \chi_{ijk}^{\text{SHG}}(2\omega;\omega,\omega)\hat{E}_j(\omega)\hat{E}_k(\omega).$$

(C2.11)

Once again, we treat the special case where $i = x$ and $jk = yz, zy$, in which case (C2.11) becomes

$$\hat{P}_x(2\omega) = \tfrac{1}{2}\varepsilon_0 \left(\chi_{xyz}^{\text{SHG}}(2\omega;\omega,\omega) + \chi_{xzy}^{\text{SHG}}(2\omega;\omega,\omega) \right) \hat{E}_y(\omega)\hat{E}_z(\omega).$$

(C2.12)

At this point it is conventional to make the replacement

$$2d_{x(yz)} = \tfrac{1}{2} \left(\chi_{xyz}^{\text{SHG}}(2\omega;\omega,\omega) + \chi_{xzy}^{\text{SHG}}(2\omega;\omega,\omega) \right)$$

(C2.13)

which leads to

$$\hat{P}_x(2\omega) = 2\varepsilon_0 d_{x(yz)}\hat{E}_y(\omega)\hat{E}_z(\omega) = 2\varepsilon_0 d_{14}\hat{E}_y(\omega)\hat{E}_z(\omega)$$

(C2.14)

where (yz) means that both yz and zy are included. The second step incorporates the numerical suffix notation that is normally introduced at this point. More details are given in Table 4.1; see also Eq. (4.10).

The factor of 2 in Eqs (C2.13) and (C2.14) is a historical artefact. It was introduced at some point in the history of the subject, probably as a way of wiping out the pre-factor of $\frac{1}{2}$ that has featured in most of the previous equations. As an example the corresponding result for $i = x$ and $jk = yy$ is

$$\hat{P}_x(2\omega) = \tfrac{1}{2}\varepsilon_0 \chi^{\text{SHG}}_{xyy}(2\omega; \omega, \omega)\hat{E}_y^2(\omega) = \varepsilon_0 d_{12}(2\omega; \omega, \omega)\hat{E}_y^2(\omega). \tag{C2.15}$$

C2.3 Note on non-collinear beams

A caveat needs to be entered at this point. It is important to understand that the distinction between ω_1 and ω_2 disappears in the case of second harmonic generation not merely because the two waves have the same frequency but *because they are derived from a single beam*. It follows that in the case of non-collinear SHG, when two separate beams are involved, albeit of the same frequency, the formulae of sum frequency generation should be used with $\omega_1 = \omega_2 = \omega$. Two distinct beams interact in this case, even if they were originally derived from the same laser source.

C2.4 The general second-order process

Further adjustments are needed if any of the participating frequencies is zero. The most inclusive form of the expression for the nonlinear polarisation is

$$\hat{P}_i(\omega_3) = \varepsilon_0 K_2(\omega_3; \omega_1, \omega_2)\sum_{jk}\chi^{(2)}_{ijk}(\omega_3; \omega_1, \omega_2)\hat{E}_j(\omega_1)\hat{E}_k(\omega_2) \tag{C2.16}$$

where $K_2(\omega_3; \omega_1, \omega_2) = \tfrac{1}{2}p_{12}\,2^{r-l}$. The three components in K_2 arise as follows:

- The initial factor of $\tfrac{1}{2}$ has its origin in the definition of the complex fields, and is the same $\tfrac{1}{2}$ that appeared in Eq. (C2.1).
- The factor p_{12} is the number of distinct permutations of ω_1 and ω_2; so $p_{12} = 2$ if the frequencies are different and 1 if they are the same. This factor is the direct result of the $\sum_p$ operation in Eq. (C2.1) and, once it is implemented, the ordering of the frequencies within the nonlinear coefficient is fixed.
- The parameters l and r are the numbers of zero frequencies to the left and right of the semicolon in the argument of the nonlinear coefficient.

Why does the number of zeros matter? The answer is that, whereas for a DC field, $E(0) = A$, for an AC field of the same amplitude, $A\cos\{\omega t + \phi\} = \tfrac{1}{2}\hat{E}(\omega)e^{i\omega t} + \text{c.c.}$ to ensure that $\left|\hat{E}(\omega)\right| = A$. As a result, the factor 2^{r-l} will appear in K_2. On the other hand, if ω_1, ω_2 and ω_3 are all non-zero, $r = l = 0$, and $K_2 = \tfrac{1}{2}p_{12}$, which links Eq. (C2.16) to Eqs (C2.6) and (C2.11).

Values of K_2, and the various components within it, are listed in Table C1 for four particular second-order processes. The key point to notice is that the values are exactly the same as those that appeared naturally in the simple treatment of Chapter 1. This is either reassuring, or a frustratingly poor return on the investment made in this appendix, depending on your point of view! The relevant equation numbers are listed in the right-hand column.

	ω_3	ω_1, ω_2	p_{12}	l	r	$K_2 = \frac{1}{2}p_{12}2^{r-l}$	Equations
Table C1 Pre-factors for second-order nonlinear processes.							
SFG	ω_3	ω_1, ω_2	2	0	0	1	(1.17) & (C2.6)
SHG	2ω	ω, ω	1	0	0	$\frac{1}{2}$	(1.8) & (C2.11)
Pockels	ω	$\omega, 0$	2	0	1	2	(1.14) & (C2.19)
OR	0	$\omega, -\omega$	2	1	0	$\frac{1}{2}$	(1.8) & (C2.17)

C2.5 Optical rectification and the Pockels effect

We now consider two further second-order examples involving DC fields.

For optical rectification (OR), $\omega_3 = 0$ and a suitable choice for the other two frequencies is $\omega_1 = \omega$ and $\omega_2 = -\omega$. Negative frequencies must be treated as distinct from positive frequencies, so $p_{12} = 2$, and since the polarisation is DC, $l = 1$ and $r = 0$. Hence $K_2 = \frac{1}{2}$, and Eq. (C2.16) becomes

$$P_i(0) = \frac{1}{2}\varepsilon_0 \sum_{jk} \chi_{ijk}^{\text{OR}}(0;\ \omega, -\omega)\hat{E}_j(\omega)\hat{E}_k^*(\omega) \qquad (C2.17)$$

where $\hat{E}_k^*(\omega) = \hat{E}_k(-\omega)$. We stress once again that the ordering of frequencies to the right of the semicolon in Eq. (C2.17) is a matter of definition, but the choice is arbitrary and we could equally have written

$$P_i(0) = \frac{1}{2}\varepsilon_0 \sum_{jk} \chi_{ijk}^{\text{OR}}(0;\ -\omega, \omega)\hat{E}_j^*(\omega)\hat{E}_k(\omega) \qquad (C2.18)$$

where the frequencies $+\omega$ and $-\omega$ have been reversed both in the coefficient and in the fields.

For the Pockels effect, $K_2 = 2$, and of the two options for the defining equation we choose

$$\hat{P}_i(\omega) = 2\varepsilon_0 \sum_{jk} \chi_{ijk}^{\text{PE}}(\omega;\ \omega, 0)\hat{E}_j(\omega)E_k(0). \qquad (C2.19)$$

C3 Third-order processes

C3.1 Summary of third-order results

Section C2.1 began with the equation for sum frequency generation. The parallel effect at third order is the process $\omega_4 = \omega_1 + \omega_2 + \omega_3$ where all four frequencies are non-zero. The defining equation, analogous to Eq. (C2.1), is

$$\hat{P}_i(\omega_4) = \frac{1}{4}\varepsilon_0 \sum_p \sum_{jkl} \chi_{ijkl}^{(3)}(\omega_4;\ \omega_1, \omega_2, \omega_3)\hat{E}_j(\omega_1)\hat{E}_k(\omega_2)\hat{E}_l(\omega_3) \qquad (C3.1)$$

	ω_3	$\omega_1,\omega_2,\omega_3$	p_{123}	l	r	$K_3 = \frac{1}{4}p_{123}\,2^{r-l}$	Equations
						Table C2 Pre-factors for third-order nonlinear processes.	
SFG	ω_4	$\omega_1,\omega_2,\omega_3$	6	0	0	$\frac{3}{2}$	(C3.1) with 6 permutations
THG	3ω	ω,ω,ω	1	0	0	$\frac{1}{4}$	(1.25)
DC SHG	2ω	$\omega,\omega,0$	3	0	1	$\frac{3}{2}$	(1.25)
DC KERR	ω	$\omega,0,0$	3	0	2	3	(1,25)
AC KERR	ω	$\omega,\omega',-\omega'$	6	0	0	$\frac{3}{2}$	(1.27)
IDRI	ω	$\omega,\omega,-\omega$	3	0	0	$\frac{3}{4}$	(1.25) & (C3.4)

where, as before, $\sum_p$ indicates that all distinct permutations of ω_1,ω_2 and ω_3 must be included. This multiplies the number of terms by 6 if all three frequencies are different, and by 3 when two are the same.

The most general third-order result, analogous to Eq. (C2.16) at second order, is

$$\hat{P}_i(\omega_4) = \varepsilon_0 K_3(\omega_4;\ \omega_1,\omega_2,\omega_3) \sum_{jkl} \chi^{(3)}_{ijkl}(\omega_4;\ \omega_1,\omega_2,\omega_3)\hat{E}_j(\omega_1)\hat{E}_k(\omega_2)\hat{E}_l(\omega_3) \quad (C3.2)$$

where $K_3(\omega_4;\ \omega_1,\omega_2,\omega_3) = \frac{1}{4}p_{123}\,2^{r-l}$ and p_{123} is the number of distinct permutations of ω_1,ω_2 and ω_3. The structure of K_3 is entirely analogous to that of K_2, which has been discussed in Section C2.4. The values of K_3 in a number of special cases are listed in Table C2 and, as for K_2, are precisely those that arose in the simple treatment of Chapter 1. We stress again that the summation $\sum_p$ in Eq. (C3.1) is an instruction to permute the frequencies to the right of the semicolon, and that the p_{123} factor in K_3 is the consequence of the permutation. As soon as p_{123} is present in the pre-factor, the ordering of the frequencies in the nonlinear coefficient is fixed, because the possible permutations have been taken into account.

C3.2 Example: intensity-dependent refractive index

As an example, we will consider intensity-dependent refractive index in detail. In this special case, Eq. (C3.1) reads

$$\hat{P}_i(\omega) = \frac{1}{4}\varepsilon_0 \sum_{jkl} \Big(\chi^{(3)}_{ijkl}(\omega;\ -\omega,\omega,\omega)\ \hat{E}_j(-\omega)\hat{E}_k(\omega)\hat{E}_l(\omega)$$
$$+ \chi^{(3)}_{ijkl}(\omega;\ \omega,-\omega,\omega)\hat{E}_j(\omega)\hat{E}_k(-\omega)\hat{E}_l(\omega) \qquad (C3.3)$$
$$+ \chi^{(3)}_{ijkl}(\omega;\ \omega,\omega,-\omega)\hat{E}_j(\omega)\ \hat{E}_k(\omega)\hat{E}_l(-\omega) \Big).$$

By invoking IPS, and juggling the indices, Eq. (C3.3) can be reduced to any of the following equivalent forms

$$(a)\ \hat{P}_i(\omega) = \tfrac{3}{4}\varepsilon_0 \sum_{jkl} \chi^{(3)}_{ijkl}(\omega;\ -\omega,\omega,\omega)\hat{E}_j(-\omega)\hat{E}_k(\omega)\hat{E}_l(\omega)$$

$$(b)\ \hat{P}_i(\omega) = \tfrac{3}{4}\varepsilon_0 \sum_{jkl} \chi^{(3)}_{ijkl}(\omega;\ \omega,-\omega,\omega)\hat{E}_j(\omega)\hat{E}_k(-\omega)\hat{E}_l(\omega) \tag{C3.4}$$

$$(c)\ \hat{P}_i(\omega) = \tfrac{3}{4}\varepsilon_0 \sum_{jkl} \chi^{(3)}_{ijkl}(\omega;\ \omega,\omega,-\omega)\hat{E}_j(\omega)\hat{E}_k(\omega)\hat{E}_l(-\omega).$$

These differ only in the position of the negative frequency in the sequence, and which one to use is entirely a matter of choice. We will adopt form (b) in which $\omega_2 = -\omega$.

In a structurally isotropic medium, it can be shown that coefficients with 60 of the 81 possible sets of *ijkl* are zero, leaving 21 non-zero combinations. These 21 sets include three of type *jjjj*, and 18 (six each) of types *jjkk*, *jkjk*, and *jkkj*. Invoking IPS and general symmetry principles shows that, within each type, all six members are equal, and that the types are also related by

$$\chi_{jjjj} = \chi_{jjkk} + \chi_{jkjk} + \chi_{jkkj} \tag{C3.5}$$

where $j \neq k$. There are therefore three independent elements.[2]

Further constraints apply for particular processes. Under form (b) in Eq. (C3.4), for example, the frequencies associated with the second and fourth indices are the same, and it follows that

$$\chi_{jkkj} = \chi_{jjkk} \neq \chi_{jkjk}. \tag{C3.6}$$

Notice that the *unequal* member of the set has been determined by the choice of form (b) in Eq. (C3.4). Had we chosen forms (a) or (c), Eq. (C3.6) would be changed.

With $i = x$, and for the case where $E_z = 0$, Eq. (C3.4b) reads

$$\hat{P}_x(\omega) = \tfrac{3}{4}\varepsilon_0 \left(\chi^{(3)}_{xxxx}\hat{E}_x(\omega)\left|\hat{E}_x(\omega)\right|^2 + \chi^{(3)}_{xyxy}\hat{E}_x^*(\omega)\hat{E}_y^2(\omega) + 2\chi^{(3)}_{xxyy}\hat{E}_x(\omega)\hat{E}_y^*(\omega)\hat{E}_y(\omega) \right)$$

$$\tag{C3.7}$$

where Eq. (C3.6) has been used. With further help from Eq. (C3.5), the equation can be written

$$\hat{P}_x(\omega) = \tfrac{3}{4}\varepsilon_0 \left(\chi^{(3)}_{xyxy}\left[\hat{E}_x^2(\omega) + \hat{E}_y^2(\omega)\right]\hat{E}_x^*(\omega) + 2\chi^{(3)}_{xxyy}\left[\left|\hat{E}_x(\omega)\right|^2 + \left|\hat{E}_y(\omega)\right|^2\right]\hat{E}_x(\omega) \right). $$

$$\tag{C3.8}$$

In the case where $\hat{E}_z \neq 0$, Eq. (C3.8) generalises to

$$\hat{P}_x(\omega) = \tfrac{3}{4}\varepsilon_0 \left(\chi^{(3)}_{xyxy}(\mathbf{E}.\mathbf{E})\hat{E}_x^* + 2\chi^{(3)}_{xxyy}(\mathbf{E}.\mathbf{E}^*)\hat{E}_x \right). \tag{C3.9}$$

---

[2] See Section 5.2.2 for details about the situation in cubic crystals.

Appendix D Non-zero d elements in non-centrosymmetric crystals

Table D1 contains information about which d elements are non-zero for each of the 21 non-centrosymmetric crystal classes. Column 3 lists the number of independent elements for each class and (in brackets) the lower number of independents when the Kleinmann symmetry condition (KSC) applies.

Detailed information is given in column 4 where semicolons separate independent elements. Elements that are equal to others under all circumstances (in value if not in sign) are so indicated, while elements grouped within brackets are identical if the KSC applies. In a few cases (e.g. class 24) $d_{14} = -d_{25}$, but both are zero under Kleinmann symmetry as indicated by the '= 0'. The rare class 29 is non-centrosymmetric, but other aspects of the symmetry force all elements of the d matrix to be zero.

Note that, in biaxial media, the mapping from the crystallographic (abc) to the physical (xyz) axes is not necessarily straightforward. It cannot be assumed that $xyz \rightarrow 123$, which many people would take for granted. For a detailed discussion, see Sections 4.4 and 4.6, as well as Chapter 2 of Dmitriev *et al.* [26].

Table D1 Non-zero d elements in non-centrosymmetric crystals.

System & class	Symmetry	Independents	Links and examples
		Biaxial	
Triclinic			
1	1	18	all elements are independent
Monoclinic			
3	2	8 (4)	(14, 25, 36); (16, 21); 22; (23, 34): e.g. NPLO
4	m	10 (6)	11; (12, 26); (13, 35); (15, 31); (24, 32): 33
Orthorhombic			
6	2 2 2	3 (1)	(14, 25, 36): e.g. barium formate
7	m m 2	5 (3)	(15, 31); (24, 32); 33
			e.g.LBO, KTP, KTA, KNbO$_3$, KB5
		Uniaxial	
Tetragonal			
9	$\bar{4}$	4 (2)	(14 = 25, 36); (15 = −24, 31 = −32)
10	4	4 (2)	(14 = −25, = 0); (15 = 24, 31 = 32); 33
11	$\bar{4}$ 2 m	2 (1)	(14 = 25, 36) e.g. KDP, ADP, CDA
12	4 2 2	1 (0)	(14 = −25, = 0) e.g. nickel sulphate

Table D1 (Continued).			
14	4 m m	3 (2)	$(15 = 24, 31 = 32)$; 33 e.g. potassium lithium niobate
Trigonal			
16	3	6 (4)	$(-11 = 12 = 26)$; $(14 = -25, = 0)$; $(15 = 24,$ $31 = 32)$ $(16 = 21 = -22)$; 33 e.g. sodium periodate
18	3 2	2 (1)	$(-11 = 12 = 26)$; $(14 = -25, = 0)$ e.g. α-quartz
19	3 m	4 (3)	$(15 = 24, 31 = 32)$; $(16 = 21 = -22)$; 33 e.g. BBO, $LiNbO_3$, proustite
Hexagonal			
21	$\bar{6}$	2	$(12 = 26 = -11)$; $(16 = 21 = -22)$
22	$\bar{6}$ 2 m	1	$(16 = 21 = -22)$ e.g. gallium selenide
23	6	4 (2)	$(14 = -25, = 0)$; $(15 = 24, 31 = 32)$; 33 e.g. Lithium iodate
24	6 2 2	1 (0)	$(14 = -25, = 0)$ e.g. β-quartz
26	6 m m	3 (2)	$(15 = 24, 31 = 32)$; 33 e.g. CdSe
Optically isotropic			
Cubic			
28	23	1	$(14 = 25 = 36)$ e.g. sodium chlorate
29	4 3 2	0	Non-centrosymmetric, but all element zero.
31	$\bar{4}$ 3 m	1	$(14 = 25 = 36)$

Appendix E   Real fields, complex fields, and the analytic signal

In Chapter 2, the real electric field was defined as the sum of a set of discrete frequency components namely

$$E(z,t) = \tfrac{1}{2}\sum_n \hat{E}_n \exp\{i\omega_n t\} + \text{c.c.} = \tfrac{1}{2}\sum_n \tilde{E}_n \exp\{i(\omega_n t - k_n z)\} + \text{c.c.} \qquad \text{(E1.1)}$$

Each term in the summation represents the component of the field in the vicinity of ω_n. The complex amplitudes $\hat{E}_n$ contain fast-varying spatial terms $\exp\{-ik_n z\}$, which are removed to form the slowly varying complex envelope function $\tilde{E}_n = \hat{E}_n \exp\{ik_n z\}$. Both $\hat{E}_n$ and $\tilde{E}_n$ vary slowly in time, or not at all.

In the case of a narrow-band signal in the vicinity of ω_0 and at a fixed point in space (say $z = 0$), we have

$$E(t) = \tfrac{1}{2}\hat{E}(t) \exp\{i\omega_0 t\} + \text{c.c.} = \tfrac{1}{2}V(t) + \text{c.c.} \qquad \text{(E1.2)}$$

where $V(t)$ is called the *analytic signal*. For a given real field, $V(t)$ is not uniquely defined, because its imaginary part can be changed at will without affecting $E(t)$ and, even if $V(t)$ were known, $\hat{E}(t)$ would not be uniquely defined either, because the choice of ω_0 is strictly arbitrary. The standard prescription for defining $V(t)$ is

$$V(t) = \int\limits_0^\infty 2\bar{E}(\omega) \exp\{i\omega t\}d\omega \qquad \text{(E1.3)}$$

where $\bar{E}(\omega)$ is the Fourier transform of $E(t)$ given by

$$\bar{E}(\omega) = \frac{1}{2\pi} \int\limits_{-\infty}^{\infty} E(t) \exp\{-i\omega t\}dt. \qquad \text{(E1.4)}$$

Notice that the definition of $V(t)$ in Eq. (E1.3) is based solely on the positive frequency components of $\bar{E}(\omega)$, and the factor of 2 has been included to ensure that $E(t) = Re\{V(t)\}$. This process involves no loss of information because the reality of $E(t)$ means that $\bar{E}(-\omega) = \bar{E}^*(\omega)$ in Eq. (E1.4).

It follows that the Fourier transform of $V(t)$ is

$$\bar{V}(\omega) = \begin{cases} \dfrac{1}{2\pi}\int\limits_{-\infty}^{\infty} V(t) \exp\{-i\omega t\}dt = 2\bar{E}(\omega) & (\omega > 0) \\ 0 & (\omega < 0) \end{cases} \qquad \text{(E1.5)}$$

which in fact comes immediately from Eq. (E1.3).

Appendix F Geometry of the grating pair

This appendix contains the detailed working leading to Eq. (6.27), which summarises the dispersive properties of the grating pair arrangement of Fig. 6.5.

The coefficient of the quadratic term in Eq. (6.26) is

$$\gamma_2 p = \frac{d^2\phi}{d\omega^2} = -\frac{d\tau_{\text{group}}}{d\omega} \tag{F1.1}$$

where p is the perpendicular distance between the gratings, and the final step follows from Eq. (6.11). To find γ_2, we first need to work out the group time delay τ_{group} of different frequency groups travelling from the point of incidence A on the first grating G1 in Fig. 6.5 to B on the second grating G2, and thence to C. The precise positions of B and C will of course depend on the frequency, but the angle $\widehat{ACB}$ remains a right angle, and all points C lie on the 'finishing line' AF. By dividing the distance ABC by c, the group time delay is easily shown to be

$$\tau_{\text{group}} = \frac{p(1 + \cos(\delta - \alpha))}{c \cos \delta} \tag{F1.2}$$

where α and δ are the angles of incidence and diffraction. Group velocity dispersion in the medium (usually air) between the gratings has been ignored. Differentiating Eq. (F1.2), we have

$$\frac{d\tau_{\text{group}}}{d\omega} = \frac{p(\sin \alpha + \sin \delta)}{c \cos^2 \delta} \frac{d\delta}{d\omega}. \tag{F1.3}$$

For first-order diffraction, the standard grating law reads

$$(\sin \alpha + \sin \delta) = \frac{\lambda}{\Lambda} = \frac{2\pi c}{\omega \Lambda} \tag{F1.4}$$

where λ is the wavelength and Λ is the ruling spacing, and differentiating this equation gives

$$\frac{d\delta}{d\omega} = -\frac{2\pi c}{\omega^2 \Lambda \cos \delta}. \tag{F1.5}$$

Substituting Eqs (F1.4) and (F1.5) into Eq. (F1.3) yields

$$\frac{d\tau_{\text{group}}}{d\omega} = -\gamma_2 p = -\frac{4\pi^2 cp}{\Lambda^2 \omega^3 \cos^3 \bar{\delta}} \tag{F1.6}$$

where $\bar{\delta}$ is the mean value of δ over the input spectrum. This is Eq. (6.27).

Appendix G The paraxial wave equation

Maxwell's wave equation in a linear isotropic medium reads

$$-\nabla^2 \mathbf{E} + \frac{n^2}{c^2} \frac{\partial^2 \mathbf{E}}{\partial t^2} = 0 \tag{G1.1}$$

which follows from Eq. (2.2) if $\mathbf{P} = \varepsilon_0 \chi^{(1)} \mathbf{E}$, and the refractive index $n = \sqrt{1 + \chi^{(1)}}$. We now look for 'beam-type' solutions by writing a component of $\mathbf{E}$ in the form

$$E = \tfrac{1}{2} U(x, y, z) \exp\{i(\omega t - kz)\} + \text{c.c.} \tag{G1.2}$$

where $k = n\omega/c$ as usual. The rapid spatial dependence in $\exp\{-ikz\}$ implies a beam travelling basically in the z-direction, while the spatial dependence of $U(x, y, z)$ is assumed to be relatively slow in comparison.

Substituting Eq. (G1.2) into Eq. (G1.1) yields

$$\frac{\partial U}{\partial z} = -\frac{i}{2k} \left(\nabla_T^2 U + \frac{\partial^2 U}{\partial z^2} \right) \tag{G1.3}$$

where the transverse Laplacian operator governing diffraction is defined by $\nabla_T^2 = \partial^2/\partial x^2 + \partial^2/\partial y^2$.

In the so-called *paraxial approximation*, the second (non-paraxial) term on the right-hand side of Eq. (G1.3) is neglected, leading to the *paraxial wave equation*

$$\frac{\partial U}{\partial z} = -\frac{i}{2k} \nabla_T^2 U. \tag{G1.4}$$

The Gaussian beam of Eq. (2.47) is an exact solution of this equation.

Appendix H   Useful formulae for numerical simulations

This appendix contains some useful formulae for anyone performing numerical simulations that involve Fourier transforms between the time and frequency domains.

Discrete Fourier transform (DFT) routines typically perform the operation

$$A_k \Leftarrow \sum_{j=0}^{N-1} A_j e^{2\pi i s j k / N} \tag{H1.1}$$

where the input complex array A_j $(j = 0, \dots, N-1)$ is replaced on output, and $s = \pm 1$ offers a choice of sign. This operation can be used to perform either part of the discrete Fourier transform pair defined by

$$E_k = \alpha \sum_{j=0}^{N-1} \bar{E}_j e^{2\pi i j k / N}$$
$$\bar{E}_j = \beta \sum_{k=0}^{N-1} E_k e^{-2\pi i j k / N} \tag{H1.2}$$

where $\alpha \beta = 1/N$. Note that two successive calls to the routine (with opposite signs s in the exponents) return the original array multiplied by the array dimension N. In the early Cooley–Tukey fast Fourier transform (FFT) algorithm [127], N was restricted to a power of 2, although a wider range of possible dimensions is available in modern FFT packages.

To use the DFT effectively, the following simple relationships are helpful. The full widths of the time and frequency windows are $T = N\delta t = 2\pi/\delta\omega$ and $\Omega = N\delta\omega = 2\pi/\delta t$, where the link between the two domains provided by the second step in these formulae is

$$\delta t \delta \omega = 2\pi/N. \tag{H1.3}$$

Consider a transform from a Gaussian profile in the time domain to the corresponding Gaussian profile in the frequency domain. We define the full widths at half maximum intensity in the time domain as $\Delta t = \Delta g_t \delta t$; the corresponding width in the frequency domain is $\Delta \omega = \Delta g_\omega \delta\omega$. The parameters Δg_t and Δg_ω are the intensity full widths in grid units. It follows that

$$\Delta t \Delta \nu = \frac{\Delta t \Delta \omega}{2\pi} = \frac{\Delta g_t \delta t \Delta g_\omega \delta\omega}{2\pi} = \frac{\Delta g_t \Delta g_\omega}{N}. \tag{H1.4}$$

The time–bandwidth product for a transform-limited pulse is $\Delta t \Delta \nu = K$ where K depends on the pulse shape; see Chapter 6. It follows that

$$\Delta g_t \Delta g_\omega = N K. \tag{H1.5}$$

This statement of the bandwidth theorem expressed in grid units comes in handy when testing computer codes.

Appendix I Useful constants

Velocity of light	$c = 2.997\,925 \times 10^8 \, \text{ms}^{-1}$
Vacuum permittivity	$\varepsilon_0 = 8.854\,188 \times 10^{-12} \, \text{Fm}^{-1} (= \text{C V}^{-1}\,\text{m}^{-1})$
Vacuum permeability	$\mu_0 (= 4\pi \times 10^{-7}) = 1.256\,637 \times 10^{-6} \, \text{H m}^{-1}$
Electron mass	$m = 9.109\,382 \times 10^{-31} \, \text{kg}$
Elementary charge	$e = 1.602\,176 \times 10^{-19} \, \text{C}$
Planck's constant	$h = 6.626\,068 \times 10^{-34} \, \text{J s}$
	$\hbar = h/2\pi = 1.054\,572 \times 10^{-34} \, \text{J s rad}^{-1}$

See Appendix A for unit conversion factors.

Answers to problems

Chapter 1

1.1 $L_{\mathrm{coh}} = \frac{1.064 \times 10^{-6}}{4 \times 0.0045} = 59.11\mu\mathrm{m}$.

1.2   Under open circuit conditions, there is no net free charge on either of the plates, so $\mathbf{D}\ (= \varepsilon\varepsilon_0\mathbf{E} + \mathbf{P}_{\mathrm{dc}}) = 0$. So, since $\mathbf{P}$ is directed upwards in the figure, $\mathbf{E} = -\mathbf{P}_{\mathrm{dc}}/(\varepsilon\varepsilon_0)$ is directed downwards, and since $|\mathbf{E}| = V_{\mathrm{open}}/d$, the answer follows immediately.

1.3   Use $2\cos A \cos B = \cos(A + B) + \cos(A - B)$.

1.4   For Eq. (1.19) to be satisfied, we require that $n_3\omega_3 \leq n_2\omega_2 + n_1\omega_1$ where $\omega_1 + \omega_2 = \omega_3$. Write $\omega_{1,2} = \frac{1}{2}(\omega_3 \mp \delta\omega)$ where $\delta\omega = \omega_2 - \omega_1$.
The condition then becomes $n_3 \leq \frac{1}{2}(n_2 + n_1) + \frac{1}{2}(n_2 - n_1)\delta$ where $\delta = \delta\omega/\omega_3 \leq 1$. But $n_3 > n_2 \geq n_1$ for normal dispersion, and it is easy to show that the two conditions are inconsistent.

1.5   Figure 1.7 (for sum frequency generation) applies to SHG if the $\mathbf{k}_1$ and $\mathbf{k}_2$ arms are of equal length. The acute angle of the triangle is $\cos^{-1}\{1.5549/1.6550\} = 20.03°$ and the angle between the two fundamental beams is twice as large.

1.6 χ_1 is dimensionless; χ_2 is in $\mathrm{m\,V^{-1}}$; and χ_3 is in $(\mathrm{m\,V^{-1}})^2$.

1.7   (a) $n + \delta n = \sqrt{1 + \chi^{(1)} + 2\chi^{(2)}A_0} = \sqrt{(1 + \chi^{(1)})\left(1 + \frac{2\chi^{(2)}A_0}{1 + \chi^{(1)}}\right)} \cong n + \frac{\chi^{(2)}A_0}{n}$.

     (b)   The effective change in $\chi^{(1)}$ is $\frac{3}{4}\chi^{(3)}A_1^2$ and the intensity is $\frac{1}{2}nc\varepsilon_0 A_1^2$. A similar procedure to that in part (a) leads to the answer.

Chapter 2

2.1 $\dfrac{dI_1}{dz} = \frac{1}{2}n_1 c\varepsilon_0 \dfrac{d|\tilde{E}_1|^2}{dz} = \frac{1}{2}n_1 c\varepsilon_0 \tilde{E}_1^* \dfrac{d\tilde{E}_1}{dz} + \mathrm{c.c.}$

$\qquad\qquad = -\frac{1}{4}\omega_1\varepsilon_0 \left[i\tilde{E}_2\tilde{E}_1^{*2}e^{-i\Delta kz} - i\tilde{E}_2^*\tilde{E}_1^2 e^{i\Delta kz} \right],$

$\qquad\qquad\qquad \dfrac{dI_2}{dz} = \frac{1}{2}n_2 c\varepsilon_0 \dfrac{d\left|\tilde{E}_2\right|^2}{dz} = \frac{1}{2}n_2 c\varepsilon_0 \tilde{E}_2^* \dfrac{d\tilde{E}_2}{dz} + \mathrm{c.c.}$

$\qquad\qquad\qquad\qquad = -\dfrac{1}{8}\omega_2\varepsilon_0 \left[i\tilde{E}_2^*\tilde{E}_1^2 e^{i\Delta kz} - i\tilde{E}_2\tilde{E}_1^{*2}e^{-i\Delta kz} \right].$

     Since $\omega_2 = 2\omega_1$, it follows that $\frac{dI_1}{dz} = -\frac{dI_2}{dz}$.

2.2 This is perhaps obvious. If the fundamental field is $A \exp\{-\alpha t^2\}$, the second harmonic field varies as $A^2 \exp\{-2\alpha t^2\}$, while the intensities vary as $A^2 \exp\{-2\alpha t^2\}$ and $A^4 \exp\{-4\alpha t^2\}$. The energies come from $\int_{-\infty}^{\infty} A^2 \exp\{-2\alpha t^2\} dt = A^2 \sqrt{\pi/2\alpha}$ and $\int_{-\infty}^{\infty} A^4 \exp\{-4\alpha t^2\} dt = A^4 \sqrt{\pi/4\alpha}$, so the power-law relationship is the same for energy as for intensity. A similar argument applies for the third harmonic.

2.3 Putting the numbers into Eq. (2.15) gives $I_2 = 132.3$ MW/cm^2. The low-depletion approximation will be slightly infringed.

2.4 In Eq. (2.13), the phase of the harmonic field is governed by the term $-ie^{i\,\Delta kz/2}$. Since $-i$ points along OA, $\widehat{AOB}$ is clearly $\Delta kz/2$. Since $\widehat{OBC} = \pi/2$, it follows that $\widehat{OCB} = \Delta kz/2$ too.

2.5 The real and imaginary parts of $E(z)$ are, respectively, $E_C + E_C \left(\frac{1-\cos\{\Delta k(z-L_{\mathrm{coh}})/2\}}{2} \right)$ and $E_C \left(\frac{\sin\{\Delta k(z-L_{\mathrm{coh}})/2\}}{2} \right)$. Take the roots of the sum of the squares to obtain the result.

2.6 You can of course slog out the answer from Eqs (2.41) and (2.42), but it is more elegant to use Eq. (2.33) to show that the derivative of $\left|\tilde{A}_2(z)\right|^2 - \left|\tilde{A}_1(z)\right|^2$ is zero. The result reflects the fact that the changes in the number of photons in the signal and idler beams is the same; see Section 2.4.2.

2.7 $\tilde{A}_2(z)e^{i\,\Delta kz/2} = \tilde{A}_2(0)\cos\Delta k'z/2 + i\left(\frac{1}{2}\Delta k\tilde{A}_2(0) - g\tilde{A}_1^*(0)\right)\left(\frac{\sin\Delta k'z/2}{\Delta k'/2}\right),$

$\tilde{A}_1(z)e^{i\,\Delta kz/2} = \tilde{A}_1(0)\cos\Delta k'z/2 + i\left(\frac{1}{2}\Delta k\tilde{A}_1(0) - g\tilde{A}_2^*(0)\right)\left(\frac{\sin\Delta k'z/2}{\Delta k'/2}\right),$

where $\Delta k'/2 = \sqrt{(\Delta k/2)^2 - g^2}$.

2.8 When $\tilde{A}_1(0) = 0$ and $g \ll \Delta k/2$.

2.9 $\tilde{A}_1(0) = 0.9i$.
When $\tilde{A}_1(0) = i\tilde{A}_2^*(0)$, both fields tend to zero at $z = \infty$ according to the equations. In practice, the fields would drop to a low value, and then grow again from noise.

2.10 This is solved by straightforward substitution.

2.11 The intersecting chords theorem gives $r^2 = (2R - \delta)\delta \cong 2R\delta$ where R is the radius of curvature of the wavefront, r is distance from the axis, and δ is the 'bulge' in the wavefront. The relationship between δ and the phase $\psi(r,\zeta) \,(= -\zeta r^2/w^2)$ is $\psi = 2\pi\delta/\lambda$. Combining these relationships, and using $z_R = \pi w_0^2/\lambda$ and $w^2 = w_0^2(1+\zeta^2)$ from Eqs (2.49) and (2.54) yields $z_R = \frac{R\zeta}{1+\zeta^2}$. The answer follows immediately.

2.12 The condition is $(2k_1 - k_2)L = \pi/2$. This leads to $L = \frac{\lambda_1}{8|n_2-n_1|} = 25.77$ mm.

2.13 The solution of any second-order ODE requires two boundary conditions, the initial value and its gradient. And it is the gradient that introduces the other field component through Eqs (2.28) and (2.29).

2.14 Starting from the origin, the arc length swept out by the right-hand side of the curve is $\int_0^\zeta \frac{d\zeta'}{1+\zeta'^2} = \tan^{-1}\zeta = \phi_{\mathrm{Gouy}}$. As $\zeta \to \infty$, the arc length goes to $\pi/2$, which is the circumference of a semicircle of radius $\frac{1}{2}$. At the same time, the phasor direction at the end of the arc is given by the phase of $(1 + i\zeta)^2 = 2\phi_{\mathrm{Gouy}}$ which is also consistent with a circular path of the same radius.

Chapter 3

3.1 For $y = 0$ and in polar form, Eq. (3.9) reads $\frac{r^2 \cos^2 \theta}{n_o^2} + \frac{r^2 \sin^2 \theta}{n_e^2} = 1$, where θ is the angle relative to the z-axis. But in the geometrical construction for $\tilde{n}_e(\theta)$, the angle θ refers to the direction of the wave vector **k**, which is perpendicular to the line defining $\tilde{n}_e(\theta)$; hence the complementary angle is needed. Equation (3.10) is therefore obtained by replacing r by $\tilde{n}_e(\theta)$, and exchanging $\cos \theta$ and $\sin \theta$.

3.2 The two allowed waves are polarised in the z-direction (index n_e) and in the y-direction (index n_o). No walk-off occurs.

3.3 Anticlockwise in the first case, and clockwise in the second.

3.4 Set $\tilde{n}(\theta) = n_y$ and $\sin^2 \theta = 1 - \cos^2 \theta$. The result follows by straightforward algebra.

3.5 The walk-off angle is $3.18°$ from Eq. (3.11).

Chapter 4

4.1 m V^{-1}.

4.2 The waves could lie in the x-y plane (perpendicular to the optic axis) with the interaction based on the d_{31} coefficient of BBO; see Table 4.4.

4.3 Unit vectors in the o-o and e-e directions are $\hat{o} = (\sin \phi, -\cos \phi, \ 0)$ and $\hat{e} = (\cos \theta \cos \phi, \ \cos \theta \sin \phi, \ -\sin \theta)$. These enable one to write

$$E_x = E_o(\sin \phi) + E_e(-\cos \theta \cos \phi) \text{ and } E_y = E_o(-\cos \phi) + E_e(-\cos \theta \sin \phi).$$

Hence the term involving both E_o and E_e in the product $E_x E_y$ (mediated by d_{36} to create P_z) is $E_o E_e(\cos \theta \cos 2\phi)$. But $P_e = P_z \sin \theta$, so the overall geometrical factor is $\frac{1}{2} \sin 2\theta \cos 2\phi$. A similar calculation to find the contribution mediated by d_{14} yields the same result.

4.4 In terms of $x'y'$ axes rotated $45°$ clockwise w.r.t. xy, we have $x = (x' + y')/\sqrt{2}$ and $y = (x' - y')/\sqrt{2}$. Substituting these results into Eq. (4.28) yields

$$x'^2(n_o^{-2} + r_{63} V/\ell) + y'^2(n_o^{-2} - r_{63} V/\ell) = 1$$

and it follows that the modified semi-axes representing the refractive indices are

$$n_o(1 \pm n_o^2 r_{63} V/\ell)^{-1/2} \cong n_o(1 \mp \tfrac{1}{2} n_o^2 r_{63} V/\ell).$$

4.5 $V_{\lambda/4} = \frac{\lambda}{4n_o^3 r_{63}} = 5279 \text{ V}$ for the parameters given.

Chapter 5

5.1 If real arithmetic is used, the polarisation along the direction of the field is simply $P = \varepsilon_0 \chi_1^{(3)} E^3$. Along the slanting axes, the field components are $E_x = rE$ and $E_y = -rE$, where $r = 1/\sqrt{2}$. It is now easy to show that $P_x = \varepsilon_0 \chi_s r^3 E^3$ and $P_y = -\varepsilon_0 \chi_s r^3 E^3$ where $\chi_s = \chi_1 + \chi_2 + \chi_3 + \chi_4$. Since $P = r(P_x - P_y)$, it follows that $P = \varepsilon_0 (\frac{1}{2}\chi_s) E^3$, which proves the result. (If complex numbers are used, the replacements $P \to \frac{1}{2}\hat{P}$ and $E \to \frac{1}{2}\hat{E}$ should be made throughout.)

5.2 If real arithmetic is used, light is circularly polarised when $E_x = E \cos \omega t$ and $E_y = \pm E \sin \omega t$. In this case,

$$P_x = \varepsilon_0 \left(\chi_1 E_x^3 + (\chi_2 + \chi_3 + \chi_4) E_y^2 E_x \right) = \varepsilon_0 \chi_1 E^3 \left(\cos^3 \omega t + \sin^2 \omega t \cos \omega t \right)$$

where $\chi_1 = \chi_2 + \chi_3 + \chi_4$ (Eq. 5.6) has been used. But $\cos^3 \omega t + \sin^2 \omega t \cos \omega t = \cos \omega t$, so there is no third harmonic component in the nonlinear polarisation. Naturally, a similar result holds for P_y.

5.3 $E(0) = (2K L_\pi)^{-1/2} = 1.066 \text{ MV m}^{-1}$; so $V_\pi = 10.66 \text{ kV}$.

5.4 Equation (C3.4b) with $i = x$ reads

$$\hat{P}_x(\omega) = \tfrac{3}{4}\varepsilon_0 \begin{pmatrix} \chi_1 \hat{E}_x \hat{E}_x^* \hat{E}_x + \chi_2 \hat{E}_x \hat{E}_y^* \hat{E}_y + \chi_2 \hat{E}_x \hat{E}_z^* \hat{E}_z \\ \chi_3 \hat{E}_y \hat{E}_x^* \hat{E}_y + \chi_3 \hat{E}_z \hat{E}_x^* \hat{E}_z + \chi_4 \hat{E}_y \hat{E}_y^* \hat{E}_x + \chi_4 \hat{E}_z \hat{E}_z^* \hat{E}_x \end{pmatrix}$$

where all relevant non-zero coefficients from Eq. (5.7) have been included. Since $\chi_1 = \chi_2 + \chi_3 + \chi_4$ from Eq. (5.6) (or Eq. C3.5), and $\chi_2 = \chi_4$ for IDRI from Eq. (C3.6), the equation can be rewritten

$$\hat{P}_x(\omega) = \tfrac{3}{4}\varepsilon_0 \begin{pmatrix} 2\chi_2 \hat{E}_x \hat{E}_x^* \hat{E}_x + 2\chi_2 \hat{E}_x \hat{E}_y^* \hat{E}_y + 2\chi_2 \hat{E}_x \hat{E}_z^* \hat{E}_z \\ +\chi_3 \hat{E}_x \hat{E}_x^* \hat{E}_x \chi_3 \hat{E}_y \hat{E}_x^* \hat{E}_y + \chi_3 \hat{E}_z \hat{E}_x^* \hat{E}_z \end{pmatrix}$$

and the first of Eqs (5.18) follows immediately. The proof of the second equation is similar.

5.5 The final term in Eq. (5.28) changes $k = \frac{n_0 \omega}{c}$ to $k = \frac{n_0 \omega}{c} + \frac{\pi n_2 I_{pk}}{\lambda}$, where Eq. (5.26) has been used. Hence

$$v_{\text{phase}} = \frac{\omega}{k} = \frac{c}{n_0}\left(1 + \frac{n_2 I_{pk}}{2n_0}\right)^{-1} \cong \frac{c}{n_0}\left(1 - \frac{n_2 I_{pk}}{2n_0}\right) = \frac{c}{n_0} - \frac{c n_2 I_{pk}}{2n_0^2}.$$

5.6 Work out $\tilde{E}_L^* \frac{\partial \tilde{E}_L}{\partial z} + \tilde{E}_L \frac{\partial \tilde{E}_L^*}{\partial z}$ and use $I_L = \frac{1}{2}n_L c \varepsilon_0 \left|\tilde{E}_L\right|^2$ and $I_S = \frac{1}{2}n_S c \varepsilon_0 \left|\tilde{E}_S\right|^2$. The result follows immediately.

5.7 The intersecting chords theorem gives $(2k_L - \Delta k)\Delta k \cong 2k_L \Delta k \cong K_A^2$. It follows that $\Delta k \cong \frac{1}{2}k_L (K_A/k_L)^2 \cong \frac{1}{2}k_L \theta^2$ and the answer follows from $L_{\text{coh}} = \pi/|\Delta k|$ and $k_L = 2\pi/\lambda_L$.

5.8 This involves some intricate algebra. Start by writing

$$\tilde{A}(z) = \tilde{A}_+ e^{i\kappa z} + \tilde{A}_- e^{-i\kappa z}$$

$$\tilde{B}(z) = \tilde{B}_+ e^{i\kappa z} + \tilde{B}_- e^{-i\kappa z}.$$

Now $\tilde{A}(0) = \tilde{A}_+ + \tilde{A}_-$, while $\tilde{B}_- = -\tilde{B}_+ e^{2i\kappa L}$ ensures that $\tilde{B}(L) = 0$. It also follows from from Eq. (5.78) that $i\kappa \tilde{A}_+ = r\tilde{B}_+$, $-i\kappa \tilde{A}_- = r\tilde{B}_-$, etc. Combining the equations leads to the answer.

5.9 The profile of the forward beam at the mirror is found by setting $z = z_m$ in Eq. (5.95). The profile of the backward wave at the mirror must be the negative of this (to ensure zero field at the mirror surface). A backward-travelling Gaussian beam centred at $z = z_m$ can be found by replacing z by $(z - 2z_m)$ in Eq. (5.95), and taking the complex conjugate (to reverse the direction). Setting $z = z_m$ in this equation gives the required profile at the mirror apart from the minus sign, which arose from the phase change at the mirror. But this corresponds simply to a π phase change of the backward beam, and so it does not affect the conclusion.

Chapter 6

6.1 $\frac{z}{c}\left(n + \omega \frac{dn}{d\omega}\right) = \frac{z}{c}\left(n + \frac{2\pi c}{\lambda} \frac{dn}{d\lambda} \frac{d\lambda}{d\omega}\right) = \frac{z}{c}\left(n - \frac{2\pi c}{\lambda} \frac{\lambda^2}{2\pi c} \frac{dn}{d\lambda}\right) = \frac{z}{c}\left(n - \lambda \frac{dn}{d\lambda}\right)$.

6.2 $\frac{d^2 v_{\text{phase}}}{d\lambda^2} = \frac{2c}{n^3}\left(\frac{dn}{d\lambda}\right)^2 - \frac{c}{n^2}\frac{d^2 n}{d\lambda^2}$, so the zeros of $\frac{d^2 v_{\text{phase}}}{d\lambda^2}$ and $\frac{d^2 n}{d\lambda^2}$ are only coincident when $\frac{dn}{d\lambda} = 0$. The suggestion is therefore generally incorrect.

6.3 (a) Since the intensity varies as $\exp\{-(t/t_0)^2\}$, it follows that $\left(\frac{\Delta t_0}{2t_0}\right)^2 = \ln 2$. The answer to part (a) follows immediately.

   (b) $\Delta v = 0.441/\Delta t = 0.441$ THz; $\Delta \lambda = \lambda^2 \Delta v/c = 1.47$ nm.

   (c) Equation (6.21) gives maximum chirp at $\xi = 1$, so $\frac{dv}{dt} = \frac{1}{4\pi t_0^2} = \frac{\ln 2}{\pi \Delta t_0^2} = 221$ GHz ps^{-1}.

6.4 If $\tau = t/t_0$, then $\operatorname{sech}^2 \tau = \frac{4}{e^{2\tau} + 2 + e^{-2\tau}} = \frac{1}{2}$ when $e^{2\tau} + e^{-2\tau} = 6$. This is equivalent to $e^{4\tau} - 6e^{2\tau} + 1 = 0$, which has the roots $e^{2\tau} = 3 \pm \sqrt{8}$. This leads to $\Delta \tau = \Delta t/t_0 = \ln\{3 \pm \sqrt{8}\}$.

6.5 34 cm.

6.6 Choose an arbitrary reference line of length s between the gratings, defined by $\delta = \delta_0$, so that $s = p/\cos \delta_0$ and $\phi = -\frac{\omega s \cos(\delta - \delta_0)}{c}$. Then

$$\tau_{\text{group}} = -\frac{d^2\varphi}{d\omega^2}\Big|_{\delta=\delta_0} = -\frac{s}{c}\left(\left(2\frac{d\delta}{d\omega} + \omega \frac{d^2\delta}{d\omega^2}\right)\sin(\delta - \delta_0)\right.$$

$$\left. + \omega\left(\frac{d\delta}{d\omega}\right)^2 \cos(\delta - \delta_0)\right)\Big|_{\delta=\delta_0} = -\frac{p\omega}{c\cos\delta_0}\left(\frac{d\delta}{d\omega}\right)^2\Big|_{\delta=\delta_0}.$$

But $\frac{d\delta}{d\omega}\Big|_{\delta=\delta_0} = -\frac{2\pi c}{\omega^2 \Lambda \cos \delta_0}$, so $\tau_{\text{group}} = -\frac{p\omega}{c\cos\delta_0}\left(\frac{2\pi c}{\omega^2 \Lambda \cos\delta_0}\right)^2 = -\frac{4\pi^2 cp}{\omega^3 \Lambda^2 \cos^3\delta_0}$.

Chapter 7

7.1 $L_{\text{spm}} = \frac{\lambda}{2\pi n_2 I_{\text{pk}}}$ so $\frac{L_{\text{spm}}}{\lambda} = \frac{1}{2\pi\, 3\times 10^{-16}\, 10^{12}} = 531$.

   Care always needs to be taken over the use of centimetres in questions of this kind. All parameters must of course be in the same units.

7.2 This is straightforward algebra.

7.3 The final term in Eq. (7.18) with $L = L_{\text{spm}}$ changes $k = \frac{n_0\omega}{c}$ to $k = \frac{n_0\omega}{c} + \frac{\pi n_2 I_{\text{pk}}}{\lambda}$. Hence

$$v_{\text{phase}} = \frac{\omega}{k} = \frac{c}{n_0}\left(1 + \frac{n_2 I_{\text{pk}}}{2n_0}\right)^{-1} \cong \frac{c}{n_0}\left(1 - \frac{n_2 I_{\text{pk}}}{2n_0}\right) = \frac{c}{n_0} - \frac{cn_2 I_{\text{pk}}}{2n_0^2}.$$

7.4 For a Gaussian intensity profile $I = I_{\text{pk}}\exp\{-\alpha t^2\}$, it is easy to show that the steepest gradient occurs at $t = \pm(2\alpha)^{-1/2}$ where $\frac{dI}{dt} = \mp\frac{e^{-1/2}\sqrt{8\ln 2}}{\Delta t} = \mp\frac{1.43}{\Delta t}$ and Δt is the intensity FWHM given by $\Delta t = \sqrt{(4\ln 2)/\alpha}$.
It follows that $\frac{L_{\text{shock}}}{L_{\text{spm}}} = \left(\frac{c\Delta t}{1.43 n_2 I_{\text{pk}}}\right) \div \left(\frac{cT}{2\pi n_2 I_{\text{pk}}}\right) \cong 4.4 \times \frac{\Delta t}{T}$ where $T = 2\pi/\omega$ is the optical period.

7.5 This question is relevant to the phenomenon of carrier-wave shocking, and to the use of two-colour pumping in high harmonic generation experiments.

7.6 The time difference is $\delta t \cong \frac{L|v_{\text{phase}}-v_{\text{group}}|}{\bar{v}^2}$ where L is the distance of travel and $\bar{v}$ is the mean velocity. Setting $\delta t = \frac{1}{4}T = \lambda/4c$ yields
$L = \frac{\lambda\bar{v}^2}{4c\delta v} \cong \frac{800\times 10^{-9}}{4}\frac{(1.92\times 10^8)^2}{3\times 10^8\ 2\times 10^6} = 12.3\,\mu\text{m}$, where values of δv and $\bar{v}$ have been estimated from Fig. 6.2.
This is a very short distance, and it shows that the carrier-envelope phase is scrambled extremely easily. The situation inside a mode-locked laser is more complicated than this, but the result shows that the slip from one transit to the next will be very large and therefore essentially random. However, this in no way affects the efficacy of the technique for CEP stabilisation described in Section 7.10.

Chapter 8

8.1 The formula $\rho_{21}^{(2)} = a_2^{(1)}a_1^{(1)*}$ is verified if the derivative of both sides can be shown to be the same. (The initial conditions match, because all three parameters are initially zero.) It is therefore necessary to show that $\dot{\rho}_{21}^{(2)} = \dot{a}_2^{(1)}a_1^{(1)*} + a_2^{(1)}\dot{a}_1^{(1)*}$.
From Eqs (8.7) and (8.15), we know that

$$\dot{\rho}_{21}^{(2)} = -i\omega_{21}\rho_{21}^{(2)} - \frac{i}{\hbar}\left(V_{20}\rho_{01}^{(1)} - \rho_{20}^{(1)}V_{01}\right)$$

$$\dot{a}_2^{(1)} = -i\omega_2 a_2^{(1)} - \frac{i}{\hbar}V_{20}a_0^{(0)}$$

$$\dot{a}_1^{(1)*} = +i\omega_1 a_1^{(1)*} + \frac{i}{\hbar}V_{01}a_0^{(0)}$$

and the answer follows immediately once it is recognised that $\rho_{20}^{(1)} = a_2^{(1)}a_0^{(0)}$, etc.

8.2 For $j \neq k$, take the complex conjugate, and satisfy yourself that the two expressions are the same. Remember that j, k, and n, n' are all dummy indices, and so the pairs can be exchanged at will. The conclusion applies equally when $j = k$, although the issue would only arise in that case if the matrix elements were complex.

8.3 The contraction to d_{14} for SFG is valid only when the Kleinmann symmetry condition applies, and in this limit the two χ coefficients become identical; see Section 4.3.

Chapter 9

9.1 This is similar to Problem 1.7. The sequence is

$$k = \frac{\omega}{c}\sqrt{1 + \chi' - i\chi''} = \frac{\omega}{c}\sqrt{(1 + \chi')\left(1 - \frac{i\chi''}{1 + \chi'}\right)} \cong \frac{n\omega}{c}\left(1 - \frac{i\chi''}{2n^2}\right)$$

where $n = \sqrt{1 + \chi'}$. The approximation is good provided $\chi'' \ll 2n^2$.

9.2 On resonance, $\tilde{\omega}_{20}^* - \omega_D = i\gamma_2$ which leads to

$$\frac{\partial \tilde{E}_L}{\partial z} = -\frac{A\omega_L \left|\tilde{E}_S\right|^2 \tilde{E}_L}{n_L} \quad \text{and} \quad \frac{\partial \tilde{E}_S}{\partial z} = \frac{A\omega_S \left|\tilde{E}_L\right|^2 \tilde{E}_S}{n_S}$$

where A is positive and contains constants. It follows that

$$\frac{n_S \tilde{E}_S^* }{\omega_S}\frac{\partial \tilde{E}_S}{\partial z} = A\left|\tilde{E}_S\right|^2 \left|\tilde{E}_L\right|^2 = -\frac{n_L \tilde{E}_L^*}{\omega_L}\frac{\partial \tilde{E}_L}{\partial z}.$$

But $\frac{\partial I_S}{\partial z} = \frac{1}{2}c\varepsilon_0 n_S \tilde{E}_S^* \frac{\partial \tilde{E}_S}{\partial z} + \text{c.c.}$ and $\frac{\partial I_L}{\partial z} = \frac{1}{2}c\varepsilon_0 n_L \tilde{E}_L^* \frac{\partial \tilde{E}_L}{\partial z} + \text{c.c.}$, where I_S and I_L are the Stokes and laser energies, respectively. So $\frac{1}{\omega_S}\frac{\partial I_S}{\partial z} = -\frac{1}{\omega_L}\frac{\partial I_L}{\partial z}$, which completes the proof.

9.3 In the first case, the lineshape function is very narrow, and the Lorentzian factor in the denominator of the integral can be neglected. The integral is then unity (because the lineshape function is normalised), and the answer follows trivially.

   In the opposite limit, the lineshape profile is much broader than the Lorentzian factor, and so $g_i(\Delta')$ can be replaced by $g_i(0)$. The integral is now $\pi g_i(0)/\tau$ and the answer follows directly.

9.4 We have $u = ab(1 - \cos \Omega_R' t)$ and $1 + w = 1 - a^2 - b^2 \cos \Omega_R' t = b^2\left(1 - \cos \Omega_R' t\right)$. Hence $\frac{u}{1+w} = \frac{a}{b} = \frac{\Delta}{\Omega_R}$ is independent of t, and the motion projected onto the u-w plane is a line slanted with respect to the 3-axis at an angle of $\tan^{-1}\{\Delta/\Omega_R\}$.

9.5 $\left.\frac{dn}{d\omega}\right|_{\omega=\omega_{10}} = -\left.\frac{dn}{d\Delta_p}\right|_{\Delta_p=0}$. The algebra is made much easier by the fact that the derivative is to be evaluated at $\Delta_p = 0$.

Chapter 10

10.1 If $U = -Ex - \frac{\alpha}{|x|}$, $\frac{dU}{dx} = -E + \frac{\alpha}{|x|^2} = 0$ when $|x| = \sqrt{\frac{\alpha}{|E|}} \sim |E|^{-1/2} \sim I^{-1/4}$, because intensity is proportional to the square of the field. While this may be an interesting exercise, the result has little bearing on what actually happens under strong-field ionisation.

10.2 Elliptically polarised light can always be reduced to two orthogonal electric field components in quadrature. However, the quasi-classical model shows that recollisions only occur in alternate quarter-cycles, so the recollision can never occur in both quadratures, let alone in both quadratures at the same time.

10.3 Consider the quarter-cycle $-\pi/2 \leq \omega t_0 < 0$ in which the field is approaching a maximum. From Eq. (10.5), the electron velocity is proportional to $V(t) = \sin \omega t_0 - \sin \omega t$ ($t > t_0$). And since $\sin \omega t_0 < 0$, $V(t)$ is a sine function with a *negative* shift. This in turn means that $\int_{t_0}^{t} V(t')dt'$, which controls the net displacement (see Eq. 10.6), is negative for all $t > t_0$, so no recollision can occur.

10.4 The factor in Eq. (10.6) is $\frac{eA}{m\omega^2} = \frac{e\lambda^2}{4\pi^2 c^2 m}\sqrt{\frac{2I}{c\varepsilon_0}} = 2.75$ nm. The peak value of X in Fig. 10.2 is about 2, so the maximum excursion is maybe 5.5 nm.

10.5 If the statement is true, it follows from trigonometry that the derivative of $E(t)$ at the time of ionisation $(= -A\omega \sin \omega t_0)$ must equal $\frac{A \cos \omega t_R - A \cos \omega t_0}{(t_R - t_0)}$. This equality is confirmed by Eq. (10.7)

10.6 The fundamental photon energy in eV is $hc/\lambda e = 1.55$ eV. The ionisation energy therefore corresponds to 15.85 harmonic units. Equation (10.10) gives 244.4 for the recollision kinetic energy in harmonic units, so the maximum harmonic is int$\{260.3\} = 260$, and the minimum wavelength is $800 \div 260.3 = 3.07$ nm. This is close to the value achieved by Chang [43], although the nominal intensity used in the classical model can hardly be expected to yield an accurate answer.

Further Reading

R.W. Boyd, *Nonlinear Optics* (3rd edn), Elsevier, Amsterdam, 2007.

A comprehensive treatment of the subject at a more advanced level, presented in a tutorial style. The third edition is in SI units; earlier editions used Gaussian units.

G.S. He and S.H. Liu, *Physics of Nonlinear Optics*, World Scientific, Singapore, 1999.

A wide-ranging treatment of nonlinear optics at an advanced level, with an especially generous coverage of third-order processes and stimulated scattering. The style is more that of a review than a tutorial text; the book contains a large number of research references.

G. Agrawal, *Nonlinear Fibre Optics*, Academic Press, New York, 1989.

A clear and accessible treatment of nonlinear optics in optical fibres.

N. Bloembergen, *Nonlinear Optics*, Benjamin, New York, 1965.

A classic by one of the leading pioneers of nonlinear optics, albeit written at a very early date.

A.C. Newell and J.V. Moloney, *Nonlinear Optics*, Addison Wesley, New York, 1992.

A full-length text on nonlinear optics with a slight bias towards aspects of the subject relevant to optical communications.

A. Yariv, *Quantum Electronics* (3rd edn), John Wiley, New York, 1988.

Chapters 14–19 contain a very useful summary of nonlinear optics.

P.N. Butcher and D. Cotter, *The Elements of Nonlinear Optics*, Cambridge University Press, Cambridge, 1990.

An authoritative account of the fundamentals of the subject.

H. Zernike and J.E. Midwinter, *Applied Nonlinear Optics*, Dover, New York, 2007 (reprint).

A valuable account of nonlinear optics as it was in the early 1970s, with a tutorial flavour and at an introductory level.

R.L. Sutherland, *Handbook of Nonlinear Optics* (2nd edn), Marcel Dekker, New York, 2003.

An indispensable handbook of nonlinear optics for the specialist.

V.G. Dmitriev, G.G. Gurzadyan and D.N. Nikogosyan, *Handbook of Nonlinear Optical Crystals* (3rd edn), Springer, Berlin, 1999.

An indispensable repository of data on nonlinear optical crystals.

References

[1] R.W. Boyd, *Nonlinear Optics* (3rd edn), Elsevier, Amsterdam, 2007.

[2] J. Kerr, Phil. Mag. (4th Series), **50** (1875) 337.

[3] F. Pockels, Abr. Gott., **39** (1894) 1.

[4] P.A. Franken, A.E. Hill, C.W. Peters, and G. Weinreich, Phys. Rev. Lett., **7** (1961) 118.

[5] P.D. Maker, R.W. Terhune, M. Nisenoff, and C.M. Savage, Phys. Rev. Lett., **8** (1962) 21.

[6] P.A. Franken and J.F. Ward, Rev. Mod. Phys., **35** (1963) 35.

[7] M. Bass, P.A. Franken, J.F. Ward, and G. Weinreich, Phys. Rev. Lett., **9** (1962) 446.

[8] M. Ebrahim-Zadeh, Phil. Trans. Roy. Soc. Lond. A, **263** (2003) 2731.

[9] J.A. Armstrong, N. Bloembergen, J. Ducuing, and P.S. Pershan, Phys. Rev., **127** (1962) 1918.

[10] J.A. Giordmaine and R.C. Miller, Phys. Rev. Lett., **14** (1965) 973.

[11] E.J. Lim, M.M. Feyer, R.L. Byer, and W.J. Kozlovsky, Electron. Lett., **25** (1989) 731.

[12] D.E. Spence, P.N. Kean, and W. Sibbett, Opt. Lett., **16** (1991) 42.

[13] P.B. Corkum and M. Krausz, Nature Physics, **3** (2007) 381.

[14] S.L. McCall and E.L. Hahn, Phys. Rev. Lett., **18** (1967) 908; S.L. McCall and E.L. Hahn, Phys. Rev. **183** (1969) 457.

[15] S.E. Harris, Physics Today, **507**, (July, 1997) 36.

[16] M. Fleischhauer, A. Imamoglu, and J.P. Marangos, Rev. Mod. Phys., **77** (2005) 633.

[17] S.E. Harris, Phys. Rev. Lett., **62** (1989) 1033; A. Imamoglu and S.E. Harris, Opt. Lett., **14** (1989) 1344.

[18] Z. Dutton, N.S. Ginsberg, C. Slowe, and L. Vestergaard Hau, Europhys. News **35** (2) (2004) 33.

[19] Nature Photonics **2** (8) (August 2008), special issue on slow light.

[20] A. Yariv, *Quantum Electronics* (3rd edn), John Wiley, New York, 1988.

[21] M. Ebrahim-Zadeh, *Optical parametric devices*, in *Handbook of Laser Technology and Applications*, IOP Publishing, Bristol, 2003, 1347; *Continuous-wave optical parametric oscillators*, in *Handbook of Optics VI,* Optical Society of America, McGraw-Hill, New York, 2010.

[22] G.H.C. New and J.F. Ward, Phys. Rev. Lett., **19** (1967) 556; J.F. Ward and G.H.C. New, Phys. Rev., **185** (1969) 57.

[23] G.C. Bjorklund, IEEE J. Quantum Electron., **QE-11** (1975) 267.

[24] J.F. Nye, *Physical Properties of Crystals*, Oxford University Press, Oxford, 1985.

[25] W.G. Cady, *Piezoelectricity* (2 vols.), Dover, New York, 1964.

[26] V.G. Dmitriev, G.G. Gurzadyan, and D.N. Nikogosyan, *Handbook of Nonlinear Optical Crystals* (3rd edn), Springer, Berlin, 1999.

[27] R.W. Boyd, *Nonlinear Optics* (3rd edn), Elsevier, Amsterdam, 2007, p. 51.

[28] A. Yariv and P. Yeh, *Optical Waves in Crystals*, Wiley Interscience, New York, 2003.

[29] M. Born and E. Wolf, *Principles of Optics* (7th edn), Cambridge University Press, Cambridge, 1999.

[30] H. Zernike and J.E. Midwinter, *Applied Nonlinear Optics*, Dover, New York, 2007, p. 64.

[31] P.D. Maker and R.W. Terhune, Phys. Rev., **137** (1965) A801.

[32] P.L. Kelley, Phys. Rev. Lett., **15** (1965) 1005.

[33] R.R. Alfano and S.L. Shapiro, Phys. Rev. Lett., **24** (1970) 592.

[34] P.G. Drazin and R.S. Johnson, *Solitons: An Introduction* (2nd edn) Cambridge University Press, Cambridge, 1989.

[35] A. Barthelemy, S. Maneuf, and C. Froehly, Opt. Commun., **55** (1985) 201.

[36] J.S. Aitchison *et al.*, Electron. Lett., **28** (1992) 1879.

[37] G. Fibich and A.L. Gaeta, Opt. Lett., **25** (2000) 335.

[38] E. J. Woodbury and W. K. Ng, Proc. IRE, **50** (1962) 2367.

[39] G.E. Eckhardt *et al.,* Phys. Rev. Lett., **9** (1962) 455.

[40] A.P. Hickman *et al.*, Phys. Rev. A, **33** (1986) 1788.

[41] G.S. McDonald *et al.*, Opt. Lett., **19** (1994) 1400.

[42] E. Takahashi *et al.*, Opt. Express, **15** (2007) 2535.

[43] P. Tournois, Opt. Commun., **140** (1997) 245.

[44] R.Y. Chiao, C.H. Townes, and B.P. Stoicheff, Phys. Rev. Lett., **12** (1964) 592.

[45] C.F. Quate, C.D.W. Wilkinson, and D.K. Winslow, Proc. IEEE, **53** (1965) 1604.

[46] A. Kobyakov, M. Sauer, and D. Chowdhury, Adv. Opt. and Phot., **2** (2010) 1.

[47] M.J. Damzen *et al.*, *Stimulated Brillouin Scattering: Fundamentals and Applications*, Taylor & Francis, London, 2003.

[48] G.S. He and S.H. Liu, *Physics of Nonlinear Optics*, World Scientific, Singapore, 1999.

[49] D.M. Pepper, Opt. Eng., **21** (1982) 156.

[50] M. Gower, *Optical Phase Conjugation*, Springer-Verlag, Berlin, 1994.

[51] A. Brignon and J-P. Huignard (Eds.), *Phase-Conjugate Laser Optics*, Wiley Interscience, New York, 2003.

[52] J.I. Gersten, Phys. Rev. A, **21** (1980) 1222.

[53] J.T. Manassah, R.R. Alfano, and M. Mustafa, Phys. Lett., A **107** (1985) 305.

[54] A.L. Gaeta, Phys. Rev. Lett., **84** (2000) 3582.

[55] I. Ilev *et al.* Appl. Opt., **35** (1996) 2548.

[56] J.M. Dudley and J.R. Taylor (Eds), *Supercontinuum Generation in Optical Fibres*, Cambridge University Press, Cambridge, 2010.

[57] E.B. Treacy, IEEE J. Quantum Electron., **QE-5** (1969) 454.

[58] O.E. Martinez, J.P. Gordon, and R.L. Fork. J. Opt. Soc. Am. A, **1** (1984) 1003.

[59] R.W. Boyd, *Nonlinear Optics* (3rd edn), Elsevier, Amsterdam, 2007, Chapter 13.

[60] T.K. Gustafson *et al.* Phys. Rev., **177** (1969) 306.

[61] G.P. Agrawal, *Nonlinear Fibre Optics* (2nd edn), Academic Press, New York, 1989, Chapter 4.

[62] L.F. Mollenauer, R.H. Stolen, and J.P. Gordon, Phys Rev. Lett., **45** (1980) 1095.

[63] L. F. Mollenauer and K. Smith, Opt. Lett., **13** (1988) 675.

[64] L.F. Mollenauer and J.P. Gordon, *Solitons in Optical Fibres: Fundamentals and Applications*, Academic Press, New York, 2006.

[65] J.R. Taylor (Ed.), *Optical Solitons –Theory and Experiment*, Cambridge University Press, Cambridge, 1992.

[66] F. DeMartini, C.H. Townes, T.K. Gustafson, and P.L. Kelley, Phys. Rev., **164** (1967) 312.

[67] G. Rosen, Phys. Rev., **139** (1965) A539.

[68] R.G. Flesch, A. Pushkarev, and J.V. Moloney, Phys. Rev. Lett., **76** (1996) 2488; L. Gilles, J.V. Moloney, and L. Vasquez, Phys. Rev. E, **60** (1999) 1051.

[69] S.B.P. Radnor *et al.*, Phys. Rev. A, **77** (2008) 033806.
[70] R.A. Fisher, P.L. Kelley, and T.K. Gustafson, Appl. Phys. Lett., **14** (1969) 140.
[71] A.M. Johnson, R.H. Stolen, and W.H. Simpson, Appl. Phys. Lett., **44** (1984) 729.
[72] J.W.G. Tisch (Imperial College Attosecond Laboratory), private communication.
[73] J. Comly and E. Garmire, Appl. Phys. Lett., **12** (1968) 7.
[74] W.H. Glenn, IEEE J. Quantum Electron., **QE- 5** (1969) 284.
[75] J.C.A. Tyrrell, P. Kinsler, and G.H.C. New, J. Mod. Opt., **52** (2005) 973.
[76] A. Dubietis *et al.*, Opt. Commun., **88** (1992) 433.
[77] I. N. Ross *et al.*, Opt. Commun., **144**, (1997) 125.
[78] D. Strickland and G. Mourou, Opt. Commun., **56** (1985) 219.
[79] I.N. Ross, P. Matousek, K. Osvay, and G.H.C. New, J. Opt. Soc. Am. B, **19** (2002) 2945.
[80] E.J. Grace, C. Tsangaris, and G.H.C. New, Opt. Commun., **261** (2006) 225.
[81] Y. Tang *et al.*, Opt. Lett., **33** (2008) 2386.
[82] C.L. Tsangaris, PhD thesis, Imperial College London, 2005.
[83] D.J. Bradley and G.H.C. New, Proc IEEE, **62** (1974) 313.
[84] E.B. Treacy, J. Appl. Phys., **42** (1971) 3848.
[85] D.J. Kane and R. Trebino, IEEE J. Quantum Electron., **QE-29** (1993) 571.
[86] C. Iaconis and I.A. Walmsley, Opt. Lett., **23** (1998) 792.
[87] Rick Trebino, *Frequency-Resolved Optical Gating: The Measurement of Ultrashort Laser Pulses*, Kluwer Academic, Dordrecht, 2000.
[88] I.A. Walmsley and V. Wong, J. Opt. Soc. Am. B, **13** (1996) 2453.
[89] I.A. Walmsley and C. Dorrer, Adv. Opt. and Photonics, **1** (2009) 308.
[90] C. Dorrer and I. Kang, J. Opt. Soc. Am. B, **25** (2008) A1.
[91] H. R. Telle *et al.*, Appl. Phys. B **69** (1999) 327.
[92] D.J. Jones *et al.*, Science, **288** (2000) 635.
[93] T. Brabec and F. Krausz, Rev. Mod. Phys., **72** (2000) 545.
[94] A. Messiah, *Quantum Mechanics,* Vol.1, Dover, New York, 2003.
[95] A. Rae, *Quantum Mechanics* (5th edn), Taylor & Francis, London, 2008, p. 131.
[96] P.N. Butcher and D. Cotter, *The Elements of Nonlinear Optics*, Cambridge University Press, Cambridge, 1990.
[97] J.F. Ward, Rev. Mod. Phys., **37** (1965) 1.
[98] M. Takatsuji, Phys. Rev. B, **2** (1970) 340.
[99] B.J. Orr and J.F. Ward, Mol. Phys., **20** (1971) 513.
[100] D.C. Hanna, M.A. Yuratich, and D. Cotter, *Nonlinear Optics of Free Atoms and Molecules*, Springer-Verlag, Berlin, 1979.
[101] L. Allen and J.H. Eberly, *Optical Resonance and Two-Level Atoms*, Dover, New York, 1988.
[102] R.P. Feynman, F.L. Vernon Jr. and R.W. Hellwarth., J. Appl. Phys., **28** (1957) 49.
[103] M.D. Crisp, Phys. Rev. A, **1** (1970) 1604.
[104] K.J. Boller, A. Imamoglu, and S.E. Harris, Phys. Rev. Lett., **66** (1991) 2593.
[105] L. Vestergaard Hau, S.E. Harris, Z. Dutton, and C. H. Behroozi, Nature, **397** (1999) 594.
[106] C. Liu, Z. Dutton, C.H. Behroozi, and L. Verstergaard Hau, Nature, **409** (2001) 490.
[107] R.L. Sandberg *et al.*, Phys. Rev. Lett., **99** (2007) 098103.
[108] E. Goulielmakis *et al.*, Science, **320** (2008) 1614
[109] S. Baker *et al.*, Science, **312** (2006) 424.
[110] A.L. Cavalieri *et al.*, Nature, **448** (2007) 1029.
[111] M. Krausz and M. Ivanov, Rev. Mod. Phys., **81** (2009) 163.
[112] J.G. Eden, Prog. Quantum Electron., **28** (2004) 197.
[113] A. McPherson *et al.* J. Opt. Soc. Am B, **4** (1987) 595.

[114] M. Ferray *et al.*, J. Phys. B, **21** (1988) L31.

[115] J.J. Macklin, J.D. Kmetec, and C.L. Gordon III, Phys. Rev. Lett., **70** (1993) 766.

[116] A. L'Huillier *et al.*, Phys. Rev. Lett., **70** (1993) 7674.

[117] Z. Chang *et al.,* Phys. Rev. Lett., **79** (1997) 2967.

[118] E. Seres, J. Seres, and C. Spielmann, Appl. Phys. Lett., **89** (2006) 18919.

[119] P.B. Corkum, Phys. Rev. Lett., **71** (1993) 1994.

[120] K.J. Schafer *et al.*, Phys. Rev. Lett., **70** (1993) 1599.

[121] M. Lewenstein *et al.*, Phys. Rev. A, **49** (1994) 2117.

[122] G.L. Yudin and M. Yu. Ivanov., Phys. Rev. A, **64** (2001) 013409.

[123] L.E. Chipperfield, private communication.

[124] L.E. Chipperfield *et al.*, Laser and Photonics Reviews, in press.

[125] G. Sansone *et al.*, Science, **314** (2006) 443.

[126] R.L. Sutherland, *Handbook of Nonlinear Optics* (2nd edn), Marcel Dekker, New York 2003.

[127] J.W. Cooley and J.W. Tukey, Math. Comput., **19** (1965) 297.

Index